METACREATION™

of Personal Realities in Space and Time

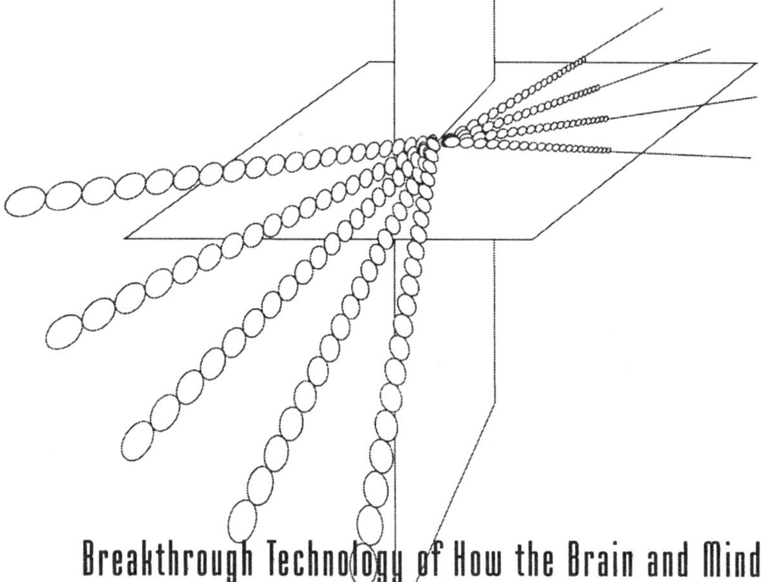

Breakthrough Technology of How the Brain and Mind Create Realities for a Prosperous and Meaningful Life

VICTOR J. GIBBONS

MetaCreation™
of Personal Realities
in Space and Time

Author: Victor J. Gibbons
Published by: MetaCreation Institute
Atlanta, Georgia

Cover Design by Mark Giles
Illustrations by Mark Giles and Victor Gibbons
First Printing: 2000
Printed in the Unites States of America

Library of Congress no. 98-092310
ISBN No. 0-9669428-0-9 (pbk)

Self Help
Psychology
Philosophy
Quantum Physics
Science
Artificial Intelligence

Price: US$ 24.95
pages: 421

This Book

Is Dedicated To
All Those Who Are Unaware
And Consequently Suffered Needlessly

Rick
I wish you
all the Best
Victor [illegible]

ACKNOWLEDGEMENTS

While the process of writing this book took a little over three years, the process of researching the contents of this book spanned a lifetime. During this time many individuals participated, either directly or indirectly, knowingly or unknowingly, in many of the life experiences upon which this book is based. To honor all of those people individually many of whom have long since passed on, would be too numerous to count. However I want to take a moment to thank and honor each and every one of these people, both living and dead, who contributed in this manner.

A special thanks and appreciation goes to Dr. Thomas Gibbons, PhD., Clinical Psychologist and Forensic Specialist, Baldwin, Michigan. Dr.Thomas Gibbons spent endless hours perfecting many of the MetaCreation™ processes during the summer workshops in Northern Michigan and in reviewing the manuscript. His wife, Dr. Louise Gibbons, PhD., Psychologist in teen counseling also made valuable contributions during those summer workshops. Next I would like to thank Dr. James Gibbons, M.D., retired surgeon and oncologist, for his work on the manuscript and participation in the workshops. I would like to also thank Dr. Joseph Gibbons, O.D. and Clinical Psychiatrist, Phoenix, Arizona for his encouragement, support and suggestions throughout this entire project.

Next I want to thank Ed and Marianne Reese of the Southern Institute of NLP for their training and support in advance NLP techniques and in Ericksonian Hypnosis. A special thanks goes to Mr. Mark Furman of Furman Research Associates for his pioneering studies in the dynamical processes of the neurological structures of the human brain as set forth in his book, *Minds in Motion, The Human Performance Technology for the next Millennium.*

Next I would like to recognize Dr. Fritjof Capra, PhD. Physics for his enlightened treatment on Quantum Physics and complexity of self organizing systems in his books, *The Tao of Physics, Turning Point* and *The Web of Life*. In like manner I would like to recognize Dr.William Bechtel, PhD. Philosophy and Dr. Adele Abrahamsen, PhD. Psychologist for their pioneering work set forth in their book, *Connectionism and the Mind,* which describes the processes of the human mind in terms of artificial intelligence as exhibited by a parallel processing computer.

Finally I would like to thank my wife Liz whose unending love and support has always been a source of encouragement and inspiration to me both throughout the writing of this book and throughout our marriage.

Victor Gibbons

FOREWORD

Within the last decade a considerable amount of research has been conducted into the neurological functions and processes of the human brain. Neuroscientists have been able to map out many of the processes and functions that create and sustain human life. They have been able to develop a considerable body of knowledge on how the neurotransmitters of the brain's neuron cells create neurological patterns giving rise to our thoughts, emotions, actions, and behaviors. At the same time, through the use of elaborate laboratory computer models, a considerable amount of research is also being conducted in the realm of artificial intelligence and robotics. These models, using sophisticated computer parallel processing techniques, simulate how the human mind acquires knowledge and how it organizes this knowledge into patterns of thought in order to create a meaningful life for us.

Yet in spite of all of these advances in scientific research there exists no comprehensive model which sets forth exactly how it is that the human brain and mind create a sense of ego personality and how it is that ego personality creates our inner and outer realities in life. What has prevented mankind from creating such a comprehensive model, in large part, is our mechanical view of our world where everything is viewed as existing outside of us in linear time and fixed space of an objective world; a world that is made up of solid objects existing in fixed three dimensional space; a world where we view ourselves as consisting of our brain, mind and body, and where we exist separate and apart from everyone else and everything else. A world view where our sense of our mind, body, ego personality, physical possessions and affairs in life constitute our physical reality in time and space. That whatever happened in our past is behind us and whatever happens in the future is out in front of us someplace. A world where we cannot access the past in order to change it and while we can plan for the future we cannot create the future simply through our thought processes alone.

Quantum Physics has changed how we perceive the creation of the material universe on the micro level of existence and at the macro level of existence. We now know that the universe does not consist solely of solid matter but the universal construct is energy. On the micro level and macro level of existence time and space are not fixed but exist relative to the creation of matter in what is referred to as the spacetime continuum of the fourth dimension. However, quantum physicists have never been able to apply quantum physics and relativity theory to the creation of realities at the quantum level of our brain and mind. Consequently, we still live out our lives in what is often referred to as the zone of the middle dimension, where time is linear, space is fixed and things, including you and me, exist in solid form in an objective world. A world over which we have very little control. A world which has become so complex that the negative effects of modern civilization are taking an increasing toll on the human condition and on the environment at an alarming rate.

All of your understandings in these respects will now change dramatically as you read this book. Paradigm shifts will start to take place in your understanding of who you are and what makes up the composite realities of your life. As you begin to apply the MetaCreation technology you will be able to create a whole new life for yourself and in the process you will be able to create a whole new you! Nothing short of your own destiny and the destiny of all mankind is now at hand. By simply reading this book your life will never be the same again!

INTRODUCTION

MetaCreation™ Technology represents a quantum breakthrough in the nature of mankind and in the nature of his realities. Never before in the annals of history has there ever been a comprehensive model that sets forth in precise detail the entire structure, functions and processes of the human brain and mind in describing human consciousness, ego personality and the many realities making up the very fabric of your life. This book, with over 50 charts, describes in considerable detail the theory and processes of what constitutes consciousness and how it operates in the life of each one of us in creating for us a sense of ego personality and our life realities. The MetaCreation process is a self help program that teaches you how to create for yourself an entirely new personality if you are not satisfied with your present one. This book also provides programs to create entirely new inner realities in life such as peace, harmony, confidence, integrity, and love, as well as outer physical realities such as material wealth, financial security, health, career goals, positive relationships and generally how to create a more enjoyable and meaningful life.

The MetaCreation processes set forth in this book are founded on scientific principles drawing from quantum physics, relativity theory, theory of self organizing systems, neuroscience, biochemistry, artificial intelligence and the latest research in the connectionist laboratory models describing the dynamics of the subconscious mind. This book explains in terms of quantum and relativity theory how the dynamic processes of the human brain and mind work to create human consciousness, the human ego personality and our many realities in life. This book describes how quantum theory is applicable to the dynamics of the analog system of the subconscious mind in organizing realities in spacetime outside linear time and fixed space of our three dimensional objective world. It shows you in detail how the human brain and mind create realities at the quantum level of the neurological structures of the mind/brain and how you can place these realities, including physical realities, anywhere you choose in the fixed space and linear time of your outer physical world.

By using the processes set forth in this book on a regular basis an individual can successfully transform his entire life from one of pain, lack, and limitation to one of purpose, happiness and prosperity. The processes in this book will show you how to find your true mission in life and how by fulfilling that mission you can find happiness and meaning in life. Ultimately the processes spelled out in this book will lead you to higher states of consciousness and expanded states of awareness. You will be able to create expanded realities from a new paradigm of higher consciousness. Ultimately this book points the way for mankind to take control of the evolutionary forces directing its future path. For the first time in the history of mankind we can, individually and collectively, take control of and direct our own destiny and the destiny of generations to come!

Table of Contents

Chapter One

How Quantum Physics Affects the Processes of the Human Brain and Mind

Quantum Physics

For over 2,700 years mankind has been looking for the basic building blocks of the entire universe including the world we live in called earth. In terms of time, manpower and money spent on finding the answer, this search goes way beyond anything you could possibly imagine that was encountered during the elusive search for the Holy Grail during the Middle Ages. Mankind has always felt that if he could just unlock the mysteries of what constitutes the basic building blocks of all of the galaxies, stars, and planets he would ultimately have the knowledge and the tools to be able to rearrange the constituent parts making up the universe and hopefully be able to build a better world for all of us.

Thales was a Greek philosopher who taught young students at his school of philosophy at his home in the City State of Ionia, Greece, back in 700 B.C. At that time a Greek school of philosophy included all of the formal studies known to mankind including science, medicine and philosophy.[1] It was Thales who theorized that if he could just discover what was later to become known as the "universal construct" he could apply this knowledge for the betterment of mankind in the construction of better roads, bridges, buildings and defenses around the walled city states of his time. He speculated that such knowledge concerning the very essence of how the universe was built and operated might even have far-reaching implications in such fields as medicine, agriculture, navigation and the techniques of ancient warfare. However, Thales quickly found that the universe did not give up its secrets easily and before his death he could only speculate that the universal construct he had been looking for all of his life was possibly locked up in the constituent parts of water. While he was incorrect in his conclusions you can easily under-

1. Leon Lederman, *The God Particle,* Dell Publishing, New York, 1994.

stand his reasoning for the entire face of the planet consists of approximately 70% water. Water also consists of approx. 70% of the composition of the entire human body.

One of Thales student's, Anaximander, observed that all of physical creation, both living organisms and inert matter, including the wild animals, birds and fish, the vegetation covering the earth, as well as the oceans, rocks and soil of the earth itself and the sky above, all seemed to exist in harmony and balance with one another. He concluded that there must be some type of essential unity underlying this seeming diversity we experienced in life. As you will see shortly he was essentially correct on this point. Like Thales he believed that there must be a basic substance out of which all things are formed. However, by the time of his death he also didn't have a clue as to what that basic substance might be.

There was a succession of philosophers following Thales and Anaximander, including Anaximenes, who believed very much like Thales and Anaximander. However, he felt that the universal construct that they all were looking for was in the basic substance of the air we breath. Not a bad conclusion for his time, but, of course, he was wide of the mark. Anaxagoras was the next one on the scene in Greece who lived in about the 5th century B.C. Anaxagoras reported that all matter comprising the universe possibly could be indefinitely divided, again and again, into smaller and smaller particles. He also believed that the origin of the universe probably came into being in an instant rather than manifesting itself slowly over great periods of time as was the current thinking of the day. Of course, he was correct in this respect. However, it took mankind another 2,000 years to prove him right. Empedocles followed Anaxagoras. It was Empedocles who theorized that it was earth, water, fire, and air that made up the essential elements. Both Plato and Aristotle accepted Empedocles' theories. Since Plato and Aristotle had large followings in the years to come these four elements were accepted by many philosophers throughout much of history as the four basic building blocks of matter.

The next Greek philosopher who tackled the problem of the universal construct was Democritus. Democritus also lived at the same time as Anaxagoras in the 5th century B.C. Democritus was influenced by the writings of Anaxagoras and concluded that the universe was made up of tiny invisible particles which he called "atomos". Atomos, when roughly translated, means "not cuttable." It was believed that the atomos was the basic indivisible particle making up all of matter. Democritus theorized that these atomos could combine with other "atomoses" to form larger elements which in turn combined to form the vast variety of different types of physical matter appearing throughout the entire universe. It was Democritus then who laid the foundation for Western Civilization's eternal quest for the elusive building blocks of all of nature. Was Democritus really on to something or did he only send Western Civilization on an extremely long and expensive wild goose chase? We shall see in just a moment.

Building upon Democritus's theories mankind had believed for a long, long time that it was the atom itself that we had been looking for in our eternal quest for the basic building block of the universe. It was the atom that constituted the basic, tiniest, most indivisible particle which made up the structure of the universe. This viewpoint persisted throughout the establishment of the Roman Empire and on into the early Christian era and even into the Middle Ages. It was during the Middle Ages that mankind believed that gold, if not the most basic element, was one of the basic elements constituting all matter. Their reasoning was that gold, which had come to be valued as a precious metal, seemed to be impervious to corrosion and decay. Fine gold filings were often used as an essential ingredient in medicine to ward off a whole host of diseases including infirmity and old age. Because gold was valued as a precious metal it became the basis for the practice of alchemy. Alchemy was the practice where the alchemist, through a process called transmutation, would try to convert elemental material such as lead into gold.[2]

Alchemy remained the dominant practice concerning the nature of the elements throughout the Middle Ages. It was not until 1661 that the Irish chemist, Robert Boyle, a member of the Royal Society of London, published a comprehensive report on the scientific analysis of the basic elements. He clearly established that there was no scientific basis for considering that earth, fire, water, and air constituted the basic elements of matter. In 1782 Antoine Lavoisier discovered that water consisted of hydrogen and oxygen. It was Lavoisier who established the scientific principle that a substance could only be considered elemental if it could not be broken down into further substances when treated chemically. Unfortunately it is scholars like Lavoisier who are often not appreciated by the masses for probing the frontiers of knowledge beyond the ordinary and the mundane. They often pay a price for such independent investigation. In this case Lavoisier was guillotined during the Reign of Terror of the French Revolution in 1794.[3] Finally we come full circle. By the early 1800s John Dalton, a professor of physical sciences in

2. David & Arnold Brody, page 54-58, *The Science Class You Wish You Had,* 1997

3. Brody, op. cit., pp. 57-58.

Manchester, England, developed a table of atomic weights of ultimate particles. It was Dalton who called these indivisible particles "atoms" in honor of Democritus. Talk about delayed recognition. It took over 2,000 years before Democritus was finally immortalized for coining the term "atom".

In the late 1800s and early 1900s subatomic particles were discovered. It was found that the atom was no longer the most indivisible unit. The atom itself consisted of electrons, protons, and neutrons. In 1912 it was Niels Bohr who believed that the theories developed by Max Planck and Einstein that stated in effect that light consisted of discrete amounts of energy called "quanta" could account for the stability of the atom. The quanta of energy in reference to light is represented by either a particle or a wave. Quantum Physics derived its name from the term "quanta". By 1998 literally hundreds of subatomic particles have been discovered. All of these subatomic particles can be grouped into two major categories, six quark groups and six lepton groups. This division of quarks and leptons have been designated the Standard Model.[4] Subatomic particles all have certain specific characteristics such as; electric charge, mass, a quanta of energy, momentum, spin, and velocity.

Well, we are at the end of the road, or are we? Scientists at the Fermi National Accelerator Laboratory in Illinois believe that quarks might be broken down even further. So I guess that we now have sub-subatomic particles and in the near future we will have sub-sub-subatomic particles. You can begin to see by now where all of this is taking us. You guessed it, more powerful accelerators and more sub-subatomic particles. Let's stop and take a breather for a minute and see what we have just learned from all of this, and then we will address what all of this has to do with the processes of the human brain and mind.

The Structure of Energy

As I have indicated subatomic particles all have certain specific properties such as, electric charge, mass, a quanta of energy, momentum, spin, and velocity. All of these properties, except mass, are readily recognizable as different forms of energy. The electrical charge is negative and positive polarized energy. A quanta of energy represents either the particle or wave configuration of light energy. Momentum, spin, and velocity all represent the dynamical movement of kinetic energy. This leaves "mass", a term often used in referring to matter and relates to an object's weight. However, it is the dynamic or active interrelationship of the other properties of the energy of the subatomic particle which constitutes its mass. Mass then relates to the weight of a particular object determined by the number of pro-

4. Ibid p. 82.

tons, electrons and neutrons in the atoms making up the object. To state it another way, the electrical charge of the atom's protons, electrons and neutrons weigh something. *Matter then at its most fundamental level consists solely of the relationships of its energy, forces, and fields.* The various combinations of these energy relationships are what form the atom's structure which is often referred to as its "structural matrix". I will be talking a lot about the energy and matrix configuration of the atom in relationship to the processes of the human brain and mind. *It is the energy of the subatomic particles that make up each atom which arrange and rearrange themselves to form an endless variety of matter found throughout our world.*[5]

Energy takes many forms. It forms the energy forces and energy fields found throughout the universe. Other types of energy you might be familiar with include heat, wind, sound, gravity, magnetism, radiation, light, radio waves, radar, and x-rays, to name just a few. Energy then can be defined as a dynamic quantity which is related to changes in conditions and states, including mental and emotional states, and changes in activity or processes. *Therefore all human thought and all other mental processes consist of some type of energy process which has an effect upon mental states, mental activity and conscious and subconscious mental processes.* Throughout this book I will be referring to various combinations of the energy and matrix images of these mental processes as they affect human thought and the conscious and subconscious processes of the mind and brain.

Electrical energy, the force you are most familiar with in your daily life, consists of polarized energy. If you take a certain amount of energy and polarize it, that is, give half a negative charge and the other half a positive charge, you have electrical energy. When you look at the atom it consists of polarized energy in the form of electricity. Its electrons are negatively charged, its protons are positively charged and while its neutrons do not carry an electrical charge on their surface they have an electrically charged interior structure. Polarized energy, by building itself up from the most basic indivisible form of the subatomic particle, forms atoms and then molecules. These, in turn, form intricate patterns of matter consisting of the 92 naturally occurring elements making up the entire universe, the earth and, yes, even you and me.

Of these 92 naturally occurring elements hydrogen, oxygen, carbon, nitrogen, and calcium make up approx 99% of the human body. The electrical matrix of an atom interacts with the electrical matrix of a second atom and in this manner chemical bonding occurs to produce an entirely new matrix that forms the basis of all matter. For

5. Brody, op. cit., p.82.

example, oxygen + carbon + nitrogen + hydrogen = amino acids and proteins. Amino acids are essential in the formation of neurotransmitters. Neurotransmitters communicate between neuron cells of the brain and therefore are an essential link in the chain creating thought. Protein, of course, is important in building body muscle and tissue. Oxygen + carbon + hydrogen + calcium = bones. Bones, of course, make up the human skeleton. Carbon + hydrogen + oxygen = carbohydrates. Carbohydrates are essential for energy. Hydrogen + oxygen = water. Water is essential in the production of blood and other bodily fluids.

To see exactly how the energy and matrix configuration of an atom combine to form matter let's briefly look at the two atoms, hydrogen and oxygen. Refer to the electric matrix of H_2O on Chart 1, at page 15, entitled "The relationships of the Electrical Matrix of Steam, Water and Ice." You will note that the molecule H_2O consists of two hydrogen atoms and one oxygen atom. However, rather than thinking that H_2O consists solely of two hydrogen atoms and one oxygen atom you have to think of it as a single molecule with unique properties all its own, properties which are very different from other atoms or molecules. That is, when two hydrogen atoms combine with one oxygen atom you do not just have two parts of one thing and one part of another thing but rather you have a single unique unit which is far more different from the three separate atoms before they combined to form H_2O. That is, hydrogen by itself is a very volatile explosive gas, and oxygen is an essential element in sustaining fire. H_2O on the other hand is neither an explosive gas nor an element in the combustion process of fire. Consequently, we didn't just mix one part oxygen with two parts of hydrogen. Instead we formed from these two hydrogen atoms and the one oxygen atom an entirely new and different single molecule whose properties have no resemblance to the three separate matrices which combine to form the single molecule H_2O. *The reason for this can be simply stated that we are dealing with energy configurations called matrices and not discrete descriptions of solid matter.*

If you view this solely in terms of various forms of matter you would easily conclude that water is simply the combination of three things, two parts hydrogen and one part oxygen and that ice is the combination of six water molecules. Whereas, in fact, the matrix configuration of water is entirely different than ice. Ice is not just six times more water. Water and ice have entirely different energy matrices and therefore constitute entirely different properties and exhibit different characteristics. When we apply heat to ice we break the

CHART 1

The Relationship of the Electrical Matrix of Steam, Water and Ice

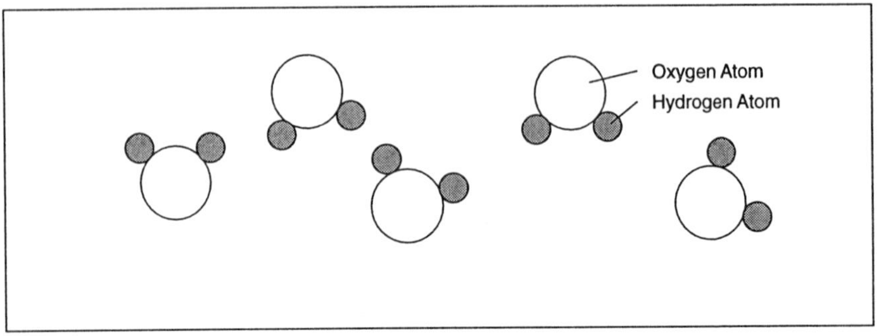

Steam Molecules

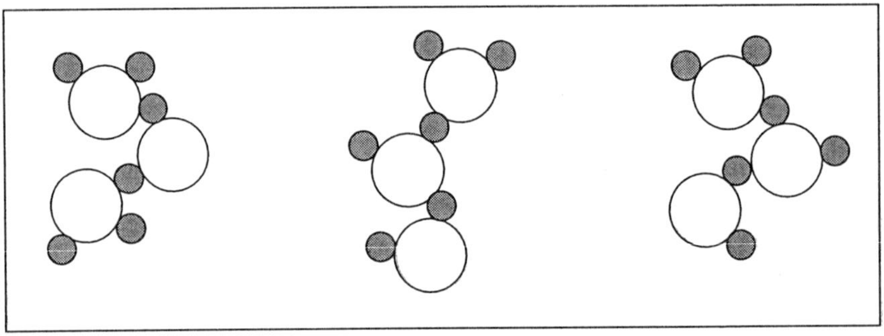

Water Molecules

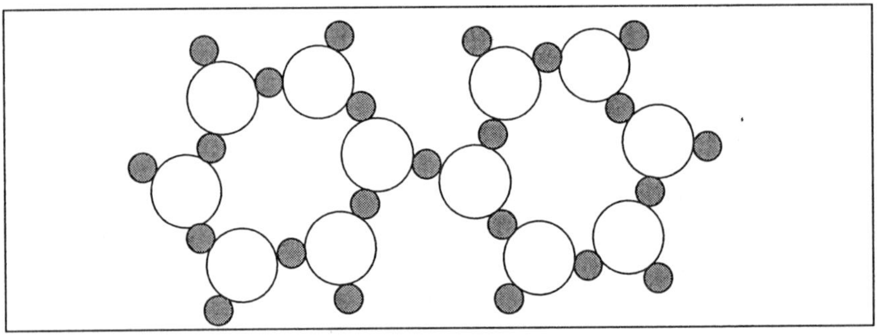

Ice Molecules

While molecules of steam, water and ice all consist of hydrogen and oxygen atoms, each consists of a single electrical matrix with unique properties all its own. Hydrogen is a volatile gas and oxygen sustains fire. Whereas H_2O forms steam, $3H_2O$ forms water and $6H_2O$ forms ice. Each has unique properties because of its electrical matrix and are not simply more of the same thing.

electrical bonds of the hexagon matrix of ice and we again have water. Water matrices really consist of a variety of combinations of H_2O molecules such as $2H_2O$, $3H_2O$, $4H_2O$[6]. By applying additional heat the water molecules break further apart to form a single H_2O molecule which has an entirely different electrical matrix which forms a gas called steam. *The electrical matrices of steam, water and ice while they all are formed out of oxygen and hydrogen atoms are actually different electrical matrices with different and unique properties all their own and are not more of the same things just in different combinations.*[7] It is O.K. to think of them in this way in your daily life. In fact, since you always get water when you melt ice and not soda pop or chocolate sauce it is reasonable to assume that ice is frozen water and that steam consists of water molecules.

However, with respect to how the energy and matrix configuration of atoms affects our brain and mind we need to view them scientifically as they are actually constituted in nature. The reason for this is that the energy and matrix configurations of the atoms organize the processes of the human brain and mind. If we viewed atoms as different aspects of solid matter only instead of energy matrices then they would precipitate out of our mental processes, so to speak, and set up overnight like a block of concrete. This, of course, is an exaggerated metaphor. However, it is made to get a point across. If, when we broke apart the matrix of a molecule and didn't get different properties from the resulting pieces but simply got smaller and smaller pieces of the same thing, then we would not be living in an electrical universe but in an Erector Set or Lego Block world. *More importantly, the brain and mind operate on primarily electrical processes and electrical matrix configurations of its atoms and molecules and not on the material aspects of solid matter in the form of chemical compositions.* The brain and mind do not organize themselves on the basis of different types of solid matter like the Erector Set or Lego Blocks. Consequently, and this is a very important point, to describe the processes of the brain and mind solely in terms of their chemical composition such as neurotransmitters, various proteins, and amino acids, is essentially describing the brain's processes in terms of matter instead of its underlying electrical and matrix processes. The brain's chemical composition is the medium which provides the brain with its dynamical energy processes. But by simply describing the processes in purely chemical terms alone is not totally accurate for it does not address the dynamics of the underlying processes. Conse-

6 David Elliot,The Electric Universe, 1996, pp. 18-24

7 Furthermore, things don't just have an electrical matrix, things are an electrical matrix.

quently, the universal construct we have been looking for is not the atom, not even subatomic particles, but energy, pure kinetic energy! The processes of the brain and mind are energy processes, more precisely mostly electrical energy processes.

The point of all this is that the processes of the brain and mind involve billions of neurological processes that are organized and function based on the energy and matrix configurations of electrochemical and electromagnetic forces and not simply on the arrangement of its chemical compositions. Forces interact on the basis of relationships. These relationships take many forms. They consist of analog patterns, associations, force fields, weights, densities, influences, attracting and repelling forces, negative and positive energies, clusters, recurrent networks, field coherence, electrical paths, trajectories, transitions, hemostatic balances, phase transitions, stable and unstable states and complimentary values. *All of these relationships are essentially electrochemical processes and not descriptions of chemical compositions or the static condition of solid matter.* Accordingly, the mind/brain is a dynamical system always in a state of motion and is not a static system. The processes of the mind/brain are fluid, active, constantly coupling and uncoupling, setting up and dissolving mental states. Thus, it is constantly in a dynamical state of mental activity even when you are sleeping. If you think of these forces as simply different forms of solid matter existing in fixed states you will not be able to get a very clear picture of how the human brain and mind work in which case our metaphor of the mind setting up like a concrete block would prevail.

One last observation before moving on. You noticed that the ice matrix did not just form in random fashion but rather it organized itself in an orderly pattern. In this case it formed a hexagon pattern. All electrical matrices form geometrical patterns. In fact geometry played an enormous role in all scientific theory since the Renaissance Ages. These types of geometrical patterns in nature repeat themselves over and over again at the subatomic level of creation of all matter.[8] It is also these types of geometrical patterns in nature that exist at the

---

8 It is these subatomic patterns arising out of the activity of subatomic particles and waves which gives chaos theory its deep unifying order underlying the appearance of surface chaos.

primary processing level of the subconscious mind.[9] That is, the subconscious mind is excellent at forming mental patterns. It does so because the incoming sensory stimuli arranges itself into an endless variety of patterns. *In fact you will see that it is the primary patterns of the mind which organizes all of the secondary patterns forming all thoughts and feelings about life.* The subconscious mind is not only good at forming patterns, it is also very good at pattern integration, pattern recognition, pattern generalization, pattern consolidation, pattern dissolution, etc. Why this is so and how these processes affect your thinking will become very clear in the following chapters. Suffice it to say at this point that we are not dealing with solid matter like a block of concrete. We are dealing with energy and chemical processes in the form of electrical energy, matrix configurations and pattern formations at the quantum level of the brain and mind.

Relativity Theory, Or Here, There and Everywhere

To begin our story of how relativity theory relates to processes of the mind we are going to again momentarily digress back in time to the Age of Enlightenment of the seventeenth century. Rene Descartes is considered the founder of the modern scientific method. Descartes set the course of all scientific thought and laid the foundation for understanding of the material universe as an objective fixed three dimensional universe operating in accordance with fixed mechanical law. The universe was viewed by Descartes as a gigantic machine which could be understood by examining its constituent parts. This emphasis on analysis of the constituent parts in understanding the whole system has led to the theory of reductionism in science whereby complexity of a system can only be understood by reducing it to its most basic constituent parts.[10]

Descartes was a mathematical genius who was able to correlate algebra and geometry into a whole new branch of math known as analytical geometry. The Cartesian view of the universe as one gigantic objective material machine dominated all fields of science and physics for the next three centuries. It was not until the early part of this twentieth century that the Cartesian mechanical view of the

9 However a word of caution. Geometry is not an inherent property of matter. Rather, geometry is simply our mental way of describing matter. The interrelationships and interconnections of the electrical charges of atoms is what describes the dynamical processes of the atom. This is not a logical relationship but rather a process relationship based on the similarity of properties which attract and repel one another as a result of their energy matrix configurations.

10 For an example of Reductionist Theory see Paul Churchland's, *The Engine of Reason, the Seat of the Soul, a Philosophical journey into the Brain,* 1996.

world started to be seriously questioned by physicists and other scientists.[11] Yet, even today, most people including many physicists and other scientists continue to view their personal world from this old Cartesian mechanical way of thinking even while at the same time acknowledging that at the quantum level of existence and at the macro level of the universe relativity and quantum theory prevail. This view is often called the "zone of the middle dimension".[12]

Isaac Newton, a contemporary of Descartes, was also a mathematical genius. He developed a whole new field of math known as differential calculus. It was Isaac Newton that developed the theory that space was fixed, absolute and uniform throughout a fixed three dimensional universe. Time also was viewed as fixed and uniformly measured throughout our world and the universe. Both time and space were viewed as absolute in the sense that they were not affected, nor could they be influenced, by any other phenomena appearing within the universe. In the Newtonian mechanical view of the world the motion of physical objects was determined by their mutual attraction to one another due to the forces of gravity. Newton's equations describing the effects of gravity by mathematical calculations became known as classical mechanics and dominated scientific thought concerning the motion of celestial bodies up until the early part of this twentieth century.[13]

Einstein's Theory of Relativity

All of this changed radically when Einstein came upon the scientific scene in the very early part of this century. Albert Einstein was born in 1879 and attended grammar school in Munich, Germany. His teachers reported, "he was mentally slow, unsociable and adrift forever in his foolish dreams".[14] When his father was asked what profession young Einstein should pursue, he replied, "It doesn't matter. . . .he'll never make a success of anything."[15] At about age of 16 when other young men were thinking about girls he was dreaming about what the world would look like if he were sitting on a light beam moving at the speed of light. He concluded that the beam of light would just oscillate back and forth without ever moving on in time. In effect at the speed of light time stood still! However, such a phenomenon was totally unheard of and contradictory to every physical law known to mankind at the time.

It wasn't until 1905 that Einstein proposed his now famous special theory of relativity. The theory of relativity is based on the prin-

11 Fritjof Capra, *The Turning Point*, 1982, pp. 60 - 65.

12 Fritz Capra, *The Tao of Physics*, p. 64.

13 Fritz Capra, *The Turning Point*, pp. 63 -67.

14 David & Arnold Brody, op.cit. p. 106.

15 Ibid page 106.

ciple that light travels at a constant velocity of 186,000 miles per second regardless of the position of the observer or the source of the light. At this speed time appears to slow down to a stationary observer observing an object moving at nearly the speed of light. For example, we will refer to twin brothers. One is sitting on the face of the earth looking at his twin on a space ship moving away from earth at nearly light speed. To the twin sitting on earth a space ship moving close to the speed of light, relative to our sitting twin, would appear to be slowing down. However, as to our space twin time would appear to be normal. Lets assume that from the perspective of the stationary twin the space trip will take approximately 10 year to go out in space and about 10 years to get back. When the space ship is moving at nearly light speed, to the stationary twin on earth, time would appear to be slowing down in order to adjust for the constant velocity of the speed of light. Given this scenario the twin remaining here on earth would be 20 years older to the traveling twin when he got back. Whereas the traveling twin would be 20 years younger then the stationary twin. This is not a perceptual thing but rather a real physical difference in their ages because of the relativistic factor of time. At these incredible speeds clocks actually slow down relative to the stationary observer, as well as heartbeats, blood flow, the aging process, the composition of the space ship and everything else moving on the space ship at nearly the speed of light. However, remember time itself would appear normal to our space traveler. It would appear to him that the space trip only took a couple of weeks.

Einstein determined that the relativistic factor of time was needed to adjust or correct the time factor because the light speed factor was always constant in relation to these two different observational frames of reference. One from the viewpoint of the stationary twin and the other of the traveling twin. If you were to add the travel time to light time on the return trip the light would appear, to the stationary twin on earth, to be traveling at about double the speed of light. However, since light always travels at the constant velocity of 186,000 miles per second and not faster or slower an adjustment has to be made in the time factor in order for the light to be observed by the stationary twin to be moving at that constant velocity. Accordingly, the factor of time itself is made relative to the observer. At these incredible speeds time is not fixed as was previously believed to be the case according to Cartesian thinking.

In relativity theory time then became a fourth coordinate existing beyond the three coordinates of three dimensional space, that is, the three coordinates of up or down, forward or backward, or sideways to the left or right. Because of relativity theory space itself can also no longer be considered as a separate coordinate. Space and time must now be considered together relative to the observer when considering extremely high speeds approaching light speed velocities. This phenomena has come to be known as the spacetime continuum of the fourth dimension. It is important to note at this juncture that you do not have to understand relativity theory. The whole point of this is that time and space are relative and are not something that exists separate and apart from us. The implications of this will become quite apparent as I develop these points shortly.

What I have been talking about so far is called the special theory of relativity. Einstein's general theory of relativity includes this spacetime continuum of the special theory and adds the factor of gravitational pull of large moving bodies in space. The general theory of relativity holds that the effect of gravity is to make the spacetime continuum curved. The rate of curvature depends on the pull of gravity generated by the mass of a body in space such as a huge massively large star. In the general theory of relativity, because of this gravitational factor, not only is space and time intricately linked to the creation of matter itself but the distribution of matter is intricately linked to the flow of time and space throughout the universe. Space and time are therefore both curved and flowing at different rates and at different degrees depending upon the distribution of matter and the gravitational pull manifesting throughout the universe.

As I said before, you do not have to understand any of this so don't worry about it. Its whole purpose is only to show that maybe we do not live solely in a three dimensional world of linear time and fixed space. Maybe it is possible for the human brain and mind to also organize realities in life outside of these fixed space and linear time references. Maybe it is we who place constraints on our ability to create realities outside of these space and time frames. Perhaps we subconsciously choose not to even consider other alternatives and instead we have created an inner mental organization where time is linear and space is fixed. We have concluded from time immemorial that this is simply the way things are when thinking about and viewing our world. It never occurred to us that there might be another way of viewing time and space in which we organize our life realities. We will consider these other possibilities in just a moment and let you be the judge.

In the fourth dimensional frame of spacetime of quantum theory, protons, electrons and neutrons involved in massive collisions in particle accelerators appear to move backwards in time as well as forward in time. This phenomena is the result of spacetime being relative in quantum physics. What is happening at the quantum level of existence and at the macro level of the universe is that what we consider to be the past, present and future in our daily day life are all happening at once and can be viewed as happening "en bloc time" or happening all within the same spacetime frame of the fourth dimension. What the physicist is seeing at the quantum level is slices of spacetime at any given moment in what is really spacetime taking place en bloc in the fourth dimension frame of the spacetime continuum. Physicists recognize that what they are experiencing en bloc in spacetime represents, when viewed by an observer at the physical level of three dimensional existence, only a slice of the fourth dimension frame at any given moment in three dimensional reality.[16] Since our five sensory receptors do not give us a sense or experience of this fourth dimension we have a very hard time mentally visualizing what this fourth dimensional phenomena would look like in our three dimensional world.

Another way of looking at the en bloc effect of the fourth dimension would be if you were to visualize what the universe would look like if you were to stand outside of the entire universe and look back at it? You would see a large spherical ball expanding out in all directions at incredible speeds with what looked like an awful lot of electrical, matrix material and other activity in its interior. It exhibits dense areas where galaxies exist and thinned out space where various gas clouds, energy forces, and fields exist. You would be viewing all of this movement en bloc. There would be no time and no space as such, just a lot of activity with varying degrees of interconnected densities. Carrying this example further, there also would be no time and no space outside the expansion of the universe. The universe is not expanding out into space at a particular interval of time. *Rather the mass of expanding matter itself creates not only the time but also the space into which it is expanding*[17] Consequently, time and space are relative and inherently connected to the organization and expansion of matter itself. The very process of the creation of matter in time and space is what makes the physical universe physical. To state this another way, if there were no matter in the universe there would be no space and no time and therefore no universe. This explanation picks up on and gives you a general feeling for Einstein's general theory of relativity.

Continuing on with our example of the expanding universe, when you are inside this enormous sphere from the perspective of standing on earth you experience yourself in a three dimensional reality where time is linear and space appears to be fixed. If you look

16 Fritjof Capra, *The Tao of Physics,* p. 185.

17 To state this concept another way is that the Big Bang did not explode out into space but created the space and time into which it exploded.

out from earth and view the expansion of the universe in all directions with a very large telescope you are viewing events taking place in a fourth dimensional frame from a three dimensional perspective. One characteristic which stands out when viewing the universe through your very large telescope is that the light you are seeing from distant galaxies and stars is arriving at the lens of your telescope from vastly different distances. Because light travels at a constant velocity you are simultaneously seeing the light from several different celestial bodies that represent considerable differences in time. This phenomena results in what you are witnessing is the light from the closest stars which are only hundreds of light years away to light from stars that are thousands and even millions of light years away. From this three dimensional perspective everything in this fourth dimension seems to be happening in a relative way and not in a fixed linear way. *Quantum theory seems to be telling us that maybe our reality, that is how you view things, is all in how you view your world. How you view your world becomes your world perspective and this forms your reality. Your reality isn't in the stuff that you are looking at. Your reality is conceived from how you look at stuff.* Why this is so and the implication it holds for us will be explored in just a moment.

Let's stop and think for a minute about what we have been discussing. If we view our world from a three dimensional frame we find that time is linear and space is fixed. However, we know from quantum physics that on the micro and macro level of existence the universe operates in the fourth dimension where time, space and matter are all interconnected and relative to one another and to the observer. We also know that on the quantum level and the macro level of the universe all matter is energy, that the universal construct is energy and not solid. If this is the case do we not then actually live in a fourth dimensional universe consisting of energy where time and space is relative to the organization of matter, instead of in a three dimensional world consisting of solid objects where time and space are fixed? Or do we live in both simultaneously? We will answer this paradox in just a moment, but first we need to look at one other aspect of the nature of matter.

The mass characteristic of quantum particles and therefore massive bodies in space are described in quantum theory terms by Einstein's famous equation $e=mc^2$. Energy equals the mass of a particle times the speed of light squared. You do not have to understand the formula. You only need to understand that this means that the mass of a particle of matter is really made up of pure energy. This should not come as any great surprise since we have already seen that matter is energy. Remember the mass of a particle is made up of its electrons, protons and neutrons, all of which are energy. However, remember the interrelationship of the energy of these subatomic particles form a matrix structure and thus the subatomic particles have mass. However the mass is latent energy locked up in the structure of the atom. Another way of stating this is that matter represents stored energy.

Mass can be described in terms of either matter or energy and therefore is really just two sides of the same thing. All mass holds energy and all energy combines to form mass. Accordingly, when subatomic particles exhibit the nature of particles you are looking at their matrix configuration. When they appear as quantum waves the subatomic particles exhibit their kinetic energy configuration. Therefore, they are either particles or waves depending upon how they are observed. Interestingly enough this same phenomena of the arrangements of quantum particles and quantum waves in forming subatomic particles is exhibited by the processes of the brain and mind. This would seem to make sense if you think about it for a minute since the subconscious mind at the quantum level of organization of mental states and patterns tracks with quantum theory as does all phenomena throughout nature at their quantum level of activity. Exactly how it does this is discussed later.

Relativity Theory and the Human Mind

The result of all this is that at the quantum level and at the macro level of the universe all matter consists of energy. Time and space are relative and are not fixed nor do they exist outside of us in fixed three dimensional frames. However, we live in a world where the speed of light going at 186,000 miles per second is not really a factor in our daily life. Space always appears through our five senses to be fixed and certainly not relative to any other thing in our environment. When Einstein proposed his special theory of relativity back in 1905, the Cartesian view of the world as consisting of a fixed three dimensional objective world made up of solid matter existing in fixed time and space prevailed. As I said, this view of our world is referred to as the zone of the middle dimension. Even today some ninety years after Einstein formulated his theories on relativity, insofar as we are concerned with everyday life, the zone of the middle dimension is by far almost a universal view of our world. Both time and space appear to be outside of us and largely outside of our control. The past is history and appears to be behind us and no longer available to us as a sequence of time which could be accessed and used in any beneficial way. Similarly the future appears from our five senses to be out there in front of us some place. While we can certainly plan for the future we cannot in any direct way through the process of our mind alone manipulate the future. Consequently, we are only left to allow for it to unfold as it always has in accordance with linear time within fixed three dimensional space.

By now you must be wondering that while quantum and relativity theory can be very interesting at times, what does all of this have to do with the internal processes of the mind, if anything at all? Well, remember the paradox I spoke about just a moment ago.You will recall we view our world from a three dimensional frame where we find that time is linear and space is fixed. However, we know from quantum physics on the micro and macro level of existence the universe operates in the fourth dimension where time, space and matter are all interconnected and relative to one another and to the observer. We also know that on the quantum level and the macro level of the universe all matter is energy. The universal construct is energy and not solid masses. If this is the case do we not then actually live in a fourth dimensional universe consisting of energy and where time and space is relative. If you stop and think about it for a minute it just seems logical that if the entire universe of which we are a part is energy and made up of relative time and space and we are in the middle of all of this, doesn't it just stands to reason that maybe we also exist in a fourth dimensional world? If we do exist in a fourth dimensional world how is it that we experience it as a fixed three dimensional reality? Is there really a middle dimension apart from the rest of the universe? If there is a three dimensional world does it consist of solid objects where time and space are fixed. Or do we live in both dimensions simultaneously?

Welcome to the Fourth Dimension!

Let's look at what is going on here for just a minute. Physicists themselves have a hard time visualizing what the fourth dimension really looks like when viewed from our normal three dimensional world. The reason for this difficulty is that the sensory stimuli we receive and interpret as our world reality gives us a sense of a fixed three dimensional world of outer objects which exist in fixed space within linear time frames. Since we receive and interpret external stimuli in this three dimensional fashion we then translate this experience into expressions of rational logical thought in order to operate effectively in this three dimensional world. Rational logical thought is simply a cognitive system of the human mind which evolved for the purpose of understanding and communicating our experiences based on this personal three dimensional frame of reference. It is then this sensory stimuli expressed by our thoughts and feelings that defines our reality. Thoughts and feelings arising within the cognitive structure of the mind are what also give us an experience of a separate self expressed as an ego personality existing separate and apart from everyone else and separate and apart from our world. It is thoughts and feelings then which gave rise to a variety of actions and behavior which in turn have allowed us historically to take effective control over, and ultimately dominate, our world environment.

This Cartesian mechanical view of life existing in an objective world has unquestionably bought mankind unbelievable advances in the last three hundred years in economic and social order, medicine, technology, fine arts, computer science, navigation, aviation, communications, development of natural resources, harnessing the atom for peaceful purposes, exploration of space, science, mathematics, engineering, and a whole host of other advances in science, arts and the humanities. At the same time such advances by mankind have also been brought at the price of over population, poverty, disease, crime, lack of social cohesiveness, wide distortions in the distribution of wealth and natural resources, child abuse, inner city slums, teenage pregnancy, breakup of the family unit, increased violence, deterioration in individual and social morals and values, regional and global warfare, and the overall deterioration of the environment which sustains and gives life to each one of us.

These seemingly insurmountable problems mankind is currently faced with however, are not a result of the Cartesian mechanical view of the world. For the Cartesian view of the world is simply symptomatic of a much more fundamental limitation in the very nature of man. *The development of the Cartesian view of the world is simply a reflection of our innermost processes of the mind of mankind. Change the inner processes of the mind and you change mankind's perspective of himself and his world reality.* How you view your world becomes your world perspective and this forms your reality. Your reality, that is, how you view things outside of yourself, is all in how you view your world. It is the cognitive structure of the mind that organizes for us a perception of our world as being separate and apart from us and not the stuff out there which makes up our outer realities in life.

Nature, unfortunately, has not endowed us with sensory receptors, nor with a process within conscious logical thought, which readily gives us a direct perceptual experience of the world as it actually exists in the fourth dimension. The reason for this is that the way the world is experienced by us is the result of how we perceive that world. Our sensory receptors not only show us what our world looks like but also confirms for us what we think that world should look like. What we mean by this will become very apparent to you in just a moment. The reason why mankind has arranged his reality as existing in conformity to the Cartesian mechanical view of the world of fixed three dimensional space in fixed linear time is that the human brain and mind have physically evolved to conceive its world in this manner. We are not only hard wired but our software programs organizing our thought processes have evolved so that we see and act in a three dimensional world. However, we are increasingly beginning to see that this Cartesian mechanical view of the world which gave rise to Western civilization, while proven beneficial historically, is now shown to be incorrect with the advent of quantum physics. Further, and of even greater importance, is the fact that we are increasingly feeling its downside negative effects which are fueling a world social order that is no longer working effectively for the betterment of all mankind.

However, in order to stem the downside negative effects of this world view we do not need to modify the components of our external reality created by such a mental organizing system. Rather, we need to address and change the internal processes creating this type of social system. The problem, and therefore the solution, does not lie within the social institutions themselves. Rather, the solution lies within the very processes of the human brain and mind. It is the human brain and mind that conceived of, and brought into reality, those social institutions. Before attempting to describe necessary changes in the internal processes of the human brain and mind in order to create a better life for ourself, and a better social system for all mankind, we need to take a historical view of how the human brain and mind first organized thoughts, feelings and actions in creating both the opportunities and the problems we are faced with today. From this perspective, hopefully, we will be able to develop entirely new ways of seeing and thinking about our life, ways in which we eliminate the downside negative effects of the old Cartesian mechanical view of our world. We will explore new ways of thinking where we conceive and bring into reality a much better world for each one of us.

An Overview of the Human Brain and Mind

We start with an overview of the organization of the human brain and mind. First of all the primary task of the human mind/brain system is to create experiences of inner and outer realities in life. It is extremely good at its task and can create an endless variety of inner and outer realities. In fact, it is the mind's facility for creating inner and outer realities that first led Rene Descartes to describe the reality of our world in Cartesian mechanical terms. We need to break the entire system down into three major functions in order to understand how the mind/brain can effectively create realities. However, keep in mind the entire mind/brain system operates as one organic whole and not as a series of three separate functions. For purposes of analysis the first functional level consists of the neurological processes of the human brain. The human brain subserves the functions of the subconscious mind which subserves the higher cognitive structures of the conscious mind. The brain's neurological processes take the incoming sensory stimuli from our external world and convert them into electrochemical impulses that form neurological networks and neuro pathways within the brain. These neuro networks and pathways form "basins of attraction" and "attractor pathways" that simply represent electrochemical patterns recognizable by the human mind. The subconscious mind organizes its analog patterns based on the similarities of the underlying energy and matrix configurations of the brain's neurological patterns which subserve the subconscious mind.

The subconscious mind operates as an analog processing system in organizing subconscious mental patterns based on the electrochemical networks created by the brain. When we say that the subconscious mind operates as an analog processing system we are simply referring to the differences between a binary processing system and an analog processing system. A binary processing system is a type of digital processing computer that most of us operate as personal home computers. Binary systems consist of numerous on-off switches and are serial in nature. They perform one function at a time. An analog system creates patterns of data or information which combine to continuously form a wide variety of patterns that make up the interior landscape of the system. Analog systems do not just process one discrete computation at a time. Analog coding differs from digital coding in that it forms interconnections and relationships between and among the various patterns formed within the system. Digital processing works in accordance with fixed programmable rules. It doesn't make exceptions to the rules. It there are to be exceptions they have to be programmed into the computer at the factory long before it is booted up.

In order to give you an even better idea of the differences between a digital system and an analog system consider the difference between a digital watch and an analog watch. A digital watch displays specific numerical digits that give you the specific time of day. A watch with both an hour hand and a minute hand is a type of analog system. It not only displays the specific time of day but it also provides you with a reference point for correlating present time with past time and future time. *The subconscious mind does not think logically, it does not organize data analytically or digitally, it organizes data in analog fashion based on similarities between patterns of information.* However, the higher order cognitive structures of conscious mind organize thought logically, that is, analytically. The conscious cognitive mind organizes information based on logical and/or causal relationships. The conscious mind utilizes deductive and inductive rules of logic in processing the language of thought into expressions of our inner and outer realities.[18]

All of this may sound a little strange and hard to grasp when you first encounter it. However, this brief description of the three functions of the human mind/brain provides the background for the following summary description of how we actually create realities in time and space. Exactly how the mind/brain creates realities will be explained in considerable detail later on. For now however let's only look at how the mind/brain system has historically created thoughts which form our belief systems framed up by the old Cartesian mechanical world view. In developing this explanation we will look at how the conscious mind utilizes the mind's cognitive structure in order to interpret the time and space patterns created by the sensory stimuli received from our outside world.[19] The cognitive structure aligns the time and space of our reality within the fixed world view of the old Cartesian model. We will also explore new ways of thinking in order to shift realities away from the old Cartesian model and align our time and space perspectives on life and our corresponding thought patterns within a new quantum theory model. This new perspective will allow you to view all of life from a much wider perspective.

As indicated earlier, linear time and fixed space are not inherent factors in the organization of energy into matter. Time and space are only relative to the sequence and placement of matter in a material universe. *In similar fashion at the quantum level of the human brain and mind, time and space are not inherent factors in the organization of thought leading to the creation of our inner and outer realities in life. Time and space are only relevant to the placement of realities in our physical three dimensional world.* As you come to experience how thought molds your perceptions of reality through the processes of your mind, you will be able to take much greater control over the time, space and the outer reality of your world. You will find

18 The subconscious mind, while it processes stimuli on an analog basis, also is capable of translating cognitive thought into analog relationships for purposes of creating inner and outer realities. In doing so, however, the subconscious mind is constrained by the analytical logic of the cognitive thought processes of the conscious mind. How the conscious and subconscious mind interrelate in creating inner and outer realities and how the constraints of the conscious mind affect the processes of the subconscious mind is explained in Chapters 9 and 10.

19 The cognitive structure is the mechanism of the mind which organizes all of our thoughts and gives us a sense of three dimensional reality. Cognitive restructuring of the mind is discussed in Chapter 10.

that if you are not satisfied with your present realities you can effectively change them to form more life sustaining and life enhancing realities. Exactly how we go about doing all of this is explained in detail in the following chapters. But first let's explore historically how we got to where we are right now.

Reality, Time and Space - The World of Ricky Mindful

To best illustrate how the mind/brain creates realities within time and space let's look at one day in the life of Ricky Mindful. Ricky turned 53 yesterday. He is a Regional Vice President of Sales for Primetex Communications, Inc., a large manufacturer of electronic and communication equipment for the office and home. While he has been very successful at what he does, recently his regional sales have been off for the last three quarters in a row. Because of this, Mac, the Executive V.P. Sales, called a special sales meeting for Monday morning to review the region's sales in general and Ricky's performance in particular. Behind his back Mac is derisively referred to as "Mac the Knife" by his sales people. Ricky knew Mac didn't get this reputation by being Mr. Nice Guy. However, Ricky put all of this out of mind last night at his birthday party. It was the best time he had in a long time. With all of his friends there he was able to really unwind. Betty Bountiful, his current squeeze, looked prettier than ever in her low cut red party dress. He drank more than he should, but he figured what the heck it was his birthday, so he just partied down until 2:00 a. m. in the morning.

He has been dating Betty, off and on, for over two years and she has increasingly been nagging at him to make a commitment. However, Ricky doesn't want to get tied down. Further if he did get married he would have a responsibility to Betty's teenage daughter, Louanne. While the kid was O.K. and her mother wasn't doing all that bad financially, he might still have the added responsibility for the kid's college education. Ricky thought to himself that under these circumstances marriage was just out of the question. Besides his folk's marriage was a total disaster. His thoughts drifted back to his early childhood memories. His Dad, Kenny, was a no good bum. His mother, Stella, was seldom home. She was always out late at night partying with her friends. He remembered those awful nights when he came home from school only to find a scribbled note on the frig telling him that his supper of either peanut butter sandwiches was in the cupboard or his macaroni TV dinner was in the oven. He felt like he was both Mom and Dad to his younger brother, Tommy, and he resented it. It's funny how time flew by and he never got the right opportunity to tell his folks exactly how he felt and now they have both passed on. Marriage, he thought to himself, "Who needs it." Beside he was happy following his own advice, "why pay for the cow when the milk is free."

With these thoughts on his mind Ricky swung both feet out of bed and onto the cool hard floor as he sat upright on his side of the bed. He looked at the digital clock on the nightstand. The large symbols on the face of the clock indicated that the time was 10:36 a.m. and the date displayed above the time in smaller symbols showed Sunday, November 23, 1997. "Damn," he thought to himself, he overslept. He and Betty were planning to attend church and the Thanksgiving dinner afterwards. He was never much for church. It was more Betty's idea than his. It's not that he didn't believe in God or anything like that. Its just that, except for the occasional business contacts he picked up from time to time, church just didn't have any meaning for him. So missing church didn't really bother him all that much. Besides he knew tomorrow's meeting with Mac was going to be a tough one and he better strategize this afternoon on how to best prepare himself for that meeting.

He could smell the coffee brewing from the kitchen. "Good," he thought to himself, Betty is already up and fixing breakfast. He could certainly use some coffee, hot and black, since he had a moderate hangover. His thoughts drifted back to last night when he and Betty got back to his townhouse after the party. She looked sexier than ever in her low cut party dress and didn't once bring up marriage or commitment. The memories of last night all came back. He could only describe them as sweet and delicious. He looked once again at the digital clock and the time was now 10:42 a.m. He had been sitting on the edge of the bed for over 6 minutes. "Damn," he thought to himself again, time really does go fast, as he eased himself off the bed and headed toward the bathroom.

This is the reality in the time and space of Sunday morning, November 23, 1997, from the perspective of Ricky Mindful. Now let's look at the same events of Sunday morning once again from the perspective of Ricky's brain and mind. As Ricky swung his feet to the hardwood floor and sat up on the edge of the bed the first thing that he noticed, besides his hangover, was the display on the digital clock indicating that the time was 10:36 a.m. and the date, Sunday, November 23 ,1997. This visual sensory input immediately oriented him in the space of his bedroom and in the time, 10:36 a.m. on Sunday morning. You would think to yourself that this is very self evident and really doesn't even need to be addressed. However, you would only think this way because we take time and space of our world for granted without ever thinking about it. The general day to day references for time and space measured by clocks and calenders have become an unconscious thing. We only become aware of time and space as they affect our lives within general frames of reference. We clearly understand that abstractly time is measured by seconds, minutes, hours, days, weeks, months, years, and decades. Therefore we do not give any conscious thought to these overall frames of reference except with respect to keeping track of the linear flow of time within such frames of reference. That's why we have clocks and calenders so we won't be late for appointments, at least not important ones.

Ricky's orientation in time and space this Sunday morning is primarily an unconscious one. However, his brain and mind actively coordinate Sunday's present time and space within the existing mental parameters of his brain and mind in order to give Ricky a sense of linear time in fixed three dimensional space. The existing mental program parameters for time and space are a historical one, one that Ricky grew up with all his life. In fact, the existing mental parameters of time and space are not only a historical one but are also a primordial one. Since mankind became self aware and started hanging around the planet communicating with one another he invented sun dials and parchment calendars to keep track of time. He also invented the wheel in order to move around in three dimensional space. So Ricky's brain and mind for a very long time had internal mental program parameters in order to keep track of the increments of time and space of his life. While he takes time and space for granted his mind and brain certainly do not!

We will briefly address how it is that Ricky's mind and brain keep track of time and space and then we will address whether or not there might be more effective ways of internally calibrating time and space. The visual stimuli of the digital clock giving Ricky the exact time and date were registered on the retinas of his eyes. The optical nerves transmitted these visual signals to his brain. More precisely, these signals were transmitted to the neural networks located in the occipital lobe area of his brain which moderates spatial and temporal dimensions in life. It is this area of the brain where these visual signals are converted by the brain's neuron cells into electrochemical impulses called neurotransmitters which form a communication link between basins of attraction made up of hundreds of thousands of neuron cells.

The electrochemical impulses of the neurotransmitters consist of atoms and molecules which, you will recall, are composed of electrical charges and matrix configurations. These atoms and molecules form the substances of the neurotransmitters which have the capacity to communicate data between cells. They can communicate data for the reason that the atoms and molecules of the neurotransmitters arrange themselves in an endless variety of substances. The configuration of these substances are recognized by the neuron cells. Based on these substances neuron cells are able to organize themselves within corresponding basins of attraction. The combined values of the basins of attraction organize interior time and space patterns within the human brain and mind by arranging themselves to adjust their current space and time patterns to assimilate new space and time coordinates coming into the system. This operation will be made much clearer to you later on in Chapter 5. In this fashion the brain keeps track of the current time and space of Ricky's activities. It does so by creating the basins' patterns which not only keeps track of specific times of day but also provides Ricky with a reference point for correlating present time in relation to past time and in relation to future time in much the same fashion as our analog watch. All of this activity, of course, is unknown to Ricky. He just knows that it is now 10:36 a.m. Sunday morning and Betty is calling him to join her for breakfast.

Just a little side note, the over consumption of alcohol last night has markedly slowed down Ricky's neurotransmitter processes. His body is working overtime to convert its alcohol content into glucose. It then converts this substance in combination with proteins along with amino acids and triglycerides stored in the body, into complex molecules which can be used by the brain in manufacturing neurotransmitters. As we have seen from our little biochemistry lesson above the brain can organize the chemical matrices of its molecules into different substances used to create new neurological substanes. However in Ricky's case his bodily systems were somewhat overwhelmed by their high blood alcohol content and, consequently, some of the excess alcohol seeped through his blood brain barrier and caused some distortions and deterioration in his neurotransmitters' level of performance this Sunday morning. The black coffee gave him an energy jolt to the systems and in this respect it helped for the short haul. However for the long run black coffee doesn't add any nutritional value to Ricky's systems.

As indicated earlier, the neurological processes of the brain subserves the mind. More precisely the brain subserves the subconscious mind. What is meant by subserving the subconscious mind is that in this case the brain's neurological space and time basins of attraction provide the subconscious mind with neurological pattern configurations. The subconscious mind uses these pattern configurations in order to make a transition between conscious cognitive thought and sensory awareness of our experiences of inner and outer realities in life. In simple terms this means that Ricky's subconscious mind translates the neurological time and space pattern configurations of his brain into a conscious awareness of the exact time and place when he awoke that Sunday morning.[20] In addition to providing Ricky with a transition between his neurological processes and his conscious cognitive thoughts, Ricky's subconscious mind also serves to arrange Ricky's conscious thoughts representing his wants and desires in life into patterns of awareness of people, places, and things which match up with his wants and desires. In this case he knows he has to prepare for that fateful meeting with Mac the Knife. It is this subconscious process of matching up desires with things and events in life which creates inner and outer realities.

Let's now briefly look at how the space and time pattern configurations of the brain's basins of attraction provide the subconscious mind with the data it needs to interpret space and time. The brain's neurological space and time basins are configured in such a way that they can be interpreted by the subconscious mind. Remember, the brain's basins of attraction are configured by the composition of the brain's neurotransmitters. These neurological pattern configurations provide the subconscious mind with the data it needs to be able to interpret time and space in terms of the energy and image patterns of the subconscious mind's analog system. In turn, it is these space and time energy and image patterns of the subconscious mind's analog system that can be translated by the conscious mind into expressions of numerical symbols. These numerical symbols can be consciously understood to represent specific time intervals existing within each day. Again in simple terms, what we are saying here is that Ricky's brain provides his subconscious mind with the data it needs to form patterns of images representing space and time. It is these space and time images of Ricky's subconscious mind that Ricky's conscious mind interprets to represent segments of times and places in Ricky's day. Details of how all this takes place in Ricky's mind will be explained later.

However the importance of all this at this juncture is that what is making up the segments of times and places experienced by Ricky

---

20 One technical point. Ricky's brain doesn't really think conscious thoughts. While it's O.K. to think in these terms in your daily life, in order to properly understand the processes of the mind and brain, it is the conscious mind that thinks thoughts. The brain only translates sensory stimuli into the electrochemical substances of its neurotransmitters. The subconscious mind also doesn't think conscious thoughts. It only organizes cognitive thought into pattern relationships in analog fashion. It is conscious mind that really is organizing our conscious thought and beliefs based on these subconscious mental patterns and relationships.

are not fixed linear time increments taking place in fixed three dimensional space existing outside of Ricky and existing separate and apart from Ricky.'s brain and mind. I am not saying by this that we do not physically live in fixed three dimensional space. However, our perceptions of that fixed three dimensional space is organized within the neurological processes of our brain and not as a result of anything outside of ourselves. True, stimuli from that outside world coming into the system are arranged to give us a perception of a fixed three dimensional world. However, we don't have to organize it in just this fashion alone as we shall see shortly.

We can also consciously organize time and space in other ways in order that we may create expanded realities in life. Fixed linear time and three dimensional space are created and thereafter are continually being organized initially at the neurological level of Ricky's basins of attraction and ultimately at the level of Ricky's subconscious and conscious mind. Exactly how time and space are organized by Ricky's brain is explained in detail in Chapter 5. How time and space are interpreted by his subconscious and conscious mind is explained in Chapter 6. Ricky, and for that matter all mankind, has, from time immemorial, organized the inner interpretations of space and time in the old Cartesian fashion. Both individually and collectively, we are totally unaware that it is our mental processes alone that are organizing our space and time and not factors residing in our outside world. In fact, if the truth be told, we actually live, and always have, in spacetime time of the fourth dimension and have never been consciously aware of it!

Creating Realities in Spacetime of the Fourth Dimension

If this is the case, you have to be thinking, why don't we experience space and time in that fashion? Your next question might be that if we did experience it what would it look like, be like, and feel like? Another question would be what are the differences between spacetime time of the fourth dimension and linear time of the third dimension? What then does the en bloc nature of spacetime really look like? The answer is simple. It is exactly as it exists in quantum spacetime and in universe spacetime. By nature it is no different on the physical plane of existence than it is in the quantum level and at the universe level of the fourth dimension. The reason for this is that spacetime is an inherent aspect in the formation of matter. It does not exist as a separate and discrete aspect of the universe exhibiting properties all its own. It only exists in relation to the process of formation of matter within the universe. It does not have a taste, smell, touch, nor are you able to see, hear or feel it, and therein lies the problem. *You just can't detect it with your five sensory receptors !*

This, of course, has been a major problem for us long before we first adopted the Cartesian way of thinking some three centuries ago. As we have seen our sensory receptors do not detect the en bloc nature of spacetime from the perspective of our three dimensional world. This fact alone still continues to baffle even quantum physicists in trying to understand and explain fourth dimensional reality in three dimensional terms. It's like seeing your two dimensional shadow in a three dimensional world. Your shadow doesn't consist as something solid and fixed but its existence depends on coordinates, the angle of the sun and your location.[21] Coordinates also have a lot to do with the organization of realities in spacetime in our three dimensional world. You will see how this is so in just a minute. Consequently, there is no conscious physical way we can personally experience spacetime as a phenomena of our outer physical world.

However this doesn't mean that we do not exist in spacetime nor that we can't take advantage of it in creating more beneficial realities for ourselves. The fact of the matter is that we can experience the en bloc nature of spacetime indirectly at the level of our subconscious mind. Once we indirectly experience it at the level of our subconscious mind we can then integrate its benefits on the conscious level of our logical mind. If we were to organize our thoughts in the spacetime of the fourth dimension we would be able to create expanded realities for ourselves and for society as a whole. Consciously orienting ourselves to the benefits of spacetime is referred to as "Conscious Analog Thinking".™ Conscious analog thinking is explained in Chapter 3 and in Appendix One.

Our subconscious mind has the ability to organize incoming data in similar fashion as energy and matter are organized on the quantum level and the universe level in spacetime of the fourth dimension. You will recall from our little quantum physics and relativity lesson earlier that at the quantum level subatomic particles organize themselves into an endless variety of matter. They are able to arrange themselves into the elements making up everything in our world including you and me. You will recall they are able to form such arrangements because of the negative and positive charge of the electrical energy and matrix configurations of the subatomic particles. Subatomic particles form atoms and atoms attract and repulse one another to form endless combinations of matter. They organize themselves to form matter notwithstanding any consideration of sequential time frames. They are not constrained in any way by cause and effect relationships of the type found within fixed three dimensional space. Similarly, on the universe level the cosmos organizes the

21 Fritz Capra, *The Tao of Physics*, p. 170.

mass of subatomic particles into a variety of celestial bodies, gasses, energy fields, and charged particles. The constraints of a three dimensional world are totally irrelevant in the quantum and universe process of converting energy into matter.[22]

On the level of the subconscious mind the same organizing principles of the quantum level and the universe level apply in all respects. Remember above we indicated that the subconscious mind functions in an analog fashion and not in digital or in analytical fashion. In this respect the subconscious mind organizes thoughts and emotions in similar fashion as matter is organized on the quantum level and on the universe level. The subconscious mind organizes incoming data by the similarities in relationships to the existing information already stored in its systems. These similarities in relationships are arranged according to the values of the energy and matrix configurations of the incoming stimuli in much the same way as the energy and matrix configurations are organized on the quantum level and on the universe level. These energy and matrix configurations are what attract and repel the various neuron cells subserving the analog patterns of the subconscious mind. It is these patterns, you will recall, that are interpreted by the conscious mind in organizing the formation of our realities.

On the subconscious level sequential frames of linear time and fixed three dimensional space also are not inherent aspects in the process of organizing realities. *They are however constraints imposed on the processes of the subconscious mind by the conscious mind.* Therefore, linear time and fixed space become major constraints to creating expanded realities in life. For linear time creates for us a sense of a past, present and a future. This organization of time and space acts as a constraint in creating the realities in life for the past is always behind us. There is nothing we can do to change the past. The future is still ahead of us and while we can plan for the future, it will, for the most part, unfold in the natural course of events. That leaves us with directing the realities of our life from present time only. In this respect linear time acts as a constraint on our ability to change the past or future through the mental processes of our mind alone. However, by intentionally adopting the en bloc nature of spacetime at the level of the subconscious mind to represent cycles of time in our life we are able to expand our ability to

22 Further, neither in the quantum process nor in the universe process of converting energy into matter is anything lost or created. The quantum process and the universe process do not destroy matter or create matter. They convert energy to matter and matter to energy to form an endless variety of material forms. However, nothing is lost in the process. Energy and matter are merely recycled.

create for ourselves expanded realities in life. We do so by accessing our subconscious analog patterns for time and space and change a negative past to a positive past and in doing so we are able to create future physical realities more to our liking. How we do all of this is explained in detail in the following chapters.

First, however, to get a sense of the metaphor of spacetime on the level of the subconscious mind you might want to view things as happening in cycles of time where realities are both created and dissolved, cycles which are interrelated with no distinct past, present or future attached to them. *Another way of expressing this concept is that on the level of the subconscious mind there is, in fact, no time, at least no linear sequential time as we consciously think of it.* Time, rather than being a separate and distinct element existing throughout the universe, is really an aspect inherent in the reality creation process itself. Time then is a part of the process of creating realities at the level of the subconscious mind. You can begin to get a better idea of spacetime by viewing it as being tied into cycles which organize and dissolve realities in life at the level of the subconscious mind. While the subconscious mind creates realities outside of linear time and fixed space it sorts for their conscious expression in linear time and fixed space of our outer physical world.[23] To graphically illustrate what we are talking about here refer to Chart 2-1 at page 39 entitled, *Illustration of Linear Time and Spacetime.* In the linear time chart, you, represented as "self", are in the middle of your time line. Time runs out from your past at the left side of the page to the present where you are in present time and space in the middle of the page. Time continues to run out toward your right where the future lies ahead of you. When viewed this way you only have control over the things happening in the small slice of time in the present, no control over things that happened in the past and not as much control as you would like over things that will happen in your future. Whereas the chart depicting the en bloc nature of spacetime of the fourth dimension of the subconscious mind shows time as a metaphor flowing in a circle. It flows from the present into the future and curves back to the

23 However, our outer realities are still experienced in linear time and in fixed three dimensional space. We need only change our conscious awareness of how it is that the subconscious mind sorts to create inner and outer realities in life. To think that we can use en bloc time to organize realities out of thin air is a misunderstanding of en bloc spacetime. The actual outer physical manifestations in life must therefore still conform to physical laws governing all of nature as we understand them. However, with respect to physical laws governing the organization of realities and not just the operation of realities spacetime is the correct aspect in the creation process of subconscious mind and not fixed linear time and fixed space.

CHART 2-1

Illustration of Linear Time and En Bloc Time

Linear Time in Three Dimensional Space

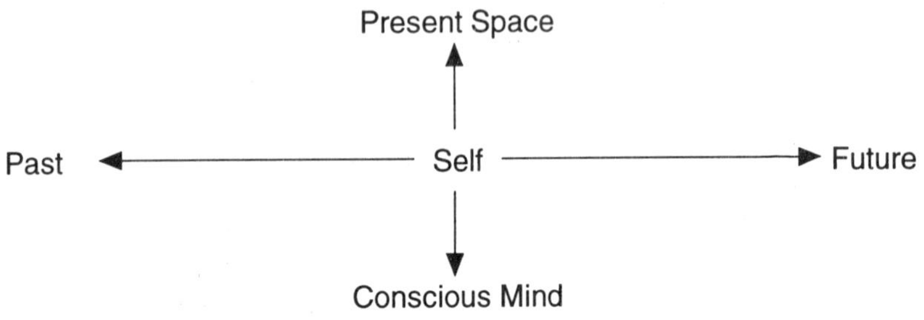

Linear Time Depicted in the Form of a Straight Line Running Out of the Past Toward the Future

En Bloc Time of the Fourth Dimension

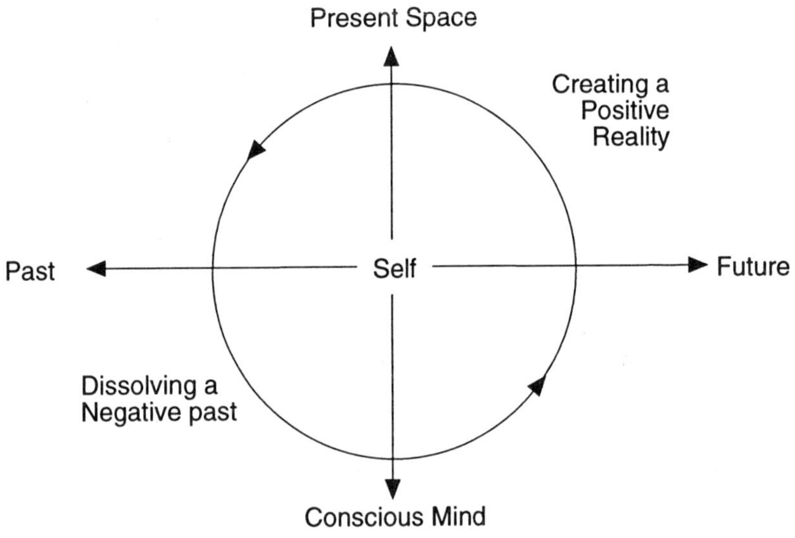

En Bloc Time Depicted as a Circular Loop Encompassing Both the Past and Future

past and swings again into the future, and on and on, creating and dissolving realities at the level of the subconscious mind. In this respect time is viewed as process. You are able to access this circle of time at any given point on the perimeter of the circle and bring about whatever changes you desire in the process of creating inner realities at the level of the subconscious mind. In this manner changing a negative past creates changes for a positive future. On the other hand, by not being able to access the past on the linear time line you are not able to change a negative past. In this case, the patterns of a negative past will continue to negatively influence future events in your life. For an example of what we are talking about here refer to Charts 2-2 and 2-3 beginning at page 41. Chart 2-2 simply places you in the middle of this time circle. K/V/A stands for kinesthetic, visual and auditory sensory stimuli. You will be hearing a lot more about these three sensory modalities throughout the remainder of this book[24]. Briefly, all three are forms of energy, the visual energy of lights waves, the sound energy of sound currents, and the pressure and feelings of the kinetic energy of motion. Pure energy will be depicted by the symbol E as in $e=mc^2$. Chart 2-2 simply illustrates that your inner and outer realities are organized at the subconscious level as a result of the action of the visual, auditory and kinesthetic stimuli of the feed forward and feed back cybernetic loops in life. Chart 2-3 illustrates the same effect of the cybernetic loops creating realities in life except that in this case the effects of the past K/V/A stimuli affect your future realities in life.

<u>How Time Relates to the Activities of Our Life</u>

To best illustrate what we are talking about we will again revisit Ricky Mindful. You will recall we left Ricky at the breakfast table drinking black coffee to relieve the effects of the hangover from last night's party. He was enjoying his danish, scrambled eggs and bacon, while Betty was explaining to him how she too certainly enjoyed the party and most of all the two of them being together. It made their evening complete. They just seem right for each other. Her artistic talents, which she was able to capitalize on in her interior design business, seem to compliment perfectly his sharp and sometimes aggressive business skills in retail sales and marketing. She felt that last night and the many other wonderful times they recently had together left little question that they clearly loved one another very deeply. She continued on in a kind and loving way that in order for their relationship to grow and mature they should start

24 There are five major sensory receptors. The other two are olfactory, the sense of smell, and gustatory, the sense of taste. While all five are important in life the first three illustrate the points addressed throughout this book.

CHART 2-2

The Effects K/V/A has on Creating the Images of Inner and Outer Reality In Present Time

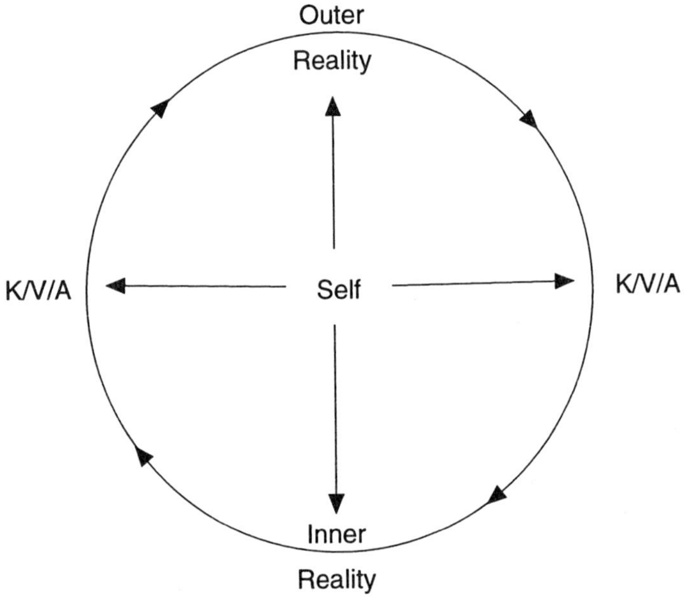

Simplified Reality Circle

CHART 2-3

How the K/V/A Images of Past Realities Affects the Formation of the K/V/A Images of Future Realities

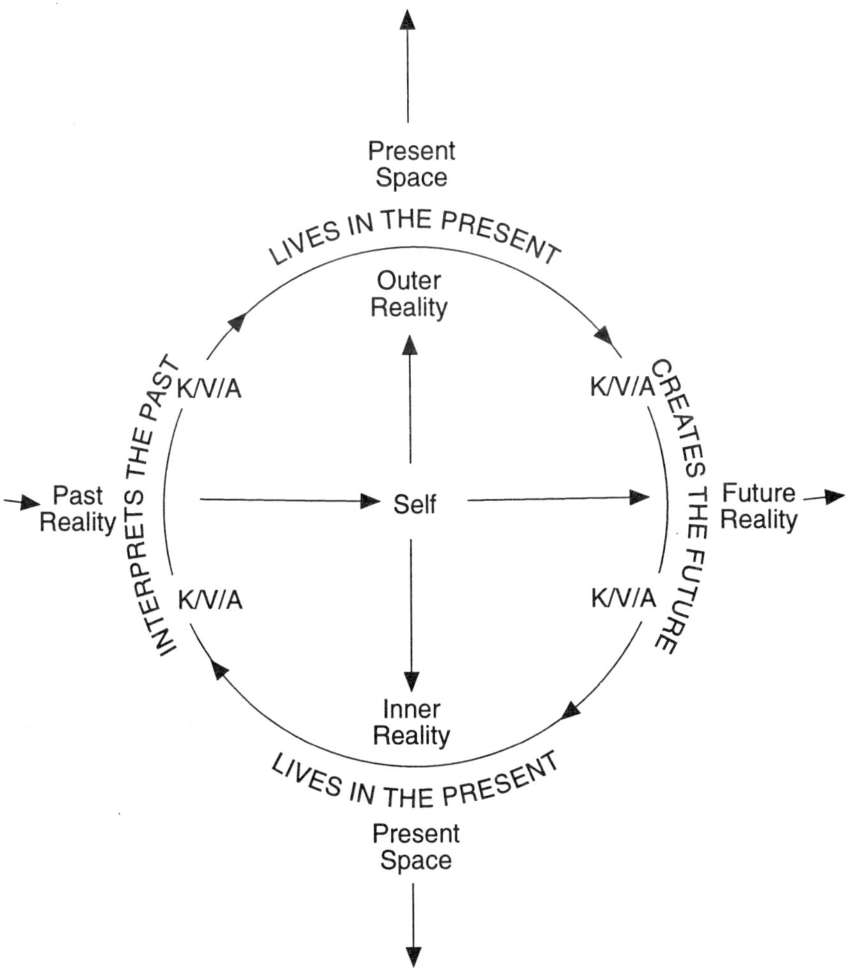

Simplified Reality Circle

thinking about making a more serious commitment. Neither one was getting any younger and now would be the right time to seriously think about taking the next step. At that moment the phone rang and Betty answered it. It was Louanne. She wanted to know if Betty was going to pick her up for her 2:00 p.m. soccer game.

As Betty talked to her daughter Ricky thought to himself, "You know, she is right". She did seem to have all the right qualities he was looking for in a women. She was bright, talented and very appealing in more ways then one. She certainly had all of the right curves and knew how to satisfy her man. Her daughter wasn't really all that bad either. In a secret way he even admired the kid. She was a survivor like her mother and could certainly hold her own in any argument with Ricky, that was for sure. Why then couldn't he just make a commitment. Betty wasn't the type of woman who was going to wait around forever. He knew she had many other admirers and didn't lack for male companionship if she wanted it. She certainly learned how to assert herself ever since she divorced Fred, her alcoholic husband. Enabler or co-dependency just wasn't in her vocabulary.

. . . .and yet he felt uneasy for some reason. He just couldn't quite consciously put his finger on it. He poured himself another cup of coffee and reached for another danish as his mind drifted off to those early years when he was left to care for Tommy. He remembered how his parent's marriage was a total disaster. He would never forget the night before Christmas many years ago when his mother never came home. Tommy, his younger brother, just couldn't understand. He kept crying, "she just doesn't care about us any more". Ricky, still a young boy at the time, tried to reassure Tommy that she really did care for them every much. She just had a lot on her mind.

He said to Tommy, "Just watch and see, she'll come home pretty soon with presents and Christmas dinner for the three of us and everything will be O.K. just like it used to be."

Secretly Ricky wondered whether Tommy was right after all. She really didn't care about them at all. If She did She sure didn't say it or show it very often. But he had to act grown up and not reveal his true feelings to Tommy. He held Tommy in his arms as they both fell asleep on the couch that Christmas Eve.

To hide their disappointment the next morning they tried to celebrate Christmas by making pancakes and Kool-Ade and they played together during the day. At 4:00 p.m. She showed up smelling of alcohol, stale cigarette smoke and needing a bath. She gave then each an unwrapped Christmas present and a turkey T.V. dinner. It wasn't the small value of the gifts that bothered Ricky so much as the fact they weren't even gift wrapped! Most of the fun was in unwrapping the gifts on Christmas morning. His gift was a model airplane kit. Tommy really loved model airplanes so Ricky gave his gift to Tommy and wished him a Merry Christmas. His mother in the meantime went upstairs, took a hot shower and went immediately to bed. . . . .

Betty was still on the telephone with Louanne. Ricky continued daydreaming. Every Christmas since Ricky and Tommy grew up they made it a point to get together and really celebrated the spirit of Christmas the way it was meant to be celebrated. A large Christmas dinner, gifts to be unwrapped under the tree, and a lot of Christmas cheer to go around. Tommy eventually got a master's degree in counseling and went on to get a PhD. degree in psychology. He had built up a very successful counseling practice in Detroit, Michigan. Ricky laughed silently to himself. He would kid Tommy, from time to time, that he really only got to be a psychotherapist in order to psychoanalyze himself and figure out what really went so terribly wrong in their early childhood years. When they were small Tommy often felt that they just weren't good enough for their parents to love them as other parents loved their kids. It seemed that thereafter, sometime during Christmas week, Tommy and Ricky would have too much Christmas cheer and would continue their late night reminiscing about those early childhood years trying to figure out what went so terribly wrong. Whatever they said, it always seemed to end up the exact same way each year like a cheap broken record. They never really got any further insight into what went wrong. Ricky would end up being very angry at his parents. Tommy would end up feeling a whole lot of self pity for himself and for Ricky. While their emotions appeared to be quite the opposite, somehow they seemed to reflect two sides of the exact same thing. While sipping his coffee Ricky rationally, if not logically, concluded that he got burned badly once before and he wasn't about to set himself up to lose all over again by marrying Betty.

He could hear Betty arguing with Louanne, something about dating and partying after the soccer game. "Teenagers are growing up all too fast these days," he thought to himself as his thoughts again drifted back to his kid brother. There was some truth to what he would kid Tommy about. There wasn't a therapy Tommy didn't like. Maybe that is what made him so successful. He certainly could empathize with his client's problems in his substance abuse and family counseling practice involving mostly AA groups, cognitive and behavioral therapies. Tommy would always call Ricky whenever there was something new in his life, for Ricky was like a father to him, the only father he really ever knew. He remembered how Tommy would excitedly call him late at night to tell him all about his latest therapy. Tommy must have seen and done it all. Ricky remembers first there was group therapy, primal scream, gestalt, abbreaction, and past lives regression. Next came Cybernetics, Neurolinguistic Programming, and Bioenergetics, followed by Zen meditation, TM and sweat lodges. Next there was therapeutic touch, energy work, and chakra and meridian alignment. Tommy next adopted vegetarianism, Hatha and Tantric Yoga, ritualistic breath work, chanting and toning as a life style. The last he heard Tommy was involved in something called Thought Field Therapy, Rapid Eye Movement, mental and emotional acupuncture, KV Disassociation, and Erickisonian Hypnosis. These last two were so vague and esoteric that no one could really understand them, Ricky thought to himself. If there was ever a new therapy Tommy would be the first to know about it. Yet with all of this Tommy still had not been able to unlock the riddle of his past.

However this time Ricky was particularly looking forward to his annual Christmas visit with Tommy and his family for he was curious to hear more about Tommy's backpacking trip last summer to Superstition Mountain near Apache Junction, Arizona. Tommy ostensively went on a solo two week backpacking trip in search of the Lost Dutchman treasure at Superstition Mountain. Unbeknownst to Ricky, however, Tommy was really going to Superstition Mountain in search of his lost childhood, not lost treasure. When Tommy called his voice was deeper and his energy and everything else about him seemed to have changed for the better. Ricky just couldn't put his finger on it. Somehow Tommy seemed more congruent, more integrated and more centered then ever before. He said that on the third day in the back country he crossed paths with an old Indian Shaman. He spent most of the trip at the Shaman's adobe lodge located in a narrow rocky canyon deep at the base of Superstition Mountain. Tommy sounded very strange on the phone that day and most of what he had to say just didn't make any sense at the time. Ricky pondered his words over and over and was looking forward to hearing more of what transpired at Superstition Mountain last summer.

As best Ricky recalled Tommy talked about how the Shaman revealed to him the secrets of the ages, the ancient wisdom of the medicine men and mother earth. He recalled that Tommy said something about transcending both time and space and being eternally present in the here and now of life. He stated that everything was one thing and that this one thing was happening right now. He said that he had finally truly internalized an objective world, and it was now his to mold and transform into whatever image he desired for himself through his thought processes alone, without any effort, doubts, or fears on his part whatsoever. He no longer had to strive after things to make a living for he now had the power to express everything he could ever want in life. He went on to state that he did not have to plan for tomorrow for, in the inner recesses of his subconscious mind, he was tomorrow, and all of the yesterdays of his life.The past was not only reflected, but truly embodied, in the future. He stated that he had at long last become the universe's metaphor expressed as whole, perfect, and complete. His every thought surpassed any belief systems that ever existed and formed the reality of whatever he sought in life. He no longer had to simply rely just on his sight, speech, thinking and feelings in order to obtain what he wanted but he actually had transcended sensory acuity. He became both the creator and the created of his world and ultimately the source of both. Tommy not only spoke about creating a positive past and a positive future but of MetaCreating™ a positive past and MetaCreating™ a positive future through metaprogramming his brain and mind. Finally he had stated that all love is truly self love. He had finally reconciled with his parents and was now able to forgive them, and yes, even love them unconditionally, even though they were never able to respond in kind during their lifetime. His life, at long last, was now made whole and he was at peace. Ricky didn't know what Tommy was talking about. But the excitement in his voice revealed a strength and conviction that he never before heard in Tommy. He was glad that Tommy was able to find some peace and reconcile with their parents. However, for his part, he still felt those deep undercurrents of hesitation and ambivalence in making a commitment to Betty, and he just didn't really know why. He was looking forward to their annual Christmas visit to hear more about what happened at Superstition Mountain.

Again let's look at Ricky's situation from the standpoint of his mind and brain which immediately sheds some light on what's going on in Ricky's life. Ricky's hesitation and ambivalence is not just a conscious logical thought thing for as we have seen Ricky truly loves Betty and sees the many advantages in making a commitment. He understood that it would be truly rewarding and fulfilling for the both of them.Yet he couldn't bring himself to make such a commitment. What is going on is that the negative stimuli from the repeated negativity he experienced in early childhood left a deep and lasting neurological impression upon him in the form of his brain's negative basins of attraction and attractor pathways. In effect this negative basin, consisting of over two hundred million negatively charged neuron cells, constitutes a very large and deep neurological negative basin containing all of his childhood neglect and trauma. This negative basin was dedicated solely to the task of continually recycling negativity, born out of his hurt and pain, throughout all of his relationships. Ricky didn't realize that this negative basin constituted over one percent of his brain's entire neurological structure and was dedicated solely to the task of organizing, maintaining, and perpetuating negativity in all of his relationships.

The negative electrochemical configuration of his childhood basin of attraction was affecting, and in a very real and physical way infecting, directly or indirectly, everything else in his life besides his relationship with Betty, and he wasn't even aware of it at all. This wasn't so much a cognitive thing, but rather the problem went much, much deeper. It not only had to do with his neglectful childhood, but his sense of self worth, self esteem, self identity and ego personality. It transcended all other areas of his life including his relationships with women in general. It was reflected in his aggressive business style and in competition with his male associates. At times it drove him to take needless risks in order to prove himself over and over again. It manifested itself when he thought about marriage, for his thoughts immediately picked up a negative neuro attractor pathway which led directly back to this negative basin of attraction of those lonely childhood years where it seemed that he was the only one that really cared about Tommy. His brain's negative basin of attraction and his subconscious mind's negative analog patterns seemed to have a mind all their own. They acted like a black hole in his life where any significant relationship, whether male or female, caused him to react either aggressively, or to withhold himself from becoming emotionally involved, for he felt that he would only be hurt all over again.

Yet, he was never able to make a direct conscious connection of how pervasive and deep this negative basin affected all areas of his life. In particular, he was unable to make a connection between his feelings about marriage to Betty and the negative feelings about marriage in general. This wasn't so much a logical cognitive thing as it was a neurological thing and a negative subconscious analog thing. Yes, he knew that his form of logic wasn't serving him very well. Yet, he felt that he was right and justifiably so. Tommy arranged for Ricky to see a counselor to sort out his thoughts and feeling about marriage to Betty. Yet all the talk in the world didn't seem to help, so he just quit seeing the counselor. Ricky's dilemma is not so much one of logical thought on the conscious cognitive level of the mind for his problems all arose from the neurological processes of his brain and the negative patterns of his subconscious mind.

The counselor suggested that Ricky look at his early childhood where he might find a solution to his current dilemma with Betty. Ricky felt that such a suggestion was a waste of time. In a curious way he was indirectly right for the problem was not primarily a cognitive one. Anybody who knew him and his parents would know that he was more than justified in being very angry for all those the years of neglect. But how do you access and discharge the negative energy of the brain and subconscious mind merely through some kind of talk therapy? It just didn't make any sense to Ricky. Why revisit painful memories in order to try to discharge them. There had to be a better way. Maybe Tommy had the answer. Hopefully he would find out at Christmas time. It never occurred to Ricky that his negative feelings manifesting in the neurological processes of his brain and subconscious mind were influencing him more than he ever realized. It even carried over into his work. That is why he was known to have a short temper, was very confrontational and overly aggressive at times.

What Ricky didn't know was that since the problem arose at the subconscious level of his brain and mind it would have been more effective to deal with the problem at those levels. Even if Ricky believed that the energy and images of the subconscious mind were the cause of his thinking about marriage, those negative incidents, and there were plenty of them, arose more than forty-five years ago. How in the world was he ever going to be able to change his past which took place more than forty-five years ago? His parents were both dead, so he wouldn't have the opportunity to even talk it out with them. He thought to himself, you would be lucky if you could just change your perception, feeling or experience of your past much less actually physically change the past itself. However, his thoughts kept coming back to his talk with Tommy. Tommy somehow seemed to be able to not only reconcile with his past, he actually seemed to transcend his past for he said that he was able to forgive his parents and himself for whatever role he played in this dysfunctional family. He said that he even loved his parents unconditionally. Something about all love being truly some form of self love. Ricky thought to himself that this Christmas was certainly going to be interesting.

Had Ricky only known about the en bloc thinking in the fourth dimension of the subconscious mind he would have known that he could have not only changed his experience of the past but he could actually physically change the past itself. For where does the past exist but in the present memories of the mind? Is this what the Shaman revealed to Tommy? Had Ricky taken the time to think about it for a moment he would have wondered exactly where does the past exist if not actually lodged in the electrochemical composition of the neurotransmitters of the synaptic junctions of over two hundred million neuron cells making up Ricky's negative childhood basin of attraction located in the neuro cortex of his brain. If his memories of the past are actually tied up in the chemical composition of these two hundred million negatively charged neuron cells making up this large negative basin of attraction then does the past truly exist in some other objectified cell, frame or register lodged in the family's past history more than forty-five years ago up there in Northern Minnesota? Or is the past actually in the present networks making up the memories of his brain and subconscious mind? If so, would you not be better off changing the negative chemical composition of the brain's basins and the negative values of the subconscious mind, rather than trying to deal with factors lying totally outside yourself and residing up in Minnesota?

Ricky wondered how in the world could you change the past? Assuming that you can change those memories in present time, do you not then change the past at least insofar as it affects your past reality of your world? If someone else remembers the past differently than you doesn't this then represent their past reality and not yours? Is it not true then that if you can possibly change your experience of the past you can also actually change the physical memories of past realities themselves? Are not memories also made up of the same neurological substances of the negative basins of attraction as the experiences? If you could change this neurological substance from its current negative form to a positive form would you not only be changing your perceptions of the past but also at the same time interpreting the past in an entirely new and different way? At the quantum level of the neurological synapsis are not the atoms and molecules of the neurotransmitter substances made up of energy and matrix configurations? Since they are all made up of energy and matrix configurations, it stands to reason if you change the energy and matrix configurations from negative to positive you change feelings, perceptions and thoughts about the past. By changing feelings, perceptions and thoughts you change the past itself for what is the past but feelings, perceptions and thoughts of what you remembered occurred as past events in life.Therefore, by actually building for yourself new basins and attractor pathways you create a new past reality. How you build new basins is addressed later on.

Tommy no longer spoke in terms of self pity but in terms of love. Tommy did not touch on the many bad times they had but only on those few good times they experienced with their folks before Dad left them at a very early age. Tommy's whole perspective seemed to have changed almost as though he were raised in a completely different loving and caring household instead of the neglectful one Ricky so vividly recalled. When Ricky spoke about the many bad times to Tommy, Tommy would not respond, but only interrupt him in a kind and gentle way, to remind him that there were also some good times. Tommy would ask Ricky if he would try to be a little more respectful toward their parents. Not understanding, Ricky would remind Tommy of how bad things really were. Tommy would only reply in a kind and loving way that they did the best that they knew how, that you have to love them, forgive them, bless them, release them, and get on with your own life.

What then is the true nature of reality? Tommy's newfound freedom based on love and forgiveness or Ricky's pain based on anger toward his parents. Is it really true that it is never too late to actually have a happy childhood? If so, Tommy seems to be a good example of this. Also what then does all of this say about the future? If en bloc thinking can effectively change the past, can it not also change the future? Can you actually create an entirely new future for yourself, not just plan for the future, but actually create a future? Can you create a future where you create a whole new personality for yourself and a life of physical abundance to make life more enjoyable, as Tommy is now saying? The answers would seem to be yes, provided your are able to let go of a negative past and take control over your mental processes in order to create a new positive future for yourself. Which, in this case, seems to be what Ricky really needs to be doing before he will ever be able to make a commitment to Betty.

That still leaves us with the question of what happened to Tommy at Superstition Mountain? Can we also reorganize the past and create for ourselves an abundant future? Is matter and energy really interchangeable on the quantum level of the mind and is time really process? Can we create for ourselves a whole new self identity as well as an abundant future in the circle of time, like the Phoenix rising from the ashes of the old to emerge with a new identity into the flow of life as represented by the circle and process of time? Finally, will Ricky be able to adopt en bloc thinking of the fourth dimension in time to clear the negative neurological basins of attraction and the mental analog patterns in order to create a positive future for himself and finally make a commitment to Betty? Will Betty wait around long enough to see whether Ricky will ever make that commitment? We will just have to wait and see.

For now I suggest that you continue to live out your life in three dimensional reality existing in sequential frames of linear time in fixed three dimensional space as you have always been doing up until now. At the same time I earnestly invite you to begin to reorganize your thinking and adopt Conscious Analog Thinking in order to create realities in the fourth dimension, where time is inherently a part of the process of creating realities in time and space of your own choice. The dimension where time exists en bloc and where matter and therefore space, exist relative to you, the observer and creator of your own life realities. The dimension were matter and energy are interchangeable and you are able to create personal realities in whatever form you want and within space frames and time frames of your own choice. In a word empower yourself by adopting Conscious Analog Thinking™ where you think fourth dimensionally and take control over your life and over your realities in spacetime.[25] Welcome to the fourth dimension!

25 Conscious Analog Thinking is described in Chapter 3.

CHAPTER 2

Complexity of Self Organizing Systems of the Human Brain and Mind

<u>Introduction to Systems Thinking</u>

The human brain and mind is a morphological, nonlinear, spacetime, multidimensional,dynamical, self organizing system which has the capability to create its own world realities in any fashion it chooses in its own time and space. This is a very fancy way of simply saying that the mind/brain system not only has the capacity for programing itself for greater realities but it can actually create entirely new programs for the creation of entirely new realities.The mind/brain has the capacity for bootstrapping itself into creating expanded realities in time frames of its own choice. It can do so even if given only partial knowledge upon which to act in creating new metaprograms. With even only partial knowledge the mind/brain has the capacity to visualize the whole and therefore it has the ability to acquire the missing pieces through its own mental processes alone. By creating a higher level of organizing synthesis the mind/brain can continue the process of further acquisition and integration of knowledge. In this fashion it is able to organize yet even greater awareness of its world realities and create even greater realities in time and space. This is what I mean when I say that the mind/brain is a self organizing, nonlinear, morphological, multidimensional, spacetime, dynamical system. It primary purpose is to create realities in time and space. *In a word, the mind of man has the capacity to take control and direct the course of human destiny given the state of current scientific knowledge and technology.*

We have seen that the old Cartesian mechanical view of the world was that space was fixed, time was linear, and everything existed as fixed objects outside of us in three dimensional space. This characterized the current thinking concerning the entire physical universe, our world and the social order up until the time of Einstein. This way of thinking served mankind very well for over three centuries. Yet Einstein was able to take this limited and partial knowledge and conceptualize it far beyond anything that could be derived from a direct logical examination of the then existing body of knowledge. He was not only able to conceptualize entirely new patterns of thought, but he was able to set forth his theories in precise mathematical formulae. He gave birth to relativity theory and quantum physics! This represents a perfect example of what is meant when I say that the human brain and mind is a morphological, self organizing system with the capacity to create expanded world realities in its own space and time.

Exactly how the mind/brain organizes itself in order to create expanded realities is addressed in this Chapter. We will first explore in general terms what complexity of self organizing systems means. We will then explore the structure of the human mind, first in terms of systems structure, then in terms of organizational functions and finally in terms of functional processes. Next we will link up the structure, functions and processes of the mind in order to understand how it is that the mind can actually organize realities for us in space and time. We will take what we learn and in the next chapter apply it in Ricky's case to see exactly how all of this works in destabilizing a negative past and how it can be used in creating entirely new positive future realities.

How Complexity Affects Your Life

Quantum Physics ushered in a whole new way of viewing ourselves and our world. Quantum physicists were initially at a loss to explain in Cartesian mechanical terms the phenomena they were encountering at the subatomic level. This entirely new quantum reality just didn't match up with anything they were familiar with in their physical world. This radical new way of viewing things did not just pose technical or scientific problems. It posed significant conceptual problems in language, thought, and the unquestioned belief in a firm reality grounded in a fixed three dimensional world. It therefore created an existential problem of immense proportions as to the very nature of our own existence in an objective world. The implications of quantum physics to this day still have not been fully grasped or appreciated by many scientists and most of the rest of us. In order to fully understand how the principles of quantum physics directly and profoundly affect all aspects of our life we start first with the complexity of self organizing systems and then we apply system analysis to the structure and processes of the human mind in order to understand just how it is that we actually create realities in life.

Systems analysis involves an understanding of the functional relationships of the properties making up the system in order to understand its intrinsic nature as a single system functioning in its environment. The properties of the whole system are derived from the organization of its components but are qualitatively different from, and greater than, the sum of the components making up the system. In similar fashion, a molecule of water is qualitatively different from a molecule of ice. The components organize themselves synergistically to create a single system that exhibits properties very different from the properties of the constituent parts making up the system.

Systems thinking emphasizes the patterns and relationships between and among the parts in order to arrive at how the system functions as a whole.[1] Systems thinking involves analyzing the functions and processes of the parts in the context of the structure, functions and processes of the whole system. In the case of the human mind/brain system we need to understand how the separate components, namely, the cognitive structures of the conscious mind, the analog system of the subconscious mind, and the neurological processes of the brain, all function together in order to arrive at some understanding of how the mind/brain system organizes incoming stimuli to create inner and outer realities in life. Once we understand how it is that the mind/brain system organizes realities we will be able to take control over these organizing processes and use them to organize more satisfying realities in our own life.

Some Basic Characteristics of System Analysis

Properties exhibited by the system are destroyed when the system is dismantled and its constituent parts are analyzed in order to understand how the system functions as a whole. Therefore, while we will be analyzing the constituent parts of the human mind/brain system keep in mind that we are doing so only for purposes of getting a complete picture of what is involved in the entire mind/brain structure in creating realities. However, I can't overemphasize the fact that no one part is superior, or in any way more essential, to any other part in the organization of the mind/brain system in creating realities. None of the parts by themselves can organize the realities we experience in life.

A second characteristic of systems analysis, which is definitely applicable to the human mind, is that a system exhibits levels of performance. The cognitive structure of the conscious and subconscious mind is organized in a hierarchy of structures. The cognitive structure is the system by which we organize ego personality and its many realities in life. The analog system of the subconscious mind, which constitutes the subconscious portion of the cognitive structure, I refer to as the subconscious or hidden primary belief systems of the conscious mind. The conscious, or known, secondary belief system constitutes the conscious portion of the cognitive structure. Horizontal shifts at the level of the cognitive structure at which we experience life are made by logical thought processes. Upward shifts between levels of the cognitive structure are much harder to achieve since such shifts require a form of creative integrative thought. Creative Integrative Thought™ entails changes in ego personality as well as changes in your inner and outer realities in life. Creative Integrative

1 For an excellent description of system analysis see Fritjof Capra's book, *The Web of Life*.

Thought™ is described in Chapter 3. Upward shifts in the cognitive level at which you are presently experiencing life are essential in reaching higher states of awareness needed to create expanded realities in life and are discussed in Chapter 9.

A third characteristic of systems theory is that the system is not static. It is always in a state of dynamical motion, moving from a state of equilibrium to disequilibrium and back to equilibrium. It always seeks a state of dynamic balance or homeostasis between states of disequilibrium taking place within tolerable limits. This is what makes the system dynamical. There is always a state of dynamic movement within its networks. Its energies and images are constantly forming and dissolving patterns and relationships to create for us a continuous conscious experience of our environment. This is what gives us a sense of a solid exterior world and an experience of a stable inner world.

The fourth aspect is that systems organize along lines of networks. It is the relationships between and among its networks that define our realities in life. Both physical and nonphysical objects making up our life are organized by these networks and are labeled consciously by the cognitive mind as our way of creating order and stability in our life. If we consciously label everything in our life we will know where to look for them when we need them. Its like tying a string around your finger. It reminds you of something specific that needs to be done. Consciously labeling everything in life is like tying a million strings as reminders of where specifically everything is located in your life. Without such a system of labeling everything there would be no way in which we could define in specific terms our physical reality. Otherwise, everything would appear to be interrelated in a series of connected networks without distinct boundaries or definitions as to their identity, purpose or location in life.

Babies first experience their world in this way. Our early ancestors had a real problem in distinguishing themselves from their environment until they were able to figure this all out. When they did finally realize that they existed apart from their environment they also figured out that they had better start working in order to feed themselves and subsequently pay the bills. What happens is that babies, like our early ancestors, become self aware.[2] Our ability to label objects in our exterior environment is what also gives rise to our body of knowledge. The study of knowledge is referred to as "epistemology". We will be looking at the process by which the mind

2 However, on the subconscious level of the mind and brain organizational processes remain as interrelated and interconnected series of networks all working synergistically in analog fashion to support conscious mind in creating reality.

creates our thoughts and belief systems that constitute our body of knowledge. It is our belief of who and what we are that gives rise to ego personality and the manner in which we organize our lives. It is the thoughts and beliefs systems of ego personality that paint the many images and emotional tones upon the framework of our life.

As an example of how we label objects and create thoughts and feelings I want you to think of a car. When you think of a car you don't just form an abstract idea of a car. Rather, you think of a specific type of car such as a Ford, Honda or Lexus. If I asked you further to get a really good idea of what a car specifically looks like you might visualize it as being a late model red two door Ford Probe fully equipped with a racing strip and white wall tires. Thus your inner visualization creates an inner image of a car which at that moment in time constitutes your inner reality of that car. Duplicate this process several hundred thousand times and you have just created for yourself a very large body of knowledge and an attitude. I say an "attitude" because you would have, in the process, also created an opinion and corresponding feelings about various aspects of that body of knowledge. That body of knowledge would not just contain cold hard facts organized in static fashion in your mind like you might find in a reference book but also would contain feelings, beliefs, ideas and opinions. We will be dealing a great deal with the processes by which we form knowledge. At the same time we will also be dealing with related emotions and feelings, all in the context of creating realities.

A fifth characteristic of complex systems is that the system can create structural changes when organizing morphological programs. Remember, morphological programming allows the system to create for itself entirely new programs to organize new and different ways in which to view and express realities in life. When it creates new programs it also has the ability to create within its networks new structures which have organizing capacity to arrange new realities based on the organization of the new programs. How these processes work will be made readily apparent to you in just a moment.

The sixth characteristic of a complex system is that it operates as an open ended nonlinear or multidimensional system. There is a constant flow of information coming into the system in the form of sensory stimuli and a constant flow going out in the form of information, behavior and actions, both in response to and in organizing its environment. This characteristic of system processing of incoming and outgoing data and information is provided by the system's cybernetic loops. The cybernetic loops operate to create self organization within the system so that the system can organize and experience its environment. The cybernetic loops, operating as part of the sensory receptor system, perform command and control functions in regulating the flow of the incoming and outgoing data. The five sensory receptors are selective in their operation and select out of the millions of sensory stimuli available at any given moment just that data that is meaningful to the system. "Meaningful data" in the context of the mind/brain system is that data that is important to the system in preserving the integrity of the system, in maintaining and enhancing its performance, and in creating realities.

For example, the system scans the landscape in order to let you know that there is a red light ahead and that you better stop in order to avoid an accident. Likewise to maintain or enhance its existence the system's visual receptors might spot relevant data on the financial page out of all of the other financial data on the page in order to allow you to make an informed decision about a stock market investment you have been considering lately. All other data that isn't immediately relevant to the system is screened out by the sensory receptors. We often think of the sensory receptors in terms of providing the system with just incoming data. However, the sensory receptors screen out more data than they let in. If the system did not have this ability to screen out most incoming data the system would immediately go into overload and literally drive you crazy.

The Structure of the Mind and Brain

With this short lesson in systems analysis we'll now explore how all this relates to the structure and processes of the mind and brain. Charts 3-1 through Chart 3-3 at pages 58-61 set forth the organizational structure of the system of the brain, the system of the subconscious mind, and the system of the cognitive structure of the conscious mind. Each chart shows the exact same thing but in a little different perspective. The reason for this is that by comparing the charts you will first get a feel for what is traditionally thought of as the separate systems of the brain, subconscious mind and conscious mind. At the same time you will begin to appreciate that these systems actually are different aspects of the one mind/brain system used in creating a meaningful life. Chart 3-1 describes 12 levels involved in the organization of reality. Level one starts with the sensory stimuli coming into the system. Data is received through the five sensory receptors. Incoming data is translated into electrochemical signals by the neurotransmitters of the brain at level 2. The neuron cells communicate with one another by their neurotransmitters. At level 3 neuron cells organize themselves into basins of attraction in accordance with the similarities of their neurotransmitters. A primary function of the brain is to regulate the autonomic regulatory systems of the

CHART 3-1

Levels of the Mind/Brain in Creating Inner and Outer Realities

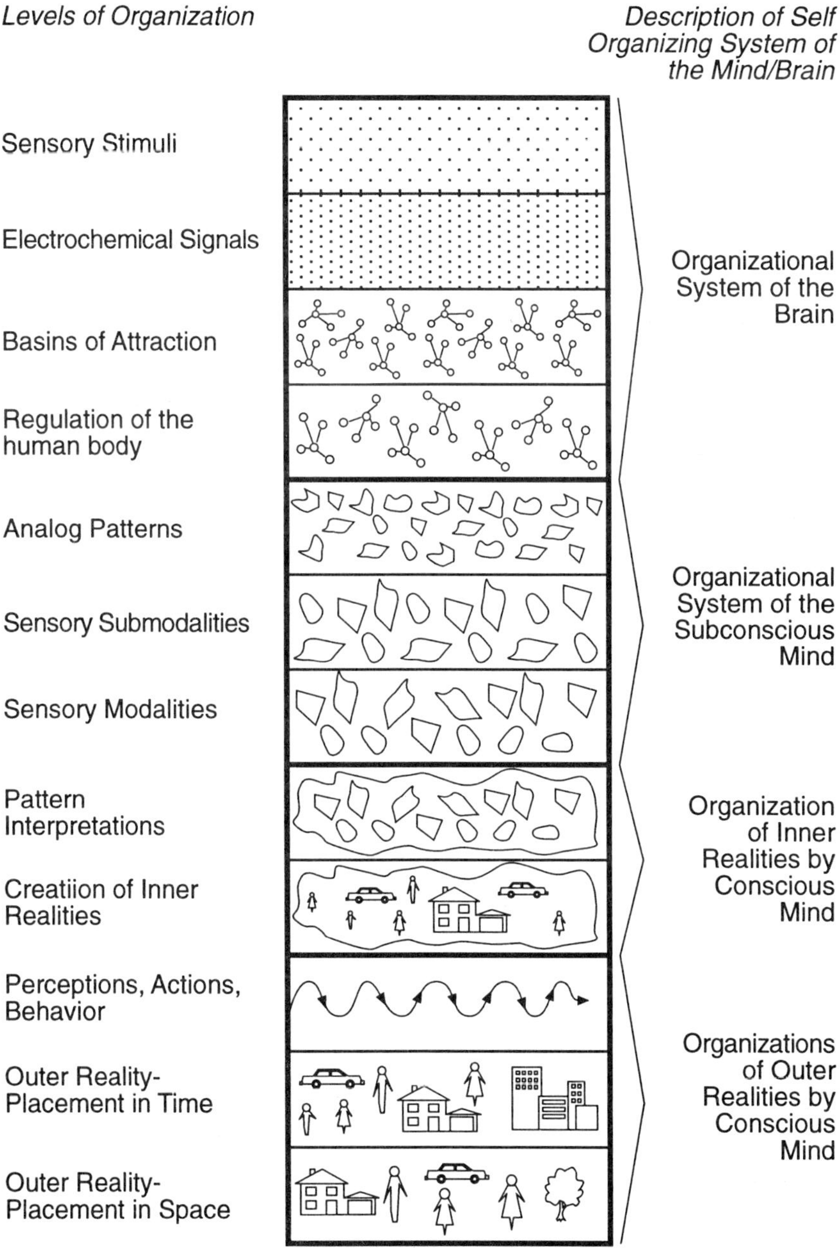

CHART 3-2

Overlapping Self Organizing Systems of the Mind and Brain Involved In the Creation of Inner and Outer Realities

Levels of Organization		*Description of Self Organizing System*
Level 1	Sensory Stimuli	Organizational System of the Brain
Level 2	Electrochemical Signals	
Level 3	Basins of Attraction	
Level 4	Regulation of the human body	
Level 5	Analog Patterns	
Level 6	Sensory Submodalities	
Level 7	Sensory Modalities	

Level 4	Regulation of the human body	Organizational System of the Subconscious Mind
Level 5	Analog Patterns	
Level 6	Sensory Submodalities	
Level 7	Sensory Modalities	
Level 8	Pattern Interpretations	

CHART 3-2 continued

Overlapping Self Organizing Systems of the Mind and Brain Involved In the Creation of Inner and Outer Realities

Levels of Organization		*Description of Self Organizing System*
Level 6	Sensory Submodalities	Organization of Inner Realities by Conscious Mind
Level 7	Sensory Modalities	
Level 8	Pattern Interpretations	
Level 9	Creation of Inner Realities	
Level 6	Sensory Submodalities	Organization of Outer Realities by Conscious Mind
Level 7	Sensory Modalities	
Level 8	Pattern Interpretations	
Level 9	Creation of Inner Realities	
Level 10	Perceptions, Actions, Behaviors	
Level 11	Outer Reality-Placement in Time	
Level 12	Outer Reality-Placement in Space	

The Chart 3-1 organization is strictly for purpose of analysis in understanding the structural organization of the systems. However in reality there is no demarcation or boundaries between the various levels of the systems. All systems are actually interrelated and interconnected with one another and work synergistically in communicating information almost simultaneously throughout the entire composite mind/brain system in creating inner and outer realities in life.

CHART 3-3

Organization of Levels 6,7,8 and 9 of the process used by the Mind in Creating Inner Realities

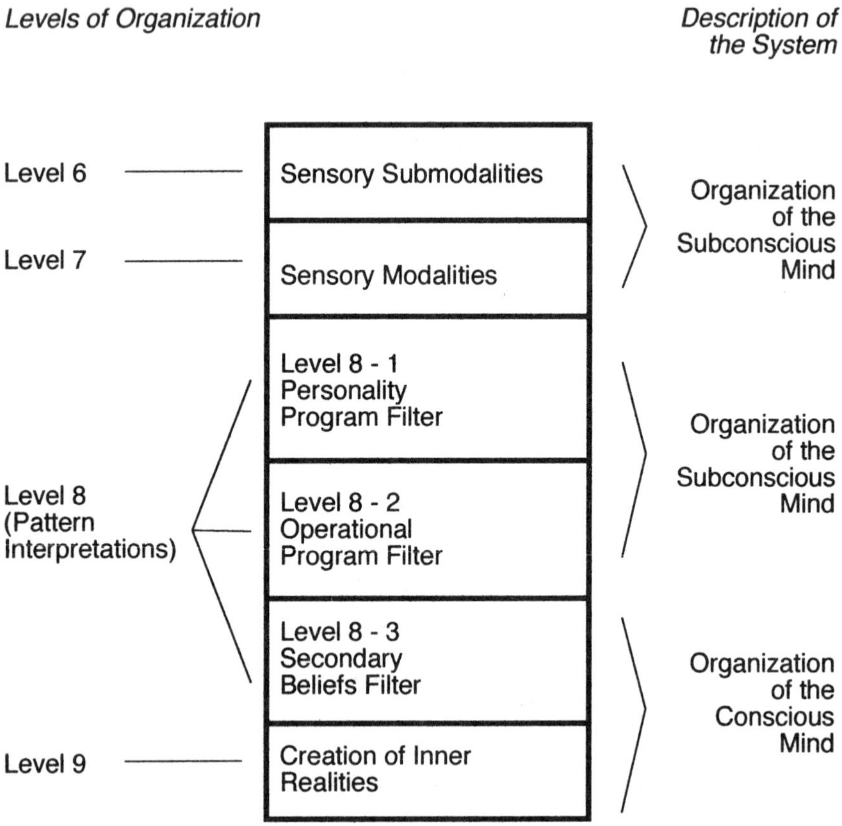

When the Mind/Brain processes sensory modalities to form the mental imagery of its pattern interpretations, the mind's subconscious primary personality and operational systems take over in the processing action to filter the mental imagery through the personality and operational cognitive structures of the mind. In this way meaningful images are framed through the processing of these cognitive filters and are then acted upon by the secondary belief systems of the conscious mind of the individual so as to form mental patterns of inner realities. It is these inner realities through our individual perceptions, actions and behavior which then forms our outer physical realities in life.

human body at level 4.[3] The systems set forth at level one through four are traditionally thought of as the structural organization of the human brain. The structure and processes of the brain are explained in detail in Chapter 5.

Level 5 consists of the analog system of the subconscious mind. The analog patterns of the subconscious mind are arranged based on the relationships of their underlying basins of attraction. Levels 6 and 7 consist of the subconscious mind's organization of the sensory submodalities and modalities based on these analog patterns. These levels are different from level 1 describing incoming sensory stimuli. Level 1 describes sensory stimuli received by the systems which is translated into electrochemical impulses by the brain. These levels actually organize basins of attraction and their corresponding analog networks of the subconscious mind to provide us with the actual sense of touch, smell, taste, sight, and hearing. The submodalities of the five sensory receptors would be: for touch - location, pressure and duration; for smell - intensity, sweet and pungent; for taste - sweet, sour; salty, or acid, for sight - distance, perspective and clarity; and for sound - loudness, location, distance and tempo. This is a summary list of the submodalities. For a more detailed list of the submodalities refer to Chart 3-A entitled, *Characteristics of the Images of the Modalities and Submodalities* located at page 63. Further the kinesthetic modality listed in Chart 3-A also includes inner and outer sensations of the kinesthetic energy of our emotions and feelings.

These submodalities and modalities are combined into a multitude of energy and images forming pattern interpretations at level 8. Levels 5, 6, and 7 all form what would commonly be referred to as the organizational system of the subconscious mind. Level 8, pattern interpretations, is a transition level which represents both subconscious and conscious processes. Level 8 represents both the subconscious primary programs of the subconscious mind and conscious patterns of thought referred to as the secondary belief system. Our conscious mind at level 8 starts to label everything in our life and assigns a value to each thing relative to other things in our life. At level 9, we consciously interpret thoughts organized at level 8 and subscribe meaning to them in accordance with their importance in creating inner realities in life. We understand and communicate all of our thoughts and feelings through the use of our native language in order to create an experience of inner realities. Accordingly, levels 8

3 While level 4 plays an extremely important role in regulating the numerous autonomic functions of the human body we do not explore in any great detail the role of level 4 in maintaining the human body for optimum health for such an analysis is beyond the scope of this book.

Chart 3-A

Characteristics of the Sensory Modalities and Submodalities

Kinesthetic

Location
Still/ moving
Pressure
Duration

Size
Intensity
Temperature
Moisture
Texture
Rhythm
Density
Weight

Visual

Perspective
Movement
Location
Distance
Clarity
Bordered
Panoramic

Color or Black & White
Moving or still
Shape
Disassociated or Associated
Brightness
3D or Flat
Foreground/background
Dim
Detail

Auditory

Distance
Location
Music/ noise
Voice (own, other, masculine,feminine
Rhythm/ beat
Tempo
Pitch high/low
Stereo
Volume
Tonality
Duration

Taste

Temperature
Texture
Intensity
Sweet,sour,acid,salty
Density
Simple/complex
Moisture

Smell

Intensity
Nature
Location
Simple/complex
Fragrance
Sweet, pungent
acid, smoky

and 9 represent the organization of our inner realities by the conscious mind. How we organize our inner realities is described in Chapters 5 and 6. Level 10 represents our perceptions, behavior and actions in our outer physical world based on our inner realities in life. Finally levels 11 and 12 represent the results of our level 10 actions in creating for ourselves the placement and timing of the organization of our outer physical realities. Levels 10, 11 and 12 represent the organization of outer realities by conscious mind.

Levels 11 and 12 are probably the most controversial. The reason for this is that today Cartesian mechanical theory still reinforces the belief that we can only create outer physical realities through creative or imaginative thought in combination with planning and physical effort on our part to brings things about or make things happen in our life. The fact of the matter is that this is the old Cartesian way of doing things and quite a primitive way at that. As indicated earlier, the subconscious mind has an almost unlimited capacity to sort endless possible combinations of people, places, and things in order to bring to our conscious awareness what it is we want to create as our outer physical realities in life. The subconscious mind can do this in time frames and space frames of our conscious choice since, as we have seen, the en bloc nature of spacetime of the fourth dimension is alive and well on the subconscious level of the mind and can organize realities on the physical plane in our three dimensional world. Precisely how we accomplish all of this is explained in detail in Chapter 10.

Chart 3-1 shows each system as being separate from all of the other systems. The reason for this organization is to show the distinct functional processes of each system. Chart 3-2 shows the same 12 levels of organization of the mind and brain but this time they are arranged as overlapping systems of the mind and brain. For instance, in Chart 3-2, levels 1 through 7 are organized as systems of the brain while levels 4 through 8 are organized as systems of the subconscious mind. The first organization represents the systems of the brain and the second organization represents the systems of the subconscious mind, yet there is considerable overlapping between these two systems. For instance, levels 6 and 7, sensory submodalities and modalities, are shown as both subconscious and conscious processes. The reason for this is that when we consciously think in terms of the submodalities and modalities they become conscious processes of the mind. However, most of the time we do not consciously think in terms of submodalities and modalities but rather in terms of whole thoughts and beliefs represented by our mental imagery. This overlapping of system levels is a more realistic portrayal of how the systems actually function for there is no real demarcation or boundaries between the various levels of the systems. All systems are interrelated and interconnected and work dymanically and synergistically in simultaneously communicating information among the various networks making up the mind/brain system in creating inner and outer realities in life.

Level 8 - Pattern Interpretations

In Chart 3-3 level 8, pattern interpretations, is broken down into three sublevels. The first sublevel is described as the Personality Program Filter, the second is described as the Operational Program Filter, and the third is described as the Secondary Beliefs Filter. The first two of these sublevels are subconscious or hidden cognitive systems of the conscious mind. At the same time they also constitute subconscious analog systems of the subconscious mind. How these two different types of processing systems interconnect and interrelate in organizing sensory imagery and information will be explained later. Don't worry if all of this isn't very clear to you at this time. The only purpose of this chapter is to merely acquaint you with a brief overall description of the structure and functions of the mind/brain system. We will describe each of these systems in much greater detail in the following chapters.

Remember how we labeled things in life in order to make sense out of the things we encounter in our daily affairs. After we labeled an awful lot of stuff we began to accumulate a rather large body of knowledge and in the process developed beliefs, feelings and attitudes about that knowledge. After awhile we started to organize that body of knowledge so that it wouldn't clutter up our life too much by creating categories of beliefs in order to pigeonhole related information. It is both this conscious and subconscious process of creating categories for containing related information that forms our conscious secondary belief systems about things in life. Secondary beliefs are set forth as level 8-3. Information we are currently using is filed by the subconscious mind in short term memory. Information we might use later on is filed in long term memory. At the same time we subconsciously assign values to these various items of knowledge based on their importance to us. Our value system reflects emotional values as well as use or utility values.

With this brief background we will now look at the three sublevels of level 8 in more detail. Level 8-1 and level 8 -2 form the personality program filter and the operational program filter that represent the subconscious or hidden cognitive structure of conscious mind. These two programs constitute the subconscious primary belief systems subserving the secondary belief systems of the conscious mind at level 8-3. The basic parameters of the primary personality and operational programs were laid down very early in life, at a time when we were not aware of their significance nor the effect they would have upon us. These primary programs became the most significant influences throughout our entire adult life.

If these hidden primary programs were of a positive nature, that is, if they were made up of positive neurological basins of attraction, and on the subconscious level, positive analog patterns, they reinforce positive mental imagery and positive behaviors in life. If they were made up of negative basins of attraction and analog systems as in Ricky's case we have a problem. The extent that a negative basin impedes you from creating a healthy and happy life for yourself depends to a large extent on the depth and width of the negative basin. The depth is measured by its severity which is measured by the intensity of its negative energy and by the vividness of the negative images of its matrix configuration.[4] The broadness is measured by how much of your personality and operational systems are adversely affected by the basin. As if this isn't bad enough, their effects unfortunately extend even further. These negative basins not only adversely affect your past and present circumstances in life but also inhibit your potential for creating happier realities in the future. They create negative cybernetic loops which are self maintaining, self reinforcing, self limiting and ultimately self defeating. Something has to be done about them, and in Ricky's case, something needs to be done really soon. Betty will not wait forever. What exactly we can and should do about them will be addressed shortly.

First let's address sublevel 8-3, the secondary beliefs filter. For purposes of describing the mind/brain system we define secondary beliefs as "all beliefs of the conscious mind". I don't care what you think about, for purposes of describing the mind/brain system, all conscious thoughts are secondary beliefs as opposed to the hidden primary belief systems. Consequently, all secondary beliefs are constrained and influenced by the hidden or subconscious primary systems although we are not usually aware of this fact. If your hidden primaries are of a positive nature you are fortunate for these positive primaries sustain, reinforce and perpetuate positive secondary beliefs in life. It is not so much your conscious beliefs about people, places and things in life that give direction and meaning to life, although they certainly contribute to your wellbeing. What is far more important is the processes of the subconscious mind that set the general tone

4 Severe childhood abuse and wartime trauma create post traumatic stress syndromes which can be so severe at times that when triggered cause a flashback effect where the individual reexperiences all of the associated submodalities and modalities of the original traumatic episode.

and direction of life. It is the subconscious processes that determine whether you will ultimately be happy or dissatisfied with life in general. This is so for it is the subconscious mind which gives physical expression to thoughts and feelings. Conscious mind doesn't create anything, it merely initiates and organizes thoughts and feelings. It is the subconscious mind that creates realities in accordance with conscious thoughts and feelings.[5]

If you are like most of us and if you have a problem you will normally think about ways of fixing it. Sometimes things works and sometimes they don't. Apart from just faulty thinking, the times that things don't work often reflect the fact that hidden negative primaries are adversely affecting the outcomes you seek in life. When this is the case you can do one of three things. First, you might do nothing at all. No one says that you have to do anything to fix it, except maybe your mother. Secondly, you could continue to think of logical alternatives for fixing the problem. If after many tries nothing seems to work, as in Ricky's case, you might try the third alternative. You might try addressing the problem on the subconscious level of the mind in order to see if a negative primary is the culprit sabotaging your efforts at finding a solution. Hopefully the methods spelled out in this book will help you clear negative hidden primaries in life. As to what needs to be done we have seen the subconscious mind operates in analog fashion and works by organizing similarities in the patterns and relationships between its inner networks and its outer realities. It does not operate on the basis of logical or analytical processing as does the conscious mind. Knowing this we can devise for ourselves a solution to the most intractable negative hidden primaries in life and clear them. We would then be able to take effective action in carrying out the intent of our conscious secondary beliefs which express our true desires in life.

The reason that we are spending so much time on how the mind and brain are structured to organize realities is that if you thoroughly understand the process you will then be able to direct the

---

5 On the face of it this seems like a radical statement but it can easily be explained. Conscious mind merely provides you with an awareness of the results of what's happening on the subconscious level of the mind. It is the subconscious mind that carries out all of the processes involved in organizing conscious thought. All processing of stimuli, all organization of thought and all organization of perceptions, feelings, actions and behavior are all organized at the subconscious level. You become consciously aware of them at the moment that they are organized at the subconscious level. However, we experience everything as though it is conscious awareness that is organizing our realities in life and not the subconscious mind.

process to work in your favor instead of against you. To begin with we do not address the negative basin directly. We do not put it under a microscope for cognitive examination. Rather, we first address the context in which the hidden primary resides. All problems in life arise within some context. They don't just spontaneously erupt into our life on their own accord. Therefore, we start by reinforcing that part of the subconscious analog system in which the problem arose. You will find that all of life ultimately is represented by the metaphor "whole, perfect and complete." We just don't experience it in this fashion because of these negative basins. It is these negative basins that give us negative experiences which is the mind's way of telling us that our life is less than whole, perfect and complete. In fact these basins often exhibit themselves as pain, lack and limitation.

Very briefly this repair work consists of reinforcing the affected area of the subconscious mind in which the negative basin resides. The deepest level of the subconscious mind consists of four prime directives in life which can best be described as: (1) your fundamental right to simply exist as a human being and a right to your own life; (2) your fundamental right to like and appreciate who and what you are; (3) your fundamental right to express yourself fully; and (4) your fundamental right to achieve whatever it is you want in life. Without the positive formation of these four fundamental prime directives early in life you potentially will encounter significant difficulty in just living and in trying to express yourself fully throughout your adult life. A negative basin, regardless of its configuration, will always impact one of these four prime directives in life. These four prime directives in life as well as all of the other primary and secondary programs of the subconscious and conscious mind are described in detail in Chapters 7 and 8.

Once we complete this repair work on the subconscious systems in which the negative basin arose we then create a positive analog. With respect to the subconscious mind, an analog is a type of mental unit which in this case is a positive network pattern similar in structure and function to the corresponding negative basin but composed of positive energy and images. For this reason it is described as a counter analog. By relaxing the conscious mind and accessing the subconscious mind we can program the subconscious to create for us the most appropriate personal power analog to destabilize and dissolve the negative basin. Our counter analog has far superior energy and imagery in its matrix configuration. We send our personal power analog™ on a mission to search out and destroy the negative basin. It is able to effectively do so because the analog similarities allow the superior positive analog to seek out the energy and imagery of the negative basin and destabilize and dissolve it. While the procedure is initiated by the conscious mind the actual creation of the power analog and the process of seeking out and destabilizing the negative analog is a subconscious one.

Because of the similarities in their biochemistry the energy and image configurations have an affinity or valance effect which attracts and binds the two analogs together and allows the superior one to merge its energy and matrix configuration with the negative energy of the lesser one and dissipate and dissolve it.[6] This process is the result of the electrical chemical binding of atoms into molecules described in Chapter One. We refer to this power analog as your "Personal Power Symbol"™ or "Personal Power Analog."™ We will be using symbols and metaphors throughout this book because this is the language of the subconscious mind.[7] For a list of the more common analog symbols recognized by the subconscious mind refer to Chart 3-B at page 70 entitled *Common Analog Symbols Recognized by the Subconscious Mind.* We then come back, if necessary, to complete whatever repair work needs to be done on the damaged portion of the subconscious analog system where the negative basin resided. In this fashion we can effectively access the most intractable negative hidden primaries in life and clear them.

The mind has enormous organizing capacity and will create such a power analog if properly programmed to do so. The analog you create in this fashion may be a universal symbol. However, it will also be personal to you in the context of destabilizing and dissolving negative basins. The reason for this is the mind will always seek completion and coherence within its systems. Negative basins are therefore an anomaly to the proper functioning of the system. Consequently, evolution of the system is on your side and against the anomaly of the negative basin. Accordingly, the power analog will destabilize and dissolve the negative basin and set the groundwork for you to organize positive realities in life. The next step is to consciously create our goals and strategies needed to fulfill our primary mission in life represented by the desires of our secondary beliefs.

---

6 In order to be perfectly correct it is the biochemistry of the basins which attract one another and it is the energy and matrix configuration of the internal patterns of the subconscious mind which match up and change the polarity of their internal values of the analog system of the subconscious mind. Just how this analog system operates is explained in Chapter 6.

7 The subconscious mind also processes cognitive logical thought. However it organizes logical thought using its analog processing systems. It is interesting to note that an analog system can easily process logical thought. It is far more difficult for a binary serial processor to process data, or the conscious mind to organize its thoughts, in analog fashion.

CHART 3-B

COMMON ANALOG SYMBOLS RECOGNIZED BY THE SUBCONSCIOUS MIND

Symbol	Symbol	Symbol	Symbol
Eagle	Lion	Hawk	Dawn
Bear	Angel	Warrior	Nightfall
White Knight	Blue Sapphire	Diamond	Anthems
Gold	White Light	North Star	Center
King	Queen	Prince	Lightning
Princess	Spirit Guide	Magician	Thunder
High Priestess	Wizard	Shaman	Voice
Medicine Man	Master	Guru	Cave
Mystical Teacher	Pilgrim	The Path	Rock
Circle	Cross	Heart	Waves
Book of Wisdom	Mantra	Sound	Snow
Spear	Arrow	Candle	Water
Torch	Zeus	Jupiter	Jester
Venus	Wolf	Phoenix	zodiac
Dragon	Potion	Chalice	Triangle
Gems	Saint	Treasure	Manuscript
Cornucopia	Armor	Secrete Place	Pen
Ring	Crown	Robe	Doorway
Kingdom	Power	Strength	Cupid
Fire	River	Ocean	Shamrock
Mountain	Majesty	Source	Rose
Creator	Butterfly	Sword	Fragrance
Statue	Tapestry	Horse	Flowers
Key	Masterpiece	Creation	Cathedral
Earth	Knowledge	Book	Beach
Staff	Emblem	Flag	Moon
Ruler	Monarch	Destiny	Space
Throne	Castle	Hand	Forest
Writing	Goblet	Cradle	Pedestal
Eye	Pyramid	Temple	Trumpet
Orb	Globe	Alchemy	Picture
Elves	Fairies	Gnome	Ancestors
Rainbow	Map	Dove	Riddle
King	Lover	Samurai	Oracle

To do so we access spacetime of the fourth dimension on the level of the subconscious mind in order to create for ourselves the most appropriate positive energy and image analogs to bring to conscious awareness those people, places and things that will manifest the specific physical realities we desire in life.

We adopt a type of creative integrative thought referred to as "Conscious Analog Thinking"™ in order to facilitate the subconscious mind to bring to conscious awareness the manifestation of expanded physical realities. This way of thinking is in addition to our daily analytical thought processes. Briefly, conscious analog thinking consists of viewing your life holistically rather than viewing life taking place solely in linear sequential fashion. You begin to sense that life takes place in the form of interconnected and interrelated series of patterns and relationships, relationships were the timing and placing of events are viewed holistically rather than just sequentially in linear time and fixed space. This conscious perception of the relationship of events in life is similar to the manner the subconscious mind organizes the patterns of its analog system. Accordingly, as you adopt conscious analog thinking you consciously assist the subconscious mind in organizing expanded realities in life. We presently do not perceive the connectiveness of things taking place in our life for they are not always logically or causally connected. Nevertheless, events represented by their pattern and relationship organizations could hold meaning for you if you become consciously aware of their significance in your life. By adopting conscious analog thinking you will begin to see meaningful patterns emerge in life which you were not aware of in the past. Such patterns can become important from the standpoint of creating expanded realities in life. For instance, if events seem to repeat themselves in your life three or more times these events may well hold a message for you. If you ignore them as being merely coincidences and nothing more you may be ignoring significant patterns in life which hold the answer to some hidden opportunity trying to reveal itself from the subconscious mind. These patterns may be the subconscious mind's way of telling you of certain events, which if acted upon, will bring about the positive realities you have been seeking. Remember, it is the subconscious mind that processes information in analog fashion as well as analytical fashion, and it might be its way of revealing to conscious awareness significant people, places and things which are important in your life.[8]

For example, prior to Tommy taking his backpacking trip to Superstition Mountain last summer he first heard from a long-time friend who lives in Scottsdale, Arizona. His friend invited Tommy to visit him and they both would look for the lost treasure of Superstition Mountain. Just before Tommy was to leave on his trip his friend called and told him that he had urgent family business back east that needed his attention immediately. Nevertheless, Tommy felt that the signals he was receiving indicated that he should go alone. He felt this way because just prior to his friend inviting him to visit him in Arizona, Tommy's travel agent called about a special airfare to Phoenix, Arizona, with a side trip to Superstition Mountain to look for lost treasure. It was a special promotional fare well below normal rates. At about the same time Lonni, the office manager, told Tommy that when she was a young girl growing up in Arizona she and her brother would often backpack into the Arizona Superstition Mountains and explore ancient indian pueblo ruins. One day she stated that she felt that Tommy should make a trip out to Superstition Mountain and that there he would find the answers to the issues he was still dealing with in his life. When the third unrelated incident concerning Superstition Mountain was brought to his attention Tommy followed his instincts and at that moment decided to take the trip. If you believed that these circumstances were just coincidental and nothing more you might just ignore them. However, the subconscious mind often communicates to us in nonlogical analog fashion ways to bring to conscious awareness those people, places, and things your conscious self needs to know in order to manifest the realities you seek in life. This is an example of conscious analog thinking. You will be hearing a whole lot more about conscious analog thinking in creating realities in life.

Back to the process of destabilizing and dissolving negative basins. We sometimes encounter states of mental disequilibrium in both the conscious and subconscious systems which is a transitory phenomena. Systems often encounter a phase state of disequilibrium when moving to a higher order of organizing synthesis. However, the systems themselves will always seek to restore a stable state

---

8 Often dreams which repeat the same pattern of events over and over may be the subconscious mind's way of telling you something important about your life. Skilled neurolinguistic practitioners and hypnotherapist often look for repetitions in a student's thought patterns and behavioral styles, and if they perceive a repeat of three or more thought patterns or behavioral styles they can effectively use such information to better understand the origin and components of the student's deeper mental structures giving rise to such thought patterns and behavioral styles.

of coherence and congruency throughout their structure and organization. By working the processes you will find over time that you will achieve higher states of organizational synthesis resulting in higher levels of consciousness, expanded states of awareness and overall optimum performance in the systems of your mind. For example, all types of obsessive, compulsive and addictive thoughts and behavior will, over time, start to destabilize and dissolve. For such thoughts and behaviors are but the reflection of negative basins operating at the subconscious level of the mind. On the other hand, positive behaviors and actions will become stabilized and reinforced over time, for this is the inherent nature of how the system works. The whole point of this book is to assist you in working in cooperation with the natural order of the operational processes of the systems instead of in opposition to them. It is often our unawareness that is keeping us in a state of confusion and therefore in states of limitation and pain. This is just another way of saying that as we learn how the systems perform we automatically start to support that which is appropriate for maximum performance in life. At the same time we stop sabotaging ourselves by indulging in self defeating thoughts and behaviors which impede the operation of our mental systems and consequently our happiness in life. Again knowing how things work on the conscious and subconscious levels we can actively participate in the organization of realities for our own good. During episodes of disequilibrium we will also know enough to continue using the processes to accelerate the positive outcomes we seek.

Finally, just a word about the physiology of the basins of attraction at the neurological level of the brain. While the network of the negative basin may have been in place for an extended period of time it is not true that the hardware itself cannot be changed. While we refer to the neurological structure as hard wired it is, in fact, soft wired and more precisely wet wired. This wet wired aspect of our brain on the quantum level is made up of neuro substances consisting of our old biochemistry friends, energy and chemical matrix configurations. These energy and matrix configurations, like all energy and matrix configurations, are in a constant state of motion. This is what makes the mental system of the conscious and subconscious mind and brain a dynamic system instead of a static system. As such their energy can be reprogrammed to a positive polarity. You don't have to change the hard wiring. You only need to change the polarity of their energy. This may take a little time but it's worth it over the long haul. Whatever amount of work it takes on your part to take complete control over the conscious and subconscious systems governing your life and convert pain, lack and limitation into peace, happiness, and prosperity will be well worth the effort.

Functional Organization of the Mind/Brain System

We reviewed the internal processes of the mind and brain in terms of their organizational structure. When we refer to structural organization we refer to the actual physiological structures of the mind/brain. We will now look at these same structures in terms of their functions. Function describes the specific role each system plays in the process of creating realities. In the next chapter we will look at the processes of each system. That is we will also look at how each system is able to process incoming stimuli to organize inner and outer realities in life.

Charts 4-1 and 4-2 at page 75-76 set forth the system functions of the organizational levels of the mind and brain just described above. Chart 4-1 describes these functions in terms of cybernetic loops showing the relationship between the various systems in creating realities. Each system constantly feeds data in the form of stimuli, energy and images back into all of the other systems of the self organizing system of the mind/brain in order to create inner and outer realities. All systems are constantly interacting and affecting the level of activity of all of the other systems. This chart illustrates the concept that positive stimuli can create positive outer realities which in turn can create positive stimuli which will create positive inner realities. Conversely, negative stimuli can create negative outer realities which in turn can create the negative stimuli which will create negative inner realities. These negative inner realities will then create the negative stimuli to reinforce a negative outer reality of life. Unless this cycle is broken these cybernetic loops will continue to create negativity on the subconscious level throughout a person's entire life and will manifest as pain, lack and limitation.

Chart 4-2 at page 76 shows the relationship between the brain in processing stimuli and the mind in processing images in organizing our experiences of inner and outer realities. This chart sets forth a series of cybernetic loops whereby the brain organizes incoming stimuli into basins of attraction which, in turn, organizes sensory submodalities and modalities. These submodalities and modalities form pattern interpretations creating the formation forming our inner realities. It is these inner realities in life which form new stimuli for the brain to process either new realities or the rearrangement of old realities. That is, our inner realities provide the subconscious mind with the patterns of energy and images forming our thoughts which, in turn, organizes our belief systems, actions and behaviors in creating our outer realities. The cybernetic activity of these loops is constantly going on all our life. It is when we consciously take control

CHART 4-1

Cybernetic Loops of the Overlapping Self Organizing System of the Mind in Creating Inner and Outer Realities

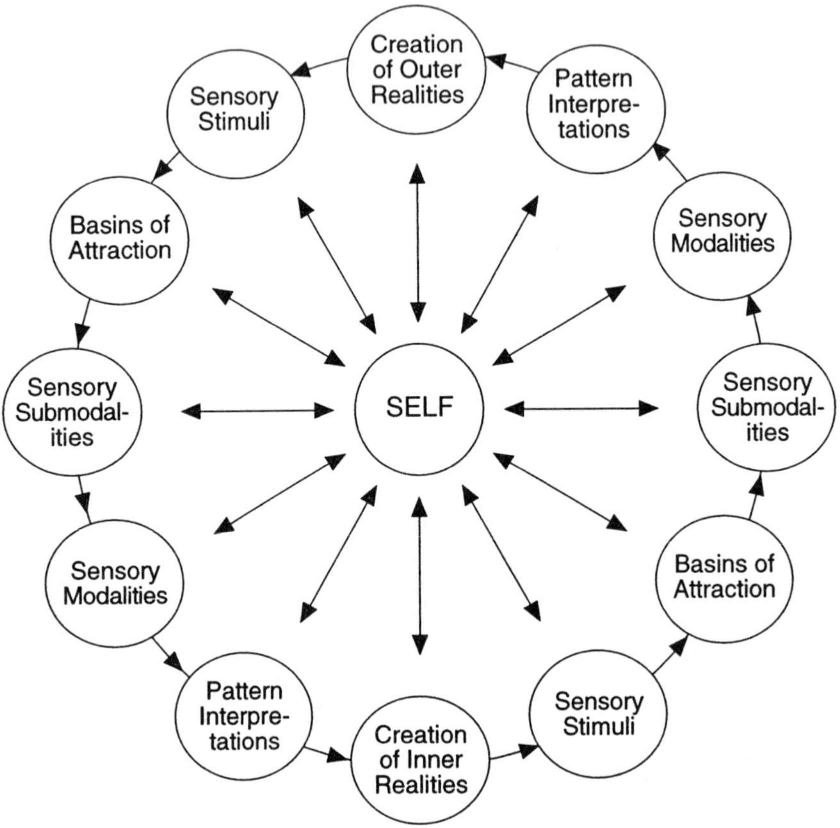

Each system constantly feeds data in the form of stimuli, energy and images back into all of the other systems of the self organizing system of the mind in order to create inner and outer realities. All systems are constantly interacting affecting the level of activity of all the other systems. This chart illustrates the concept that negative stimuli can create negative outer realities which in turn can create the negative stimuli which will create negative inner realities. These negative inner realities will then create the negative stimuli to reinforce the negative outer reality of life. Unless this cycle is broken it will continue on creating negativity on the subconscious level throughout a person's entire life, and will manifest in his outer life as pain, lack or limitation.

CHART 4-2

Comparisons and Distinctions Between the Structure and Function of the Human Brain and the Human Mind

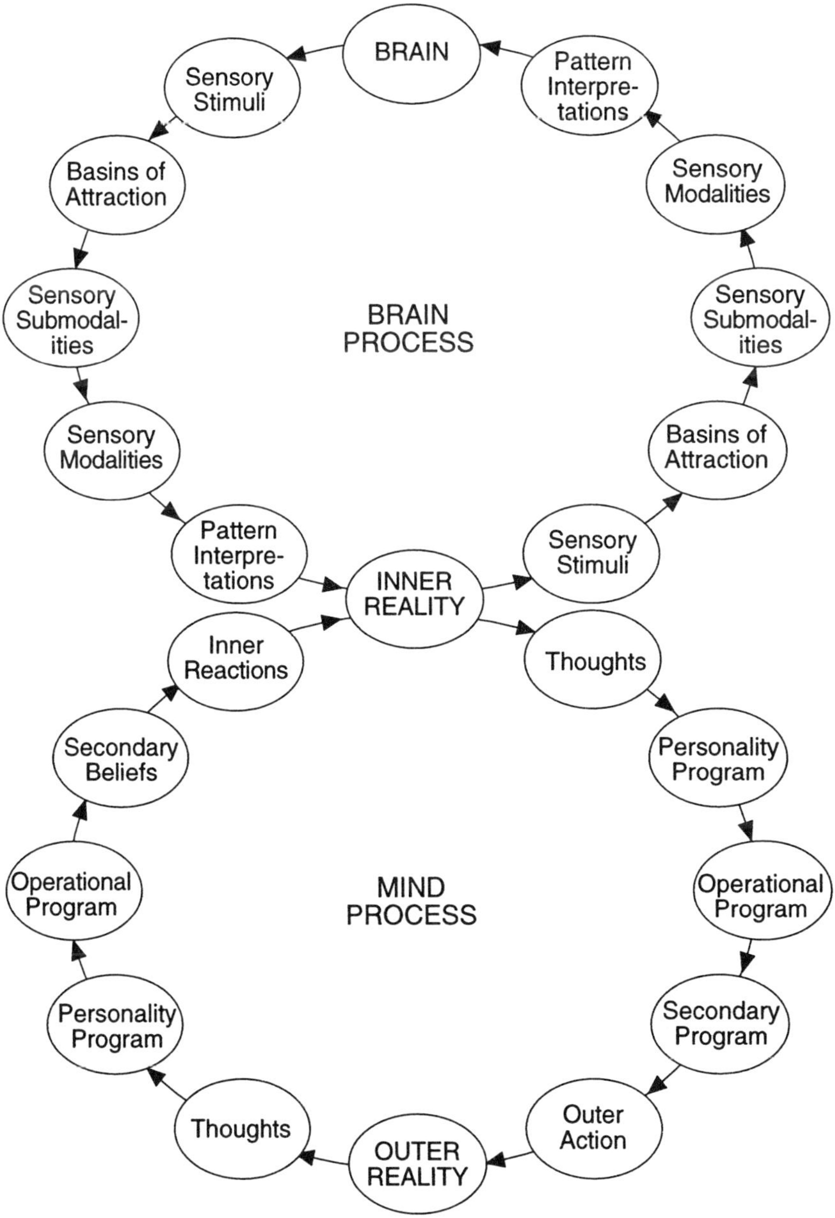

of this process at any given stage in its organization of realities that we truly become self actualized. It is incumbent upon us, rather than others or circumstances in life, to ultimately take control over our own thoughts, emotions and behavior in order to effectively take control over the processes creating our life.

Chart 5, *Comparative Analysis Between the Brain and the Mind,* at page 78 sets forth a comprehensive list of items describing the differences between the mind and brain in terms of structures, functions and processes. A significant feature of this chart, as already explained, is that the brain subserves the mind and functions to convert incoming stimuli into neurological formats that the mind can understand and process in the chain leading to the creation of realities in life. Highlights include the fact that the brain always seeks states of equilibrium whereas the mind can live with conflicting thoughts which create cognitive dissonance for long periods of time. The brain can simultaneously process literally thousands of bits of information whereas the conscious mind can only handle a very limited number of variables at any given time. The quantity of items the conscious mind can organize meaningfully is usually placed at 7 items give or take 2.[9] However, in the final analysis, both systems, the mind and brain, hold unlimited latent capacity for the organization of an endless variety of inner and outer realities in life if only accessed and properly programmed to do so.

Summary on the Relationships of the Mind/brain System

Chart 6, at page 80, *"Dynamical Relationships between the Conscious Mind, the Subconscious Mind and the Human Brain"*, represents the final chart in this series where we compare the structure and function of the mind and brain. This is probably one of the more important charts in this book. If you understand this chart you will have a pretty good grasp of the nature of the mind and brain and how they work together to create the realities you experience each day. This chart sets forth in summary form the relationships among the conscious mind, the subconscious mind, and the brain in terms of structure, function, and process. Viewing the differences of each system side by side gives you a very clear picture of both their similarities as well as their differences. Since there are so many items covered by this chart I will first give you an overall assessment of the chart and then describe some of the more important items in a little more detail.

Considerable confusion arises in discussions concerning the proper way to describe the structure, function and process of the mind and brain. The most common error you hear is that the brain

9 See Ed & Marianne Reese's NLP Practitioner Workbook, 1996.

CHART 5
Comparative Analysis Between the Brain and the Mind

Description of the Brain	Description of the Mind
1. Processes sensory stimuli	Interprets mental Imagery
2. The Subconscious provides the command and control functions for the Brain	Cognitive Structures provides the command and control functions for the Conscious Mind
3. The Brain is contained within the cerebral cavity	The processes of the Mind influences, and is influenced by, its environment
4. Does not process cognitive thought	Processes cognitive thought
5. Does not make volitional decisions	Processes volitional decisions
6. Processes both conscious and subconscious stimuli	Processes both conscious and subconscious thoughts and emotions
7. Organizes data into Basins of Attraction	Organizes mental imagery into meaningful thoughts and emotions
8. Basins subserve analog processes	Creates thoughts in response to Cognitive Structures
9. Processes sensory stimuli unrestricted by time & space	Creates thoughts within linear time and fixed space
10. Medium neurological basins of attraction	Communicates primarily in linguistic patterns
11. Does not make value judgments	Makes value judgments
12. Can process almost unlimited amounts of data simultaneously	Can consciously organize only a very limited amount of data at any given time
13. Constantly seeks equilibrium in its many phase transitions	Can adjust to conflict in thoughts and emotions for extended periods of time
14. Can coordinate the processing of a wide range of possibilities at any given time	Limited in its capacity to coordinate a very large number of possibilities at any given time
15. Does not process goals or purposes in life	Creates goals and purposes to give meaning to life.
16. Responses to stimuli of physical reality	Creates thought in order to affect physical reality
17. Deals in specific data bits to create electrochemical stimuli	Can deal in abstractions and generalizations when creating thought
18. When processing data it operates outside both time and space	When creating thought it is restricted by space, time and a physical world.
19. Provides for the automatic functions of the human body	Makes conscious decisions concerning the human body
20. Has large capacity for expansion in processing	Has considerable untapped potential for expanding conscious awareness
21. Has not reached its potential	Has not reached its potential

thinks. It clearly does not originate or process cognitive thought directly. It simply organizes incoming stimuli into the electrochemical substances of its neurotransmitters. It is these neurotransmitters that allow the neuron cells to communicate with one another. Using the energy and matrix configurations of its neurotransmitters the brain organizes similarities between neuron cells into what we have described as basins of attraction and attractor pathways. The point here is simply that the brain only converts incoming stimuli into electrochemical impulses which are then organized into neurological basins and pathways. This is the only thing the brain does with respect to the chain in the process of organizing thought. It does its job extremely well. It just does not organize thought directly and never will. That is just not its job. It already has enough to do.

Up until now we have not addressed each item in every chart. It was left to the reader to review the information in the charts. However, with respect to this Chart we will be addressing most of the items for, as I just said, an overall understanding of this chart will give you a pretty good understanding of the theory and techniques concerning the mind and brain described in the following chapters.

With respect to the first column "System Structure", we have already seen that the conscious mind uses a system of cognitive structures to organize conscious thought. Conscious thought is organized analytically or logically. Cognitive structures are covered in Chapter 9. As I have indicated, the subconscious mind uses analog processing to organize information. The specific type of processing system used by the subconscious mind is similar to a parallel distributed processing system".[10] We refer to a parallel distributed processing system as a "PDP". What a parallel distributed processing system is and how it functions is discussed in Chapter 6. As indicated earlier, a PDP could be described as a type of parallel computer instead of a binary serial computer. It has no central processor, unlike our binary serial processor of our PC home computer. The operating units in the PDP system operate on an analog basis where each unit within the system communicates with its neighboring units. Each unit has a specific quality that is referred to as the unit's weight or value. The unit's value simply consists of the energy and its matrix configuration. Depending on the positive or negative polarity of the unit's energy the units either attract or repel other units within the system. In this way the units form patterns which in turn form networks of relationships based upon the attraction and repelling effect of the individual unit's values. PDP's do not follow fixed programmable rules as do binary serial processors but can be taught to produce the most

10 For an excellent discussion on PDPs see William Bechtel & Adele Abrahamsen's book, *Connectionism and the Mind.*

CHART 6

Dynamic Relationships of the Conscious Mind, Subconscious Mind and the Brain

	System Structure	Processing System	Conscious State	Space Orientation	Time Orientation
Conscious Mind	Cognitive Structure	Logical Thought	Consciousness state	Three Dimensional Space	Linear Time
Subconscious Mind	Analog System	Analog Processing	Subconscious state	Non-Local Space	NonLlinear Time
Brain (Neuro Network)	Neuro Networks	Electro-Chemical Processing	Nonconscious state	Non-Local Space	NonLlinear Time

	Program Format	System Constraints	Program Structure	Program Parameters	System Orientation
Conscious Mind	Semantic Language	Logical Hard Constraints	Intellectual Thought Processes	Directed by Belief	Primarily Outer Oriented
Subconscious Mind	Abstract Symbols and Metaphors	Weighted Soft Constraints	Pattern Integration Process	Directed by Shifts in Weights	Primarily Inner Directed
Brain (Neuro Network)	Basin Configurations	Electro-Chemical Constraints	Neurological Pattern Process	Directed by Shifts in Neuro-transmitters	Exclusively Internal Process

CHART 6 continued

Dynamic Relationships of the Conscious Mind, Subconscious Mind and the Brain

	Parameter Constraints	Environmental Relationship	Prime Directive	Self Awareness	Process Orientation
Conscious Mind	Personality Constraint	Operates Independantly of Environment	Survival-propogation of the species	Clearly Defined Sense of Self	Central Processor-Self Awareness
Subconscious Mind	Subconscious Weight Constraints	Inter-dependent Relationship with Environment	Program Process Completion	Non-Local Operating Reference	No Central Processor
Brain (Neuro Network)	Neuro-transmitter Stimulus Constraint	Inter-dependent Relationship with Environment	Program Process Completion	Non-Local Operating Reference	No Central Processor

	Time/Space Orientation	Maturation Mode	Volitional Capacity	Causal Relationships	Cognitive Abilities
Conscious Mind	Local	Primarily Adult	Volitional Will	Cause and Effect Connection	Thought Relationships
Subconscious Mind	Global	Primarily Child	Automatic Response	Interrelated/Interconnected	Analog Relationships
Brain (Neuro Network)	Global	Adult and Child	Automatic Response	Interrelated/Interconnected	Neuro Network Relationships

CHART 6 continued

Dynamic Relationships of the Conscious Mind, Subconscious Mind and the Brain

	Spatial Orientation	Causal Orientation	Logical Orientation	Programmable Potential	Input/Output Capacity
Conscious Mind	Limited/Linear Sequential	Causal Constraints	Logical Constraints	Unlimited Morphological Potential	Potential to Expand Input/Output Source
Subconscious Mind	Holistic	Unlimited Processing Potential	Unlimited Capacity Potential	Unlimited Morphological Potential	Potential to Expand Input/Output Source
Brain (Neuro Network)	Holistic	Unlimited Network Capacity	Unlimited Network Capacity	Unlimited Neurological Potential	Expansion of Neurological Capacity

	Expansion of Capacity	Expansion of Awareness	Creations of Inner Reality	Creations of Outer Reality	Evolution of the System
Conscious Mind	Unlimited	Unlimited	Excess Program Capacity	Excess Program Capacity	Toward Completion
Subconscious Mind	Unlimited	Automatic	Excess Program Capacity	Interrelated/Interconnected	Toward Completion
Brain (Neuro Network)	Unlimited	Automatic	Neurological Support Capacity	Neurological Support Capacity	Toward Completion

appropriate output results given the program parameters and the capacity constraints within the system.

Once the initial weights of the units are programmed into the system, the system can take over and reorganize and reprogram itself. It is able to change the values of its units in order to obtain the level of performance needed to reach optimum results. In this respect a PDP system exhibits a type of machine acquisition and organization of data which could be characterized as a type of machine learning. Therefore, these laboratory models provide invaluable information concerning the processes involved in the formation of thought itself at the level of the subconscious mind. PDP's are still in their infancy, however, a considerable body of knowledge has already been gathered from the experimental laboratory models in existence today. We review the similarities between these PDP models and the processes of the subconscious mind in organizing thought in Chapter 6.

The next columns, Processing Systems, Conscious States, Space Orientation and Time Orientation, are all self explanatory based upon what we have already covered. The Program Format item for the conscious mind is shown as "semantic language". This simply means that we use meaningful language as the form in which we formulate thoughts and communicate to others what we are thinking about. We have seen that the subconscious mind uses the language of abstract symbols and metaphors to organize its patterns and relationships at the subconscious level. The brain on the other hand organizes its electrochemical impulses in the formation of its basins of attraction and attractor pathways. System Constraints form some type of constraint upon the operation of the mind and brain in creating realities. We have already seen that the way we view time and space act as a constraint on mentally creating a life for ourself. Other major constraints include ego personality, logical rational thought, and the primary personality and operational systems of the subconscious mind. The cumulative effect of these constraints is to cut us off from the vast latent capacity of both the mind and brain to create expanded realities far beyond anything we could possibly envision for ourselves. How we modify, and in certain instances eliminate, these constraints in order to access and use more of the potential capacity of the mind is discussed in Chapter 9.

I do not want to leave you with the impression that these constraints are all bad. They serve a beneficial purpose. They ground us in the space and time of a three dimensional world where we are comfortable with labeling everything in order to create for ourselves an orderly and meaningful life. If we stripped away all of these constraints all at once we would literally go into sensory overload which would drive us crazy. Therefore, we selectively determine which constraints to loosen, and to what degree they should be loosened, in order to have the best of both worlds. That is, we deal extensively with time and space of a three dimensional world since this is a major constraint. However, we only deal selectively and partially with the negative or self limiting constraints on ego personality and the operational systems of the subconscious mind since the positive aspects of both systems serve a very useful purpose in life.

The remainder of the columns are self explanatory with the exception of the next few items. The column described as "Prime Directive" at page 81, on the conscious level is represented as survival and propagation of the species. This was probably of greater importance in earlier times than it is in modern society. However, as to the subconscious mind and the brain the prime directive is simply program completion. This fact is very useful to know since we will be taking advantage of it in creating desirable realities for ourselves. As to the item described as "Self Awareness", the conscious self has a clearly defined image of self whereas the subconscious mind and brain do not operate from a position of a self but operate on the basis of what we refer to as "nonlocal operating references". Nonlocal operating references are an inherent characteristic of an analog system. As you will see later, changes within the system can propagate themselves almost instantaneously and multidimensionally throughout the entire system. Consequently, as to the next item "Process Orientation" the conscious mind operates similar to a central processor consisting of self awareness. The subconscious mind and brain, as we have already seen, do not have a central processor and operate on the basis of nonlocal operating references within their systems. We deal with consciousness and self awareness in Chapter 4.

With respect to the item described as "Input/Output Capacity" at page 82, when we state that all three systems have the potential to take control and expand the input and output source we mean that we have the capacity to take control, not only over our outer realities in life, but also over the effects the incoming stimuli has upon our operating systems in creating those realities in the first place. The last item "Evolution of the System", like the evolution of life itself, is always toward completion. We have unlimited capacity to direct our own destiny. We need only to understand and take complete control over the systems that create our life realities, and in this way we take control over our own destiny.

This is a brief summary of the comparative aspects of the conscious mind, subconscious mind and brain. You will want to refer to this chart from time to time for each item on this chart will again be viewed in several different contexts in the following chapters. The point of this chapter is to give you a summary overview of self organization systems theory and of the entire structure and function of the mind/brain before we delve a little deeper into their organizational processes starting with the next chapter.

CHAPTER - 3

A Description of the Processing Systems of the Human Brain and Mind

Introduction to the Processes of the Mind

First, we will look at how we actually create inner and outer realities in the present time and space of our existing three dimensional world. We will then look at how we can effectively take control over the creative processes of our mind by expanding our cognitive structure in order to create expanded realities in life. The cognitive structure is the programmable framework of the conscious and subconscious mind in which we organize all of our thoughts and behaviors. Thoughts and behaviors never originate outside of our cognitive structure. Physically, of course, we experience life in the time and space of a fixed three dimensional world. The reason for this is that sensory data is received and organized by the brain and mind to give us an experience of a physical world which exists in fixed time and space and existing outside of ourselves, and largely outside the direct control of our thought processes alone. As you will recall, this is often referred to as living in the zone of the middle dimension. This is the case even though thought, all thought, originates unrestricted by time and space.

Our operating frame of reference has always been that our thoughts, arising within our existing cognitive structure, will always conform to our perceptions of an outside objective world in fixed time and space. *This way of viewing our world has persisted and has become entrenched in Western civilization as not only the proper way, but the only way, of viewing our world realities and our relationship to them.* Unfortunately this view of our world leaves us with very little control over that world. We nevertheless persist in seeing ourselves as living out our lives in the zone of the middle dimension in spite of the fact that quantum physics has shown clearly that the universe and everything in it, including you and me, exist within spacetime of the fourth dimension.

Therefore, our existing cognitive structure needs to be expanded from a perception of an objective world lying outside of ourselves and largely outside of our control, to a perception of a world where we control the process of its creation through our very own thought processes unrestricted by the appearances of time, space and physical reality. Since our perceived relationship to an objective world arises from our own self-limiting cognitive structure and not as a result of anything outside of ourselves, our cognitive structure can be expanded to allow us to take greater control over our physical world. We expand our cognitive structure by adopting Conscious Analog Thinking™ which is explained below. As we expand the cognitive structure by adopting conscious analog thinking we create expanded realities in life.

Paradigm shifts in your thought processes start to take place when you begin to understand that subconscious mind can place the images of your thoughts anywhere in time and anywhere in space. The subconscious mind, operating outside of time and space, has the capacity to sort for endless possibilities to bring to conscious awareness those people, places and things you desire within the time and space frames you choose. However, subconscious mind will only act on beliefs accepted as valid by the conscious mind. Therefore, a considerable number of charts have been included in this chapter to graphically illustrate how we create realities. These have been included because if you don't understand and believe in the process it will never work for you. The subconscious mind will only act on what the conscious mind perceives to be true. Once you fully grasp how the mind operates and not just intellectually understand it, you can begin to allow the subconscious mind to sort endless possibilities and bring to full conscious awareness the many manifestations you seek in life.

To understand how this process works we will start with a brief explanation of how the subconscious mind organizes realities. We are always organizing inner and outer realities in life whether we are aware it or not. In the last chapter we indicated that the conscious mind merely provides us with an awareness of the results of what is happening on the subconscious level of the mind. You will recall that it is the subconscious mind that carries out all of the processes involved in organizing conscious thought and our perceptions, feelings, and behavior. It is our perceptions of reality that is derived from our cognitive thoughts processes. *We perceive an objective world because we believe it to be an objective world for the subconscious mind will only create realities to conform to conscious thought.*. If you think that you really only live in an objective world, you do in fact, only live in an objective world! How the subconscious mind organizes objective realities for the conscious mind is explained in the next section of this chapter. Also explained is how you can expand the cognitive structure and change the old Cartesian way of thinking of an objective world existing in the zone of the middle dimension to thought processes which creates realities holistically in spacetime.

<u>The Process of Creating Personal Realities</u>

Charts 7-1 and 7-2 at pages 89-90 deal with the organization of the various mental processes involved in the creation of life's realities. The left side of Chart 7-1 refers to the systems involved in organizing realities and the right side refers to the processes involved in organizing realities. At the bottom of the chart sensory stimuli received by the brain are converted into the electrochemical substances of its neurotransmitters. Remember, the electrochemical configurations of these neurotransmitters attract neuron cells with similar patterns to form basins of attraction and attractor pathways. Basins of attraction subserve the processes of the subconscious mind. Energy and matrix configurations of the basins provide the subconscious mind with the values of its patterns which form the subconscious analog system. These patterns form mental imagery which the conscious mind interprets in order to give meaning to our environment. It does so by labeling these patterns so that we might identify the many physical representations in our environment. Meanings created from these mental images also form the substance of our inner realities. Remember I indicated that if we continue to label everything in our environment we would eventually end up with a considerable body of knowledge and an attitude. These patterns of mental imagery are what constitute that considerable body of knowledge.

Finally, our inner reality acts as a platform upon which we create thoughts and beliefs about ourselves and our world. The subconscious mind sorts endless possibilities to bring to conscious awareness those people, places and things to organize an expression of what that exterior reality should look like. What it ends up actually looking like is based on our inner realities of what we think it ought to look like. The intention of our thoughts programs the subconscious mind to find the most appropriate expression of the exterior reality that matches up with our patterns of inner reality. This is the mechanism that the conscious and subconscious minds use in creating inner and outer realities.

However, we are not consciously aware that it is the many processes taking place in our head that are actually creating for us a personal awareness of our inner and outer realities. We erroneously conclude that it is self awareness that is creating all of these realities that we are experiencing in our daily life. Worse yet, we even conclude, again erroneously, that when things go wrong in our life it must be that things in our outside world are the cause of our problems. We think bad things happen as a result of: other people are trying to do us harm; no one really cares about us; no one really wants to help us; circumstances in life are against us; we just happen to be born into the wrong family, at the wrong place, at the wrong time; we are just plain unlucky; we were born under the wrong astrological sign; or we

CHART 7-1

The Organization of the Mental Processes of the Mind Involved in the Creation of Life's Realities

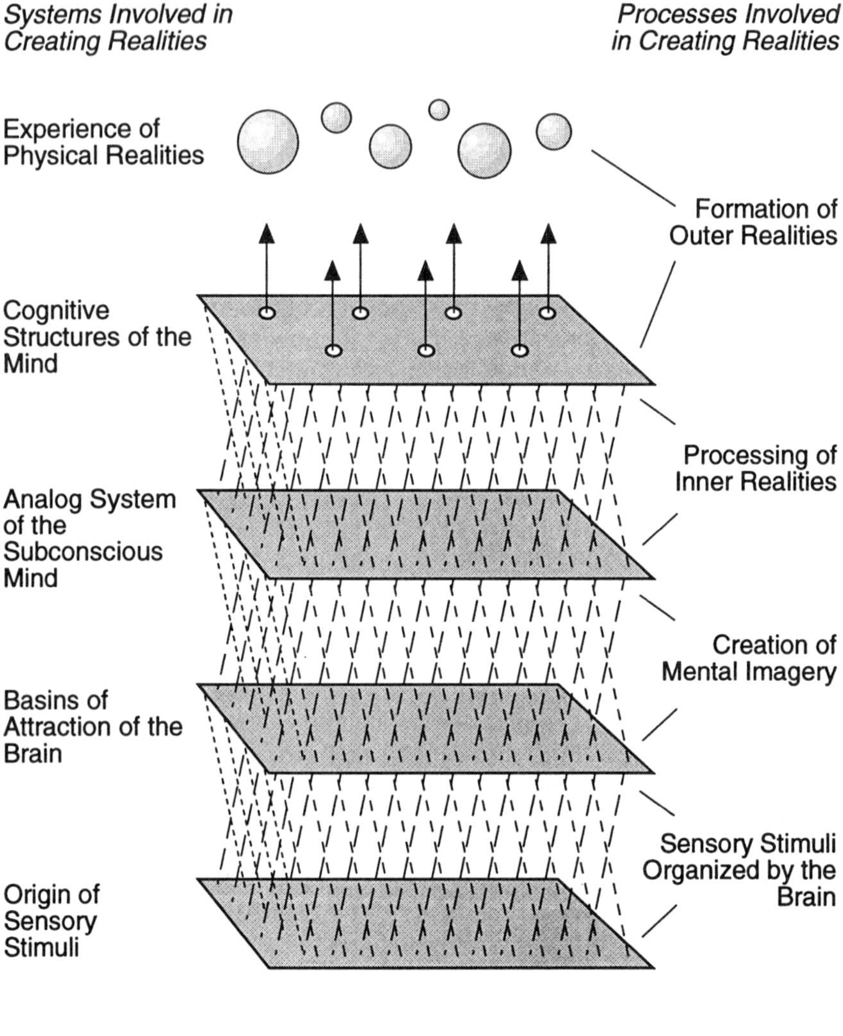

The sensory stimuli of the Brain creates electrical chemical patterns creating Basins of Attraction. The analog system of the subconscious mind creates mental imagery based on the arrangement of the Basins of Attraction. The cognitive structure then creates patterns from the mental imagery of the analog system. The individual Self creates Inner Realities from these Pattern Interpretations. The Subconscious Mind then brings to personal awareness those places, people and things needed in order to manifest the desired outer physical realities in the life of the individual.

CHART 7-2

Cybernetic Feedback Loops within the Spatial Field of Outer Reality Feeding Data back into the Operating Systems of the Mind

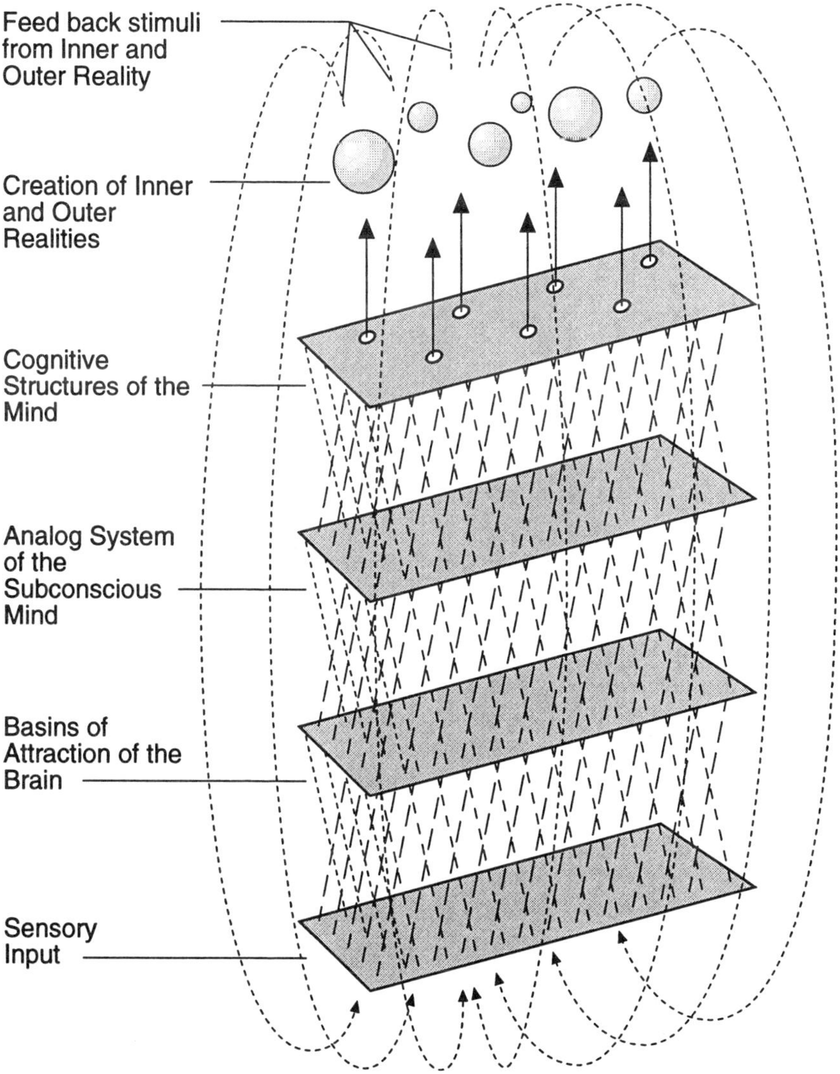

Sensory feedback from the environment and its outer realities feeds data back to the brain to either positively or negatively affect the brains neurological processes. The brain then creates either positive or negative signals to the mind which then modifies its mental imagery which is interpreted by self awareness that it is experiencing either a positive or negative inner and outer environment. Self Awareness then takes appropriate actiion in response to its interpretation of its environment are just experiencing bad karma; or any combination of the foregoing. It seldom, if ever, occurs to us that the origin of our troubles lies deep within the operating systems of our brain and mind. It's much easier to blame others, or the circumstances in life, for our problems than it is to take responsibility for our problems and seek their solutions within our own minds.

Chart 7-2 depicts the same process that is set forth in Chart 7-1. However, it adds the feedback back effect of the cybernetic loops operating in the spatial field of our outer reality. These cybernetic loops feed data back into the operating systems of the subconscious mind. Sensory data from the environment is fed back into the brain to affect, either positively or negatively, the brains neurological processes. If the brain's basins of attraction are positive then the incoming stimuli will be interpreted positively. However, if the basins are negative the incoming stimuli will be interpreted negatively. Depending upon the energy of its basins the brain will then send either positive or negative signals to the subconscious mind. The subconscious mind will modify the patterns of mental imagery in a positive or negative fashion depending upon how it interprets the energy of our basins of attraction. The conscious cognitive structure of the mind will then interpret such patterns to determine whether we live in either a positive or negative environment. Self awareness then takes whatever action it perceives to be appropriate depending upon how it interprets its environment.

For instance, the social amenity of a warm handshake comes from the historical practice of keeping one's hands in clear sight when meeting up with a stranger as a gesture of friendship or at least meaning him no harm. It was believed that if your hands were hidden in your coat pockets you might be holding a knife or pistol and meant to do harm. Unfortunately, quite often we misinterpret incoming signals from our environment and read neutral signals or even positive ones as being negative. As a result we often take inappropriate, defensive, or even hostile action based on a misreading of the signals. This isn't bad karma; it is just a bad response to life.

Charts 8-1, 8-2 and 8-3 at pages 92-94 show how easy it is to misinterpret incoming stimuli based on whether we have a positive or negative disposition in life. Chart 8-1 is a simplified cybernetic loop showing that it is the incoming stimuli from our environment that form the subconscious patterns that give rise to the formation of our thoughts and feelings in response to our environment. These thoughts and feelings create for us an experience of our environment. In turn, our experiences create additional stimuli which create further thoughts and feelings. This process goes on and on creating

CHART 8-1

Simple Cybernetic Loop in The Creation of Reality

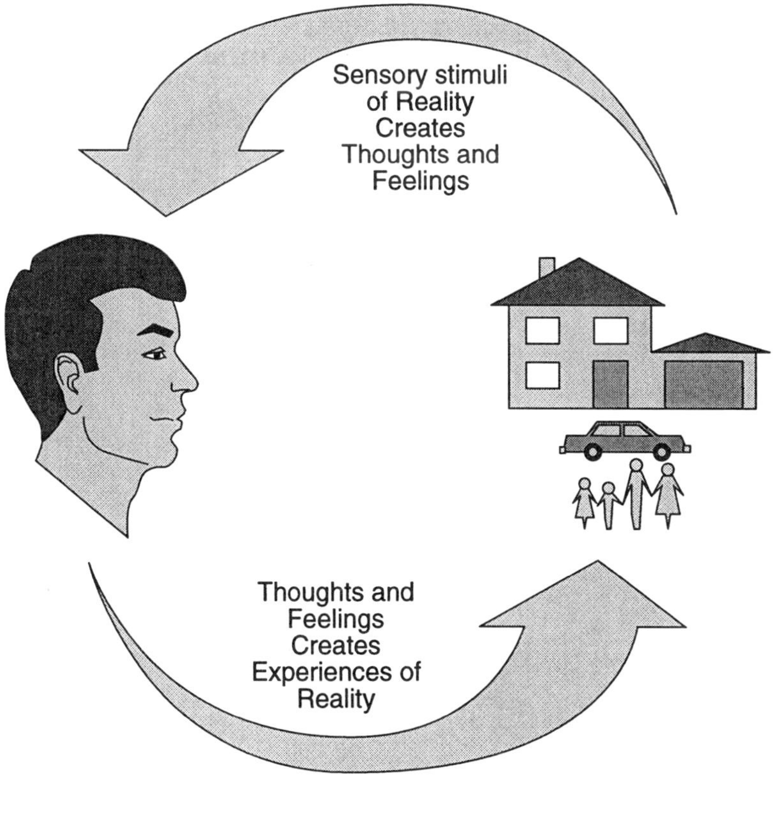

CHART 8-2

How reality affects the operational systems of ego personality, how ego personality affects the brain's basins and the subconscious analog systems and how these internal systems affect our experiences of reality

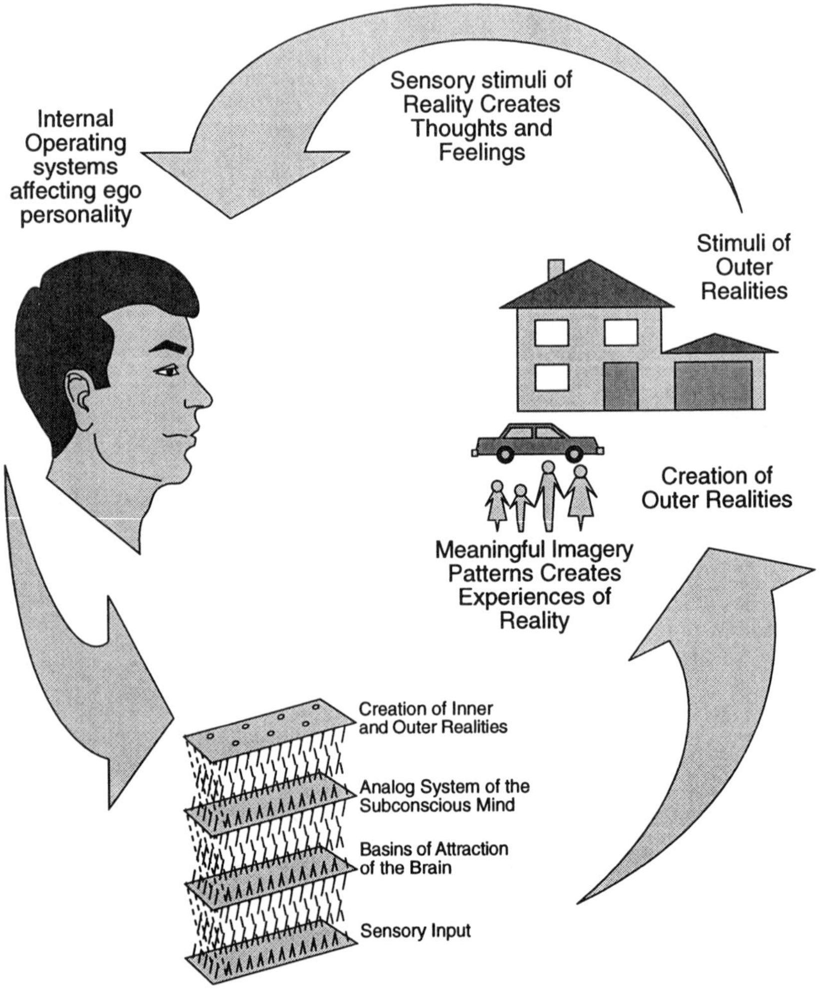

The feed back affect of the cybernetic loops feeds stimuli from outer reality into the internal operating systems of the ego personality and affects, either positively or negatively, ego personality and its operational systems. Ego personality and its operational systems then create either positive or negative meaningful patterns of mental imagery which, in turn, creates thoughts and feelings which ego personality interprets as positive or negative experiences of inner and outer reality.

CHART 8-3

Cybernetic Loops depicting the effect Stimuli have in Creating Positive and Negative Realities

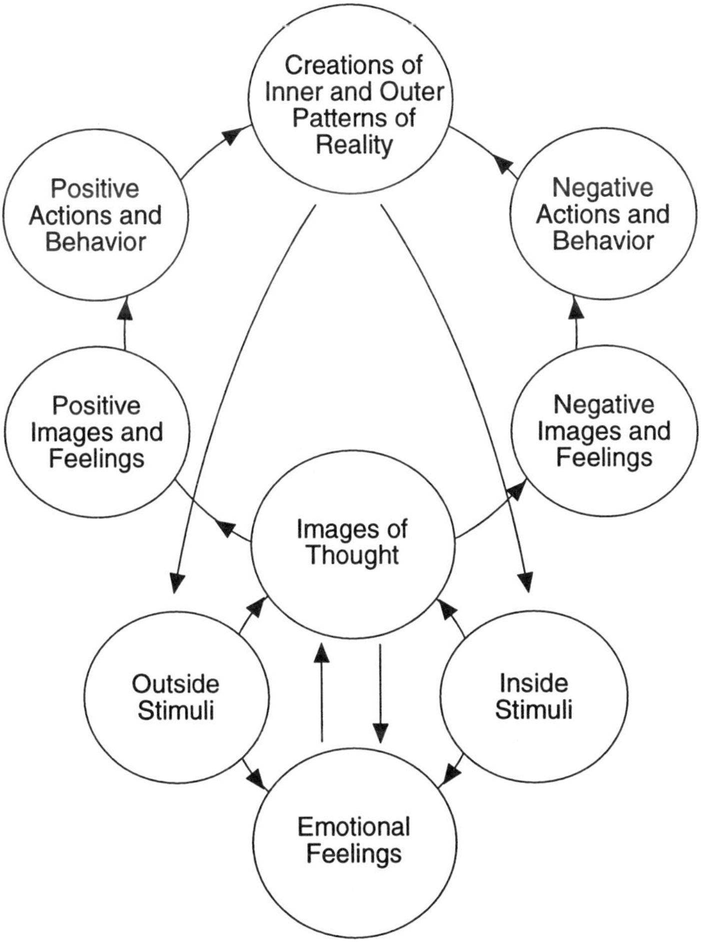

and dissolving inner and outer realities throughout our entire lifetime.

Chart 8-2 is similar to Chart 8-1 except that it factors into the cybernetic loop the effects ego personality has on the creation of inner and outer realities. In this chart the cybernetic loops feed stimuli from our environment into the internal operating systems of the subconscious mind and, either positively or negatively, affect ego personality and its operational systems. Our personality and operational systems then create thoughts and feelings that ego personality interprets as meaningful patterns of mental imagery. It is ego personality's interpretation of our mental imagery that creates our experiences of inner and outer reality. However, if ego personality and our operational systems are primarily negative the incoming stimuli will be interpreted in a negative fashion giving rise to further negativity in our environment. How we change our negative primary programming to positive programming is addressed in Chapter 8. How we expand the parameters of ego personality in order to break these negative cycles is explained in Chapters 9 and 10.

Chart 8-3 illustrates in a little different fashion the effects stimuli have in creating positive and negative realities in life. This then gives you a brief overview of how the conscious mind, the subconscious mind and the brain create experiences of inner and outer realities. We will now address how we might use these processes of the mind and brain to organize more positive realities for our greater enjoyment. We need to expand our present analytical way of thinking to also include conscious analog thinking in order to create more positive realities in life. Chart 9, at page 96 entitled *Conscious Analog Thinking,* provides a summary description of conscious analog thinking. How we adopt conscious analog thinking is addressed in the next section and is explored in much greater detail in Appendix One entitled *Conscious Analog Thinking.*

<u>Conscious Analog Thinking</u>

In this section we explore how analog thinking actually works in assisting you in dissolving past negative realities and in creating new expanded future realities. In order to do so we will again revisit different ways of interpreting both space and time. Chart 10-1 at page 98 depicts our inner mental construction of space. The analog system of the subconscious mind forms our impressions of spatial depth. We perceive the placement of physical objects in the three dimensional space of our world. Everything in our physical world takes place in three degrees of space, that is, backward and forward, up and down, and left and right. Spatial dimensions are organized by the analog systems of the mind and brain. The analog system of the subconscious mind creates an experience of our field of three

CHART 9
CONSCIOUS ANALOG THINKING

The following sets forth in comparative fashion the differences between the processes of conscious analytical thinking in fixed space and linear time of an objective three dimensional world and conscious analog thinking which conceptualizes time and space in the en bloc spacetime of the fourth dimension:

Comparative Analysis in terms of the Conscious and Subconscious Mind

1. Analytical thinking primarily follows logical or causal connections. Analog thinking primarily organizes similarities between broad patterns and relationship in life that are not necessarily logically or causally connected but rather are related by reason of the closeness of the similarities between or among the many patterns experienced in life.
2. Analytical thinking often involves reductionist methods of understanding things. That is, the whole or entire picture of a given situation is understood from an analysis of its parts. Analog thinking often starts with an understanding of the whole or the big picture in life and then discerns the smaller patterns and relationships as they relate to the whole.
3. Analytical thinking relies primarily on the mental and physical efforts of the individual to bring into manifestation those things desired in life. Analog thinking uses very little effort but rather relies on the unlimited expressions of the subconscious mind to organize themselves into manifestations desired in life.
4. Analytical thinking relies primarily on conscious awareness to attain what is needed in life. Analog thinking relies primarily on the subconscious processes of the mind to provide an awareness of what is needed in life.
5. Analytical thinking usually sets specific goals in life and formulates plans and strategies to accomplish those goals. Analog thinking primarily relies on the unfoldment of life itself in order to give meaning and expression to life.
6. Analytical thinking primarily relies on the intellect to solve problems and to set the agenda for life. Analog thinking primarily relies on instinct and intuition to provide guidance and direction concerning the affairs of life.
7. Analytical thinking often involves taking control, exerting discipline and using power over self, others and the environment. Analog thinking involves letting go and releasing control and the hold we have on the events and things of this life and relies on the higher energies and images of the conscious and subconscious mind for direction and guidance in life.

dimensional space and an experience of physical objects existing in that space. By processing data in this manner our sensory receptors give us experiences of three dimensional space. The upper diagram on the left in Chart 10-1 shows what appears to be a solid sphere and a solid cone in three dimensional space. The diagram on the right shows the exact same circular objects. This time, however, they are shaded with simple lines instead of shaded surfaces resulting in our visual receptors interpreting these objects as being flat two dimensional discs. Yet both sets of objects are two dimensional figures. Our visual receptors interpret them differently because we have trained our subconscious mind to give us an experience of depth in order for us to function effectively in a three dimensional environment. If we were to interpret everything as being flat, and we had no depth perception, we would be running stop lights and bumping into everybody and everything. So processing everything in three dimensional spatial frames help immensely in maneuvering through the physical and mental obstacles of life.

The bottom diagram is a classical diagram we have all seen at one time or another. It gives us both a background and a foreground interpretation. What we see depends on our visual interpretation of whether the objects appear in the foreground on in the background. When we focus on the background we see a chalice. When we focus on the foreground we see two faces in profile looking at each other. The point of this chart is to help you understand that it is our perceptual mental processes of how we view our world that give us a three dimensional view of reality. This is not to imply that there is anything wrong in perceiving our world in this fashion for, in fact, it serves us very well for it helps us keep track of everything in our world. However, the point I am making here is that the world out there isn't the one organizing our experiences of it this way. We ourselves organize our experiences of it this way. Consequently, we do not have to keep on believing that we only live in an objectified world existing outside of us and largely outside our control. We do not have to organize it only in this fashion. We can start to take control over how we experience our world and mentally organize more desirable experiences of our world through our thought processes alone. How we accomplish this is the subject of this Chapter.

In similar fashion Chart 10-2 at page 99 depicts how we mentally construct linear time frames in ordering our life in sequential fashion in our three dimensional environment. Again it is our visual receptors that provide us with not only spatial dimensions but also sequential time dimensions. The little stick man at the upper left of the diagram has just left the past behind him and is in the present moment

CHART 10-1

The analog systems of the subconscious mind create for us an experience of the field of three dimensional space and a sense of physical objects existing in that three dimensional space

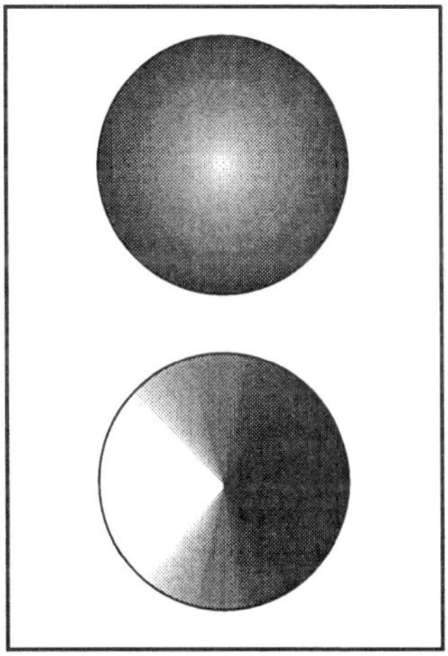

Our visual interpretation creates for us sense of a sphere and a cone

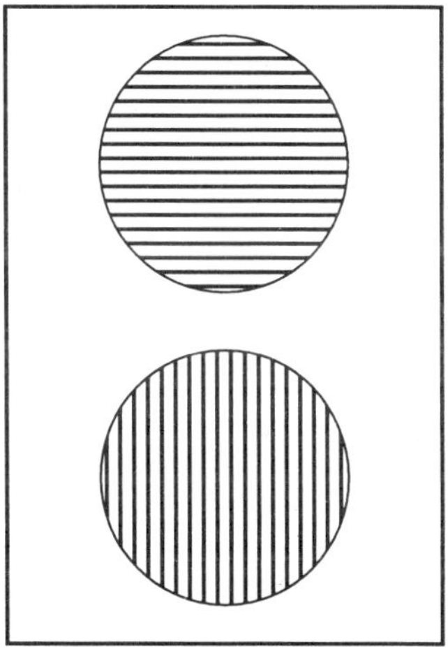

Our visual interpretation creates for us a sense of flat shaded surfaces

Visual interpretation of either foreground or background frames of reference

CHART 10-2

Mental Representations of Time and Space in a Three Dimensional World

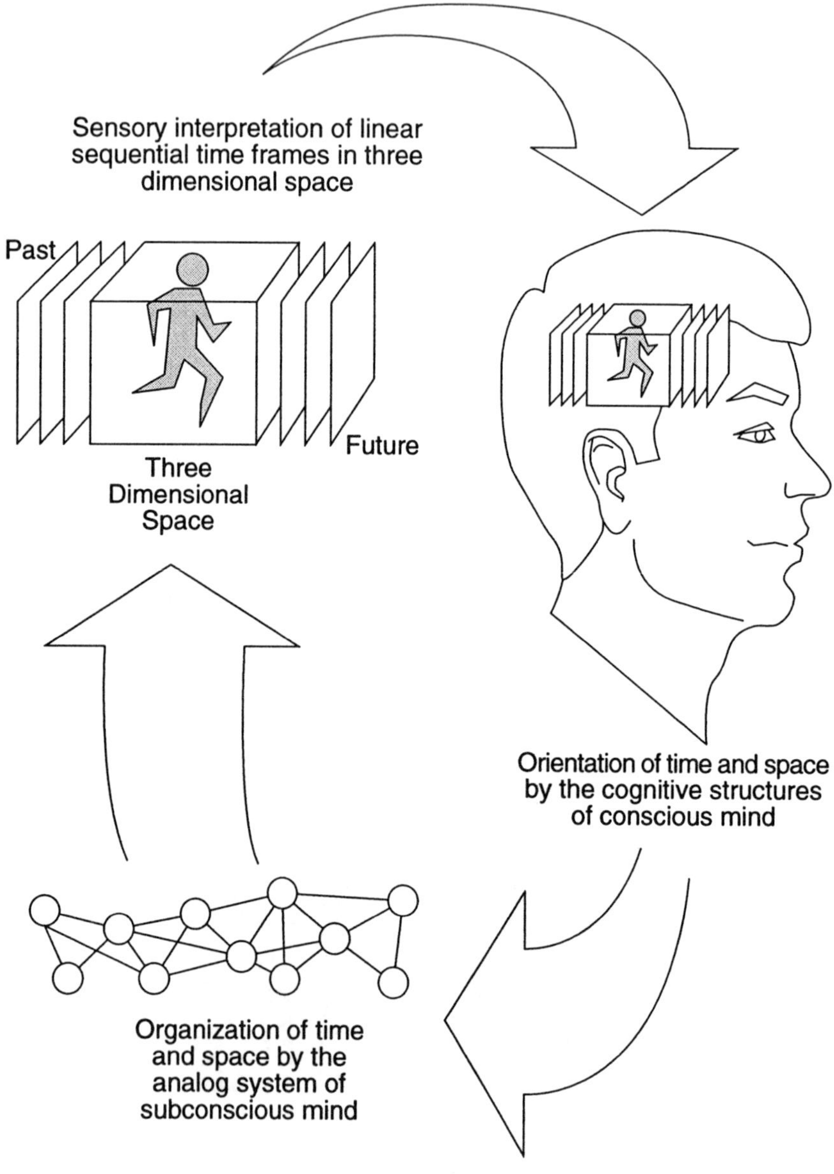

CHART 10-3

Creation of future realities in space and time of our three dimensional world

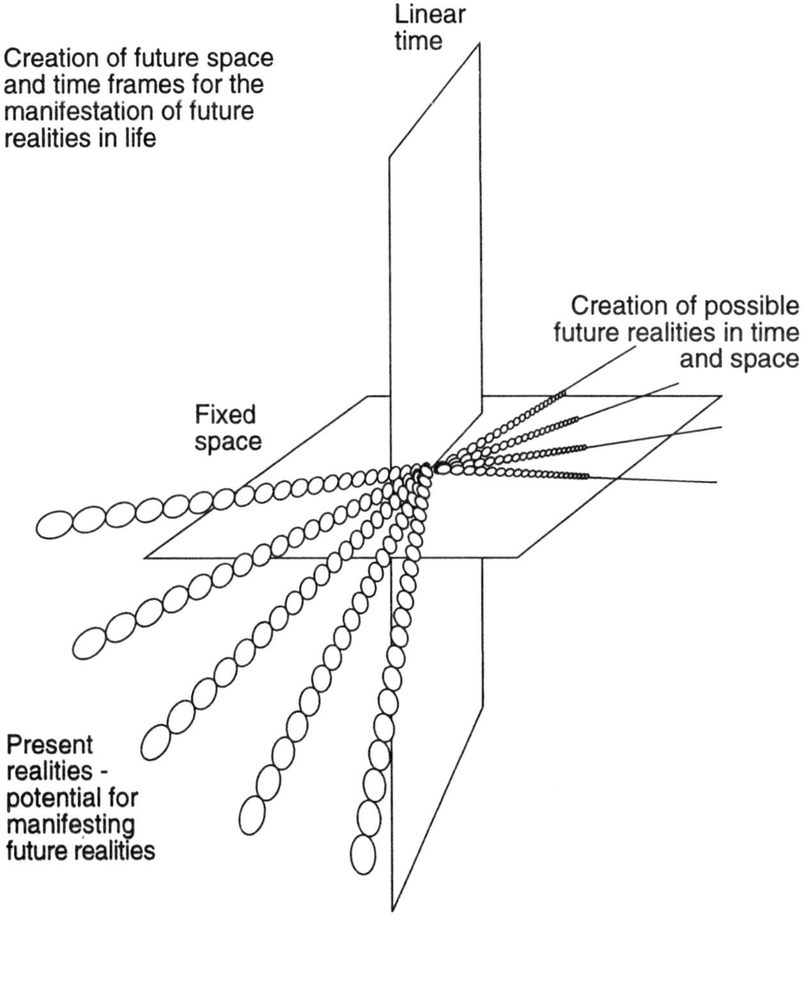

SPACETIME

The subconscious mind can search for almost infinite possibilities of people, places and things in spacetime in order to bring to conscious mind an awareness of future realities in the linear time frames and fixed space frames you desire for the manifestation of your future physical realities in life.

running head long toward tomorrow. It is the cognitive structure of the conscious mind and the analog systems of the subconscious that support and perpetuate this three dimensional view of our world taking place in linear sequential time frames.

When we add fourth dimensional reality to the three dimensional world we create the same effect as when we viewed our two dimensional shadow in three dimensional space. You remember the question is not what is the exact size of your shadow, but rather what are the coordinates to locate the size and shape of your shadow at any given time in three dimensional space. Similarly the same applies when viewing three dimensional space from the fourth dimension. We are not subconsciously prevented from creating and placing realities in time and space frames of our own choice as you would normally conclude from a three dimensional perspective of our world. We do not somehow distort three dimensional space when viewing our world from the fourth dimension; we merely expand our range of possibilities in creating realities for ourself in three dimensional space. We just need to keep in mind that when we subconsciously create realities from the spacetime of the fourth dimension we need to consider the coordinates for the timing and placing of realities in three dimensional space. What is happening is that you are subconsciously creating realities outside of linear time and three dimensional space. What this means is that you are simply organizing realities at the subconscious level in the fourth dimension. When you create realities from the fourth dimension you need to know specifically when and where to place realities in your three dimensional world. The energy and matrix configuration of your thoughts will program the subconscious mind to sort for expressions of your creations in the time and space frames you desire. The subconscious mind not only sorts for the manifestation of the particular reality sought but also sorts for the timing and placement of those realities in your physical three dimensional world.

For an abstract depiction of spacetime refer to Chart 10-3 at page 100. In the lower left we are creating present realities in life. The creation of our present realities will affect, either positively or negatively, our future realities, which are running out to the right of the Chart. All of this activity is taking place in spacetime although we are not consciously aware of it. The subconscious mind can search for almost infinite possibilities of people, places and things in spacetime in order to bring to conscious mind an awareness of future realities in linear time frames and fixed space frames you desire for the manifestation of your future physical realities in life. As you become aware of how the creation of present realities affect the creation of future realities in spacetime you can start to take effective control over the process by which we create realities in life and create for yourself positive future realities in time frames and space frames of your choice.

In Chart 10-2 we left our little stick man running out from yesterday and heading toward tomorrow. However, unbeknownst to him tomorrow never arrives when viewed from the subconscious level of the fourth dimension. He will always be in the present moment while heading toward tomorrow. By now you are probably thinking to yourself, "Well, if today is Sunday, I will be in tomorrow which is Monday when 24 hours have elapsed, so you are clearly wrong in your analysis about time and space." Your statement is correct when you consciously view life from the third dimension. It is incorrect when viewed subconsciously from the fourth dimension. What you have done is labeled Monday as being tomorrow. But you are really always just in the present moment when viewed from the fourth dimension. It is neither the future nor is it the past. Again your response to all of this might be, "Well, the clock and calendar say Monday is tomorrow and my body is aging, day by day, so I know for a fact that Monday is tomorrow." True the calendar says Monday is tomorrow, and in fact your body is aging, day by day, for sure. However, all such phenomena are consciously taking place in a three dimensional frame of reference because we have come to believe that this is the only world in which we live. *And more importantly we have come to believe that this is the only real way in which physical reality can exist.* We believe there is no other way of interpreting time and space. We do not exist in any other dimension.Consequently, it never occurs to us that maybe we also live out our life in the fourth dimension. We would never have found this out if the quantum physicists did not discover spacetime of the fourth dimension existing at the quantum level and at the universe level of all existence. If you were to view consciously your life from the perspective of the fourth dimension you would come to realize that everything is happening right now and that there really are no tomorrows or yesterdays. Remember in the fourth dimension your three dimensional world is merely a small slice out of the spacetime of the fourth dimension. In this context that small slice looks in our present moment in time that yesterday was the day before today and that tomorrow is the day following today. From this perspective in life your are absolutely right. *However, it is from the perspective of spacetime of the fourth dimension that constitutes the level at which we subconsciously create reality.*

Refer again to Chart 10-3 at page 100 to get a better visual impression of spacetime of the fourth dimension. As we said the vertical plane represents linear time and the horizontal plane represents three dimensional space. The circles in the lower left corner of the chart represent the realities we are organizing in present time. The fixed point in the center of the chart represents a future point in time and space where we desire to manifest our future realities. It is the intentionality of your conscious thoughts and the organization of the analog patterns of your subconscious programs that determine the exact nature of your future realities. Future realities are shown as various possibilities running out to the right of the chart. *All of this activity is taking place in spacetime.* If you visualize the center point in the chart as present time and the lower left realities as existing in past time you can understand how it is that in the spacetime of the subconscious mind you change a negative past to a positive past through the intentionality of your thoughts. This is possible at the level of the subconscious mind because space and time are not factors in changing a negative past to a positive past. By changing a negative past you are able to create a positive future shown as the future realities running out to the right side of the chart.

Therefore from the standpoint of creating for ourselves a life, the fourth dimension is by far the most important dimension for it forms our realities in life. From the standpoint of merely living out our life it takes place in our three dimensional world. I ask you which perspective is of greater importance to you? Is it more important to actually create a life or merely live a life? If you want to actually create a life for yourself start thinking fourth dimensionally. If you merely want to live out a life continue to think three dimensionally.

Do not misinterpret this seemingly abstract philosophical discussion on time and space for it holds very real consequences in ordering your daily affairs. For the real point of all of this is that we will consciously continue to order our life in a three dimensional world for that is the way we perceive our world. This gives order and stability to our life. However, by also adopting a type of fourth dimensional conscious analog thinking simply expands the range of possibilities we have in creating realities and placing them in our three dimensional world. For as I have repeatedly said the subconscious mind operates outside of time and space as we know it and can just as easily create awareness of realities for yesterday as well as for tomorrow.[1] This opens up entirely new possibilities for each one of us in clearing up unfinished business in the past and for creating entirely new realities for ourselves in the future. How exactly we go about doing this is best explained by the following example.

---

1 This is just another way of saying that the subconscious mind organizes realities in the fourth dimension outside linear time and fixed space.

How Ricky Deals with His Past Childhood - Or How to Clean Up a Messy Personal History

We will again revisit Ricky, Tommy, and Betty in order to give you a clearer picture of exactly how you can access and change a negative past in order to create a positive future. You will recall we left Ricky at the kitchen table talking to Betty Sunday morning November 23, 1997. The following day Ricky arrived early at the office. While he felt he was prepared for the special regional sales meeting that Mac called for that afternoon, he still felt a little nervous about answering Mac's questions about why his sales have fallen off these last three quarters. It was hard enough to keep your cool under very difficult circumstances, but Mac's relentless and harsh style of questioning and probing made matters even worse.

Ricky was visually relieved when Terri, his secretary, buzzed him and told him that she just got a call from corporate headquarters. Mac suffered a mild reoccurrence of his bleeding ulcer and was rushed to the hospital Sunday night for emergency surgery early Monday morning. Mac's executive assistant, Cheryl, came on the phone to warn Ricky that before Mac left for the hospital he instructed her to tell Ricky that the meeting would be rescheduled for the 23rd of January. By then the year end sales figures would be in and they better be good, otherwise Ricky's neck was on the chopping block, not to say anything about his career. No wonder Mac was rather unaffectionately called "Mac the Knife" behind his back by his subordinates. Ricky felt that he had a 60 day reprieve and that he better act fast to save both his skin and his career. He didn't even worry about a year end promotion or raise at this critical time. He did hear that Corporate was thinking about consolidating his eastern region with the western region of his counterpart, A.Tad Tiredly. Prior to this crisis Ricky was hoping that he would be in line for the Executive V.P. position over the consolidated regions. Now things weren't looking all that good. At that moment Ricky's secretary buzzed him to pick up on line two in his office. It was Tommy.

Tommy was as excited as ever. He told Ricky that he was getting excellent results from something called Conscious Analog Thinking™ in the fourth dimension and MetaCreation™ and that he wanted to share it all with Ricky as soon as possible. Since there wasn't much time Tommy figured that it would be O.K. with Ricky that he had gone ahead and purchased airline tickets. He, his wife Claire, and his two teenaged sons, Philip and Andy, were flying to Atlanta that night to be with Ricky, Betty and Louanne for the Thanksgiving Holidays.

Ricky picked Tommy and his family up at the airport late that night and they all went back to Ricky's place to have a late dinner with

Betty and Louanne. After Betty and Louanne went home and Tommy's family were all in bed, Tommy convinced Ricky that he should seriously consider trying out the new analog thinking strategies. He said it might help Ricky sort out many of the life issues he was dealing with lately. Maybe he might even find the answers concerning his relationship with Betty. Since Ricky's place was so crowded with holiday guests the two of them decided on the spot to leave early the next morning for an impromptu trip to the north Georgia mountains for the next two days before Thanksgiving. They set out early the next morning as planned and had breakfast by 9:30 a.m. in Helen, Georgia. Helen, Georgia, is a quaint Alpine village deep in the north Georgia mountains. As luck would have it Tommy was able to rent that special cabin in the mountains his family had used on many occasions in the past.

After unpacking their bags they hiked up to Anna Ruby Falls and then to Dukes Creek Falls where the mountain scenery was absolutely beautiful. The fall foliage was gone but the evergreen forest, the rocky cliffs and waterfalls had a special beauty and serenity all their own. They had a late lunch in one of the Bavarian restaurants in downtown Helen and after some holiday shopping it was evening before they knew it. They went back to their cabin and had a light snack for dinner and Tommy started a fire in the large fireplace in the living room to take the evening chill off the cabin. Tommy and Ricky's' conversation drifted back to those lonely days in Hutchens, Minnesota, where they grew up.

Tommy stated to Ricky that he could again revisit Hutchens and maybe find some of the answers he has been seeking these past years. Ricky was puzzled by Tommy's comment since they where in Helen, Georgia, over 1,400 miles from Hutchens. Further, those childhood events took place well over 45 years ago. Tommy explained that time and space were relative in the spacetime of the subconscious mind and that Ricky could time travel anywhere in the world and access the past as well as the future. If he wasn't satisfied with what he found he could make the appropriate changes in order to make his life more satisfying and meaningful. This sounded like a really good idea to Ricky and he stated he was ready to go, but how in the world would they ever go about doing it?

Tommy told Ricky to just lean back in his overstuffed armchair and close his eyes. Tommy proceeded to count backwards from ten to one and at each count Ricky became more and more relaxed. Finally at the count of one Tommy suggested to Ricky that he visualize in his mind's eye that the two of them were back at the Atlanta airport and that Tommy had just chartered a Lear Jet to take them back to Hutchens, Minnesota. However, this was a very special Lear Jet and it had aboard an altimeter which not only could tell how high they were flying but could also time travel and take them back in space and time. As Ricky drifted off further into a deeper state of relaxation, he could visualize flying back through the years. First he experienced when he was hired as a young outside salesman for Primetex Communications. Then he saw in his mind's eye his two year stint in the army infantry and before that his college days when he earned his MBA degree at Georgia State University.

Next he went back to his high school days where he was the quarterback for the Hutchens High Rollers football team. Finally back through his grammar school years to the little house on Donley street where he grew up. He found himself sitting on the couch in their living room next to Tommy who was sound asleep by his side. He felt so lonely, tired and frightened for it was Christmas Eve and his mother wasn't there to spend Christmas Eve with them.There wasn't a Christmas tree or gifts in the room which was mostly bare except for a few pieces of old secondhand furniture. The house was rather cold and there wasn't even any supper for Tommy and him except a half loaf of stale bread, an old jar of peanut butter and only one can of soda pop to be split between the two of them. As he sat there on the couch he could hear very faintly in the far off distance Tommy's voice saying, "Just look up to your left Ricky and slightly above your head and you will see what will appear to be your own Analog Power Symbol™. Your own Analog Power Symbol™ will take the form of a person, an object, an abstract symbol, a voice, a master teacher or other abstract form. It might also have a color, a light, a voice, or a vibration to it. Whatever form it takes it will be your own analog power symbol, and it will be imbued with considerable power, strength and wisdom sufficient to answer any question or solve any problem you might have in life. It will give you whatever it is you are seeking in order to give full expression and meaning to your life".

At first Ricky didn't see anything at all in the direction Tommy suggested. Then it started out as a very small speck of light about ten feet from where he was sitting in an otherwise rather dark and gloomy room. The speck of light started to grow and expand until its brilliant light filled the entire room. The light then started to take form. First there appeared a shape of an outline of what looked to be a person. As the figure grew more distinct Ricky could make out the figure of what appeared to be an angelic radiant being dressed in a pure white robe with a garland sash around her waist and a garland wreath around her head. Her hair was a brilliant golden color which hung down below her shoulders. She appeared to be young and very beautiful.

Ricky sat there in utter amazement. He finally said, "Who are you?" The angelic being replied that she represented the spacetime of Ricky's life. She was Ricky's Christmas past and Christmas future and all the days in between that filled the entire span of his life She said she represented both his hopes and dreams and his disappointments and failures in life. She was an expression of all of his thoughts and feelings about anything and everything that was, is, and would ever be in his life. He then asked, "Are you real or just a figment of my imagination?" She replied, "Are your thoughts real, are your emotions real, is your breath real, is the wind real, is the child's laughter real, are the currents of the ocean real, are the clouds up above real, are the energy forces and fields of the entire universe real, is your sight, hearing, taste, smell, and feelings real? For I am as real as all of these things, and even more real, for I am all of the energy and images that will ever exist in time and space making up every thought and every feeling you will ever have in life. As such I represent the energy and images of all physical reality both in the past and yet to come for I represent the source of all of these things where they all originate at the very quantum level of your subconscious mind."

His final question was, 'Why are you here?" She replied, "I am here because you have been earnestly seeking me for these last few years. It is just that you have been looking in all the wrong places. For I am not found in the power, might, and glory of the physical plane. Rather than looking outside for your answers you need only look deep within and there you will find that which you have been so desperately seeking all of your life. I am here now to answer any question you might have concerning your life. I am here to give you whatever you seek in order that you might be able to unlock the riddle of your past and to truly find the deeper meaning of your life."

Ricky thought for just a moment and then tears came to his eyes. "I always wanted a traditional Christmas with Tommy and Mom with a decorated Christmas tree and gift wrapped presents and holiday decorations and a large turkey dinner and. . . . . ." At that very instant the room lit up and was decorated with garlands over each window and the doorway. A large wreath with a big red bow hung over the fireplace which came to life with a crackling warm wood fire. Red and white poinsettias plants were everywhere and in the corner by the fireplace stood a large silver Christmas tree. It was decorated with the most beautiful ornaments he could think of, of every color, size and shape imaginable, with an angel on the top just like the one he was talking to. Below the tree were numerous presents beautifully gifted wrapped with large bows and ribbons of silver, gold, red, yellow and deep blue.

Ricky thought to himself that this was an absolutely perfect Christmas with just one exception. If only Mom could be here to enjoy all of this with him and Tommy. With that thought there was a knock at the door. Ricky quickly answered it and there stood his mother. He had not seen her in many, many years. She look as tired, hurt and pained as ever. The deep lines of all of those painful years were still deeply etched into her face. However, she managed a smile. She didn't smell of alcohol. She was clean and neat and she was wearing her best winter coat and dress. Even though very worn they looked beautiful on her. Her arms were laden down with large grocery sacks. One had a fully roasted turkey, gibbet gravy, dressing and cranberry sauce. The other sack had a pumpkin and a pecan pie and containers of creamed peas and carrots, corn bread and steaming mashed potatoes.

They quickly set everything out on the table and Ricky's mother turned to him and said, "Ricky I am so very, very sorry for having neglected you and Tommy all those years while you were growing up. I truly loved both of you. But when Dad left us I was out of money, I was about to lose this house, I didn't have a job and the only way I knew to stop the pain and misery in my life was to drown my sorrows with whisky. If you can only understand that it was not your fault that . . . ."

Ricky interrupted her in mid-sentence and took her in his arms and embraced her. His eyes were filled with tears of love and forgiveness. He stated, "I finally, at last, truly understand. You really did love us after all. You did the best you knew how. It must have been very hard for you after Dad left us. Had I only been more aware of what you were going through at the time I could have helped you more instead of complaining and making your life miserable. Can you truly forgive me"?

"Of course, I do", she said, "but it is not your fault. It never was".

Ricky was so pleased that he could finally forgive and love his mother after years of hating and ridiculing her for how he perceived she treated him and Tommy as they were growing up. They sat down and ate their Christmas dinner together as a real family. They were finally, at long last, truly reunited after all these years of separation brought on by his own hate and animosity toward her. "This was truly a beautiful and memorial Christmas" he thought to himself.

As he sat there on the couch savoring these fond moments with his mother on this Christmas Eve the angel turned to him and asked, "What did you truly learn from all of this?"

Ricky thought that this was a rather curious question to be asking at a time like this. Didn't this angel just witness the reunion of a son with his mother after all of those hate filled years? Didn't this angel grasp the true significance of this important event in his life? Doesn't this angel even understand the very nature of love? "What's wrong with this angel anyway," Ricky thought to himself? Then in an instant he recalled what Tommy said upon his return from Superstition Mountain. It simultaneously struck him like a lightening bolt that shot through his mind and body alike. It literally tore through his very being to the depth of his own soul. "Oh my God, " Ricky cried out! "All love starts with self love". With that he fell to the floor and wept uncontrollably for what seemed like hours. "It is I who I hated all these years, it is I who felt unloved, it is I whom I was ridiculing, it is I who felt ashamed, It is I who never measured up, It is I whom I blamed, it is I who was guilty, it is I who felt like a failure in life, it is I who was unable to freely give of myself to Betty, it is I who had to keep proving myself over and over again, it is I that I could never really forgive, never forgive, never forgive. . . . . . ."

"All love starts with self love" kept washing over him and pouring into him from every direction. It seeped into every pore in his body. It penetrated to the subconscious levels of his mind, to the neurological level of his brain and to every cell in his body. "So this was the true message from the angel" he thought to himself. At that very instant his life was transformed forever. His entire being instantly shifted, like the shifting of the tectonic plates deep within the earth. They shifted from negative to positive, at the very quantum level of his mind where the energy and images of his mind formed the realities of his world. His whole world would never be the same again!

He slowly rose from the floor. He sat down on the couch next to Tommy who remained sleeping. For how long this experience lasted he just couldn't tell. The Christmas scene before him started to fade and he and Tommy where back again in the gloomy living room all by themselves. However, the insights and feelings of love still remained with him as strong and as vivid as the moment he first felt them. The angel too started to fade out of view but before she left entirely she called out to Ricky "I will always be with you to give you guidance and direction in life. When you need me you will now know where to look in order to find me". With that the angel was gone. Ricky could hear Tommy's voice in the far off distance telling him that it was time to take the magical time travel Lear Jet back to the north Georgia mountains. They left the house of their birth on Donley Street in Hutchens, Minnesota, and headed through space and time back to where they started their trip in the cabin deep in the mountains of north Georgia.

Ricky could hear Tommy faintly in the background saying "Ricky I'm going to count backward from three to one and at the count of one you will open your eyes. Three, you are back sitting in your chair next to the fireplace in our cabin. Two, you can feel the heat from the fire, smell the wood burning and you can feel yourself back inside your body. One, open your eyes, One, open your eyes." With that Ricky opened his eyes, stretched, looked around for a moment as though he was reorienting himself to his surroundings. He then got up and embraced his brother Tommy and declared "Tommy, I truly love you as my brother. I thank you for giving me back my life." Then he said, "and oh, by the way, your new analog way of thinking in the fourth dimension really does work!" Tommy replied that they had the rest of the night and all of tomorrow and tomorrow night to work on some of the other MetaCreation™ techniques he had recently developed. Particularly the techniques where you can actually create for yourself the type of future realities you desire in life. However, Ricky interrupted him and said that he found out what he had come for and if it was O.K. with Tommy he would like to head back early in the morning. He wanted to buy that ring Betty had been looking at for the last several months, the one with the American Classic two carat brilliant blue white diamond circled by ten smaller diamonds. It would make the perfect gift as he proposed to Betty on Thanksgiving Day!

Tommy and Ricky made it back to Atlanta by mid-morning the next day, and instead of going straight to Ricky's place they stopped to purchase the ring. Ricky stated that he wanted to pick up some items for the dinner Thursday. Specifically he wanted a large 20 lb. turkey, gibbet gravy, dressing and cranberry sauce, pumpkin and pecan pie, creamed peas and carrots, corn bread and mashed potatoes. Tommy thought that this was rather curious since, as a bachelor, Ricky ate out most of the time and never gave much thought to grocery shopping or cooking. But since his selections sounded pretty good Tommy just said, "O.K., let's do it." They stopped by the neighborhood supermarket and got the items Ricky had on his shopping list before going on to his place.

On Thanksgiving Day Ricky and Tommy watched two Bowl Games that morning. Louanne, Philip and Andy played Atari video games in the office. Betty and Claire were in the kitchen putting the finishing touches on the Thanksgiving dinner. By 2:00 p.m. they were all very hungry and sat down to eat. But before they started Ricky said that he wanted to give the blessing. Betty thought to herself that was rather peculiar. She heard Ricky curse a person or two in the past, particularly his mother, but he never blessed anyone before. Ricky had them join hands as he give the blessing, "Lord bless this gathering here today. Bless this meal which we are about to receive. Bless my Mom and Dad and keep them in your grace. Bless my brother Tommy and his family who have given me so much to be thankful for these last few days. And Lord, bless Betty and Louanne for having come into my life. They have brought me happiness and joy which I never before experienced. And Lord lastly bless this commitment I am about to make to Betty." With that he reached into his pocket and pulled out the ring and slipped it onto Betty's finger and said, "Betty will you marry me?" . . . . . . a surprised Betty leaned over and gave Ricky a big hug and kiss and said, "Yes, I surely will"!

How the Mind Processes Realities

Let's stop for a minute and take a look at what happened to Ricky at the level of his subconscious mind. The techniques Tommy used concerning the relaxation exercise and the power symbol will be dealt with in considerably more detail in later chapters. Except for the following comment, we will not deal with them here. The purpose of the relaxation exercise was to quiet the conscious mind in order to access the subconscious mind. It is the subconscious mind where we create realities in life. We have seen that conscious cognitive thought doesn't create anything. All processing and all organization of realities take place on the level of the subconscious mind. Cognitive analytical thought is not a very good medium in which to work if your intent is to create new and different realities in life. It is excellent for thinking in logical terms and for keeping track of the many activities in our daily life. However, it is a very poor medium to use when it comes to dissolving old patterns in life and for creating new positive patterns to replace the old ones. As we have seen the reason for this is that the conscious mind has considerable difficulty in organizing thoughts in analog fashion. It just isn't constituted in this fashion. We have seen how a binary serial system has trouble with analog processing, but an analog system can easily handle serial processing. Therefore, we access and utilize the analog system of the subconscious mind in creating entirely new patterns of realities in life.

Charts 11-1 through 11-3 at pages 112-114 illustrate how changes take place at the analog system of Ricky's subconscious mind. Chart 11-1 sets forth a simplified diagram of the basic operational systems of the mind in creating patterns of realities in the present space and time of our world. You will note that we put you, represented as "Self", at the center of the concentric circles. Self represents the cognitive structure of the conscious mind. It is from our cognitive structure that we form our thoughts and beliefs about life. Our thoughts and beliefs are filtered through our personality and

CHART 11-1

Processes of the Mind in Creating Realities in the Time and Space of our World

How we conceive our world and bring our awareness of it into existence as our personal reality

Present Time

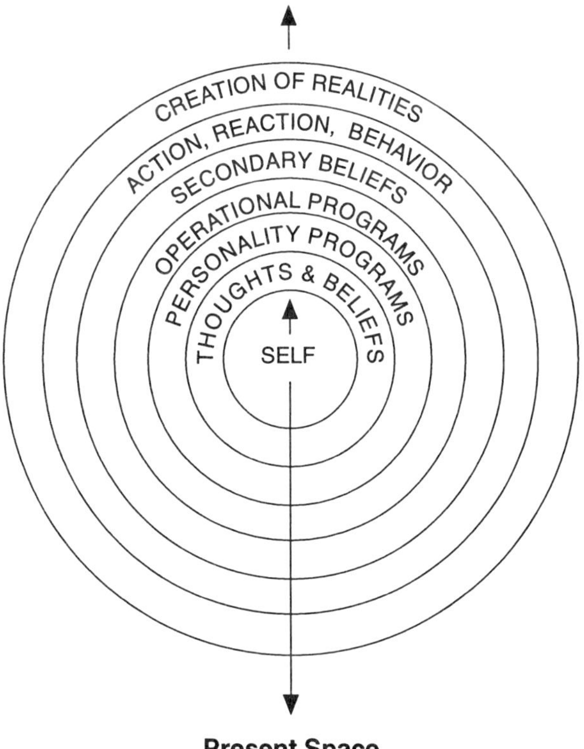

Present Space

Simplified illustration of the basic operational systems of the mind in creating realities in the present space and time of our world.

CHART 11-2

Operational processes of the Mind in Past, Present and Future Time and Space

This chart illustrates how a negative past can create a negative self image, negative operational programs, and a negative future. Conversely, a positive past can reinforce a positive self image and create a positive future.

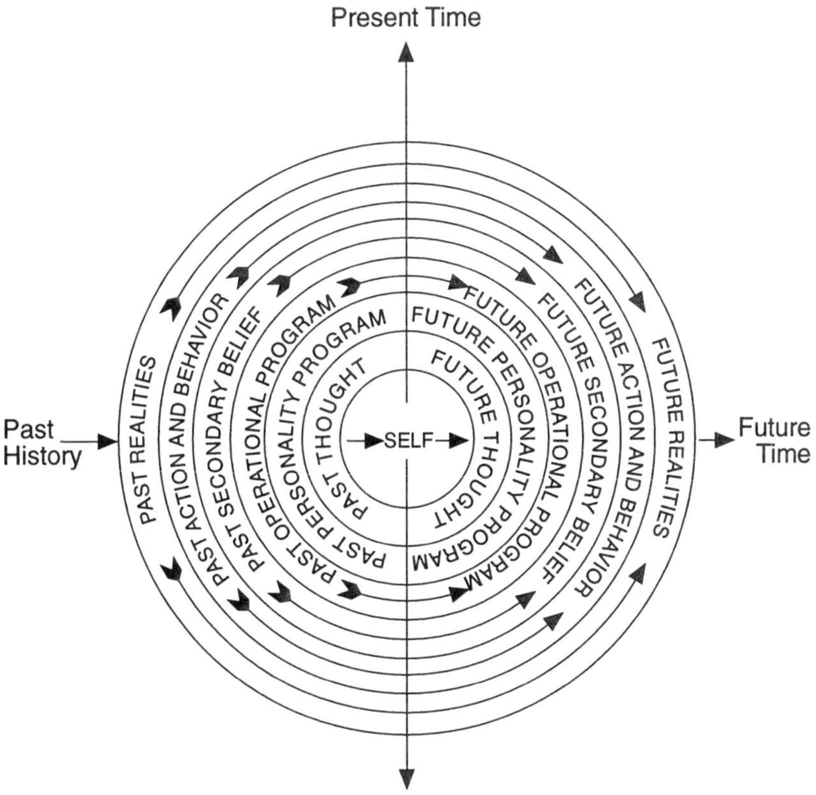

Past stimuli programed our past personality and its operational programs which in turn created secondary beliefs that created actions and behavior that formed the basis of our past realities. This past reality affects our present action and behavior in response to new stimuli. Our response either modifies or reinforces our secondary beliefs, either positively or negatively, which modifies or reinforces our personality and operational programs which then either reinforces the past realities or creates new realities. If we view our world as objective and outside of us and the past as fixed in time & space we lose control over changing our internal programs & over the creation of new positive inner and outer realities for our future.

CHART 11 - 3

How the Cognitive Structure of the Mind Changes the Effects a Negative Past has on its Ability to Create a Positive Future

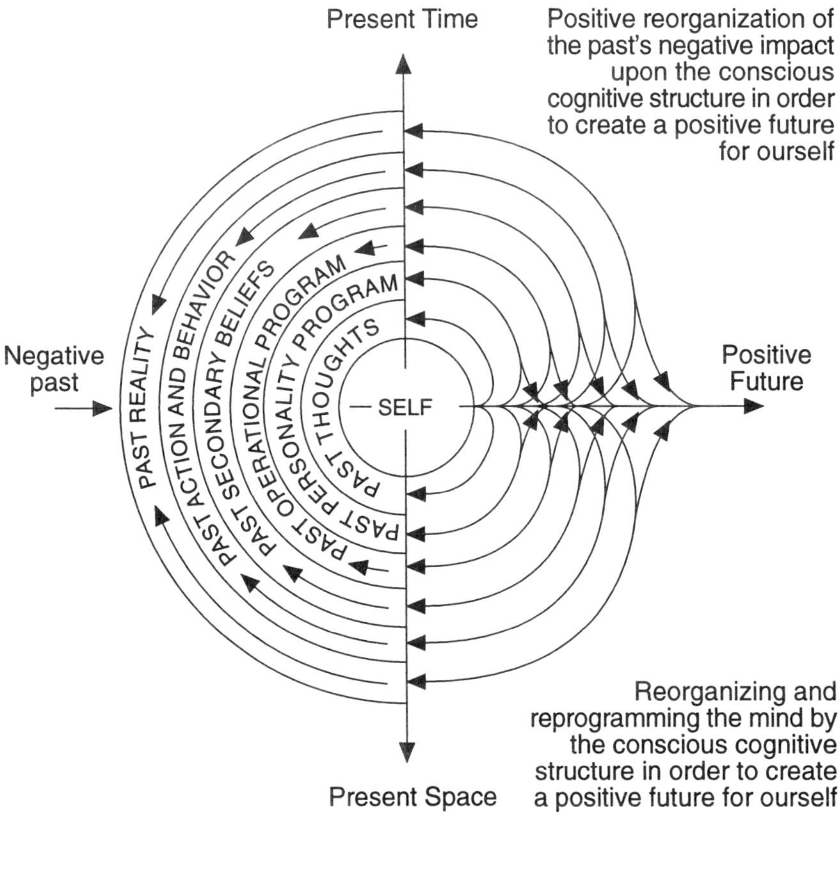

This Chart illustrates how the Cognitive Structure can reorganize and reprogram the mind in order to eliminate the detrimental effects a negative past has on its internal mental programs in order to be able to create a positive more supportive future for itself.

operational programs of the subconscious mind represented by the next two concentric circles. They, in turn, form conscious secondary thoughts and beliefs which form your specific life style represented by the next circle. We take action based on these secondary thoughts and beliefs in an effort to give form to our outer physical realities in life, which is represented by the last circle. The vertical line drawn through the circles represents that all of this activity takes place in the present time and space of your life.

Chart 11-2 is very similar to Chart 11-1 except that it shows how events of the past directly affect the formation of future events in life. This is illustrated by the activity taking place along the horizontal time line representing both our past and future. It intersects the vertical line of the present time and space of our life. The concentric circles illustrate how our past secondary beliefs, personality and operational programs, and past actions, all affect, either positively or negatively, our future actions, personality and operational programs, and our future beliefs about life. These, in turn, impact either positively or negatively, the organization of our future realities in life. Therefore, if you encountered a negative past in the form of neglect or abuse, as was the case with Ricky, these early childhood programs will adversely affect many areas and activities of your future adult life.

For instance, Ricky's neglectful childhood affected his first prime directive in life "I like who I am". We briefly discussed the four subconscious prime directives in life described as; (1) "I like who I am", (2) "I have a right to a life", (3) "I can express myself fully", and (4) "I can achieve whatever I want." These four prime directives set the mental and emotional tone and direction of all of our other thoughts and feelings about ourself and our world. If these early subconscious prime directives in life are negatively impacted they will negatively affect our conscious secondary beliefs about ourselves and our life. However, when things don't go right in life because these prime directives are defective we often do not know what is the cause of our problems. We are unaware of how these negative directives impact our daily affairs.

Chart 12 at page 116 illustrates how one negative prime directive can propagate it's negative energy and images throughout the entire analog system of our ego personality. It can adversely affect our conscious thoughts and beliefs and the organization of our inner and outer realities in life. In Ricky's case he unconsciously blamed himself for the neglect he experienced in early life. Since his mother never showed much love or affection toward him or his brother he unconsciously concluded that there was something basically wrong with him. This adversely set the basis for the negative

CHART 12

Schematic Diagram of How a Dysfunctional Primary can Propagate its Negative Energy and Images Throughout the Cognitive Structure of Ego Personality

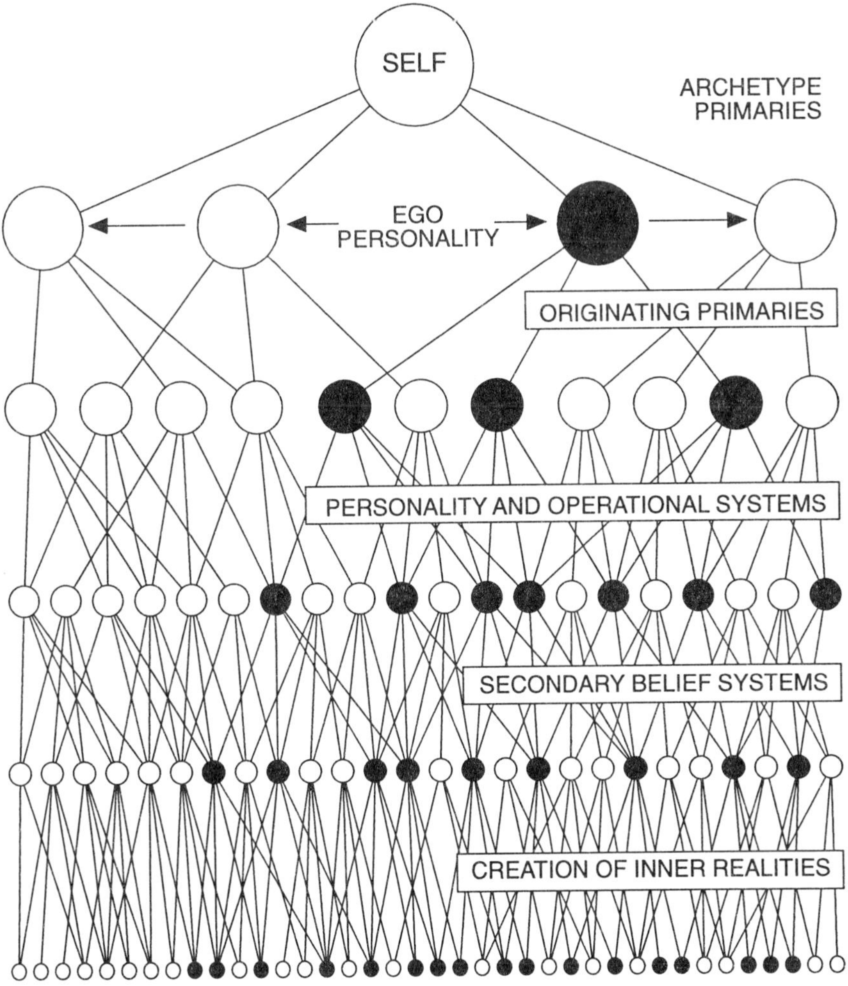

energy and images of the patterns of the subconscious analog system of his mind. This left Ricky with feelings of inadequacy, low self esteem, and low self worth. Consciously he knew that he felt this way but he did not know why he felt this way. He would become very angry at his mother and others who were important in his life. He would also become overly aggressive and competitive in business to prove himself in an effort to overcome these deep feelings of unworthiness and inadequacy. Yet, he never understood in any conscious or meaningful way that this negativity in his life was the result of feeling unloved and unaccepted at the deepest levels of his subconscious mind. Consequently, he could not open himself up to intimacy and unconditionally love Betty since he could not unconditionally love himself.

While in the deep relaxed state the power symbol of his subconscious mind brought to him the realization that his problems in life stemmed from the lack of self love not from the fact that he didn't love his mother. Not loving his mother was symptomatic of a much deeper problem. He finally learned to deal with his early programs and not with the early programmers in his life. Getting to this deeper level was triggered by unconsciously recalling the statement Tommy made earlier that "all love is self love." It is only at the level of the subconscious mind that you can change these early childhood prime directives in life. They are seldom changed at the conscious cognitive level of the mind for they didn't originate there. Nevertheless, we far too often try to talk ourself into a good mood or try to convince ourself that there is really nothing wrong with us and that everything in life is really O.K. Then when we encounter circumstances in life as simple as a perceived slight by someone we know, our reaction mirrors a negative prime directive in life which finds a direct pathway leading right back to the corresponding negative basin of attraction on the neurological level of the brain. We are then right back to square one, either feeling sorry for ourself, or angry at the other person, both of which are inappropriate responses in life. We will often resort to inappropriate thoughts and actions to satisfy our negative feelings such as excessive drinking, eating, drugs, obsessive and compulsive sex, or other obsessive thoughts or compulsive behavior. When the propagation effect of a negative prime directive is severe we duplicate these negative thoughts, feelings, and behaviors about ourself and our world, over and over again, in many areas of our life. Life often becomes just too painful and some of us resort to excessive drinking to ease the mental and emotional pain as was the case with Ricky's mom.

Conscious Analog Thinking™ can bring about significant

changes by altering negative patterns into positive patterns in life. Chart 11-3, at page114, illustrates the positive effects of analog thinking. Since linear time and fixed space are not factors at the level of the subconscious mind you can go back in time to repair the damage at a prime directive and go forward in time and create new more life supporting and life enhancing realities for yourself. The past is not outside your control when you access spacetime of the subconscious mind. You can effectively go back in time and reprogram the affected areas of the prime directives in life through the creation of your very own Personal Power Symbol.™ Remember the subconscious mind has unlimited potential for the creation of abstract power symbols and for creating inner realities as well as sorting for experiences of outer realities in life. Therefore, you need only program the subconscious for a power symbol. In Ricky's case his subconscious mind created the Christmas angel as his power symbol to destabilize and dissolve the negative analogs which were causing the problems he was encountering in life. The subconscious mind created the Christmas angel as Ricky's power analog since Ricky could easily identify with the Christmas angel. He and Tommy always tried to recreate a happy family Christmas each year in a conscious effort to make up for all of those neglectful childhood Christmases. Refer back to Chart 3-B at page 70 for a list of common analog symbols recognized by the subconscious mind.

Once your prime directives are cleared and reprogrammed to positive energy and images they will no longer sabotage you from creating positive future realities. They will actually support the creation of positive future realities in life. Consequently, we first deal with a negative past before tackling the problem of how we create newly expanded future realities for ourselves.[2] Chart 11-3 illustrates how the conscious cognitive structure can reorganize and reprogram the mind in order to eliminate the detrimental effects a negative past has on your internal mental programs. You will then be able to create a positive more supportive future life for yourself. Chart 13, at page 119, shows that not all changes are as spontaneous and dramatic as Ricky's changes. Nevertheless, over time significant changes do take place at the deepest level of the subconscious mind, the level where our prime directives originate the energy and images of our life.

Chart 13 shows that significant changes, particularly changes of a characterlogical nature, often take repeated applications of your analog power symbol in order to clear all residual negativity that might

2 You do not have to clear all negativity in the primary systems before creating future positive outer realities in life. Both processes can go on simultaneously in life.

CHART 13

How the mind organizes newly adapted stimuli by reordering its cognitive structure to destabilize older negative energy and images in order to assimilate the introduction of newly created positive energy and images

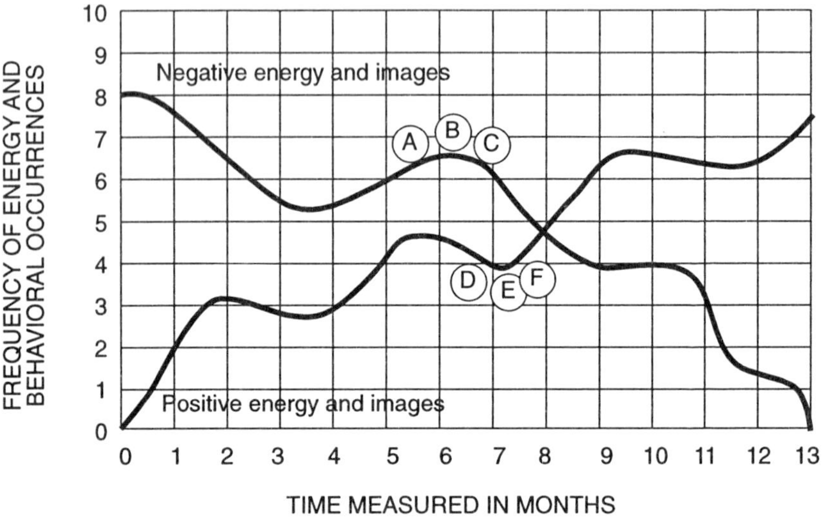

The horizontal line of the above chart represents time intervals measured in months. The vertical line measures the frequency of both positive and negative occurrences over time. The references to the symbols, A, B and C in the sixth month represents the occurrence of several negative occurrences. If viewed at this point only on the chart you might conclude that the negative energy and the resulting negative behavior is not being effectively removed from the affected cognitive structure. However, when viewed over an extended period of time you would see that the overall trend is downward. Similarly with respect to the symbols D, E and F on the line representing the installation of the newly installed positive energy. It would appear that at this point in time there is no experience of positive occurrences in the system. However again when viewed over time the overall trend is upward. These results are typical of large capacity parallel processing systems such as the mental systems of the mind and represent examples of the principle of soft constraints and soft reinforcements. The above chart illustrates that the cognitive structures of the mind operate on the principle of soft constraints and soft reinforcements and not hard computational rules. By this we mean that in order to remove negative images and replace them with newly created positive energy, images and positive behavioral traits quite often takes time. This time period can extend over several months in the case of deep seated negative deficits in the primary levels of the mind. During this period of time the destabilization of negative energy, because of the operating principle of soft constraints, does not dissipate over time on a straight line basis.

remain in the prime directives of the subconscious mind.[3] This chart illustrates that at any given time in the process there might be periods when it seems that little progress is being made. However, this is not the case for the reason that negative analogs are not arranged in a clean-cut one issue format. As we have seen these primary analogs are intertwined patterns and relationships. As you clear one aspect other aspects appear which also need to be cleared. However, repeated applications of the positive power symbol will eventually clear all negative energy affecting a prime directive.

In our illustration Tommy took Ricky back in time to his early childhood. In cases of severe abuse or trauma, however, it is not necessary to revisit painful past memories of traumatic episodes in life.The abstract power symbol will work just as effectively in clearing a negative primary analog resulting from early childhood trauma and abuse without accessing the imagery and emotions involved in the trauma. The reason for this is that the subconscious mind is not a cognitive binary serial type system but an analog type system. Its language is abstract symbols and metaphors. It is very comfortable using abstract power symbols to do the work of clearing any type of negativity lodged in its systems. Remember we are doing subconscious analog energy work. We are not doing conscious cognitive work and, therefore, we do not need to reexperience the trauma of past abusive events in life. If and when cognitive recognition and integration is needed it will come about naturally of its own accord without the attendant pain of the earlier trauma in life. It will surface when you are ready to deal with it on the conscious cognitive level. However, it is certainly not necessary in working the subconscious analog processes in spacetime of the subconscious mind.

Summary on Analog Thinking

Conscious Analog Thinking™ becomes a way of life and not just a process to realize greater realities in life. It is a type of life style where you start to discreate a negative past and start to create a positive future through the processes of your mind alone. You will find that pain, lack and limitation on the material plane in life start to dissipate and dissolve as the energy and images of their origin are removed from the subconscious mind. As you start to slow down and relax the rhythms of your life your mental and emotional energy becomes more aligned, congruent and cohesive. You find that you

3 Characterlogical changes consist of changes dealing with systemic problems in the very nature of your ego personality such as self identity issues; sexual identity and sexual orientation issues; personality issues; operational and competency issues; issues dealing with your right to be seen and heard; and issues dealing with defining and expressing thoughts and feelings of who you are and your purpose in life.

are able to remain centered even at those tense moments when you are in the midst of conflicts in life. You find, over time, that you do not seek the emotional images of violence and destruction often portrayed in magazines, movies and entertainment. You begin to seek more harmonious and edifying circumstances in life. Instead of being cut off from everyone else and from everything else in your physical world you begin to sense a connection to everyone and everything in your life and in your surroundings. You come to know your truth and you are able to speak your truth honestly and directly, but with compassion toward others.

You begin to seek greater intimacy, companionship, and love toward self and others. Life becomes more beautiful, edifying, uplifting and enjoyable. You feel centered and congruent within yourself and with all of nature. You ultimately find love, both love of self, of others, and of nature. Out of such love comes expressions of kindness, consideration, compassion and support for others. You find your path and direction in life. Life becomes more meaningful and you begin to unfold in a natural way. You will feel content within yourself. You do not grasp for things of this world. You do not seek the many and varied physical and sensual pleasures and thrills of this world in order to find meaning or fulfillment.

You are mastering the creation processes and you are able to create those things you desire for yourself. You find your true joy and happiness in the simple things of the spirit, such as hope for the future, laughter in the moment, expressions of love, joy and beauty in all that you see and do.You experience the joy of the moment, in being with those you truly love, in simply watching a sunset or in taking an early morning walk. You begin to get a sense in a very personal way what the universe metaphor of "whole, perfect and complete" truly means. You are content for your life is unfolding just as it should. You come to know the quietness of contentment. You are being made whole and, at long last, you are at peace with you yourself, with others in your life and with your world.

This brings us to the question of who is it that is creating and discreating all of these realities in life? Up until now I have been referring to the conscious cognitive structure of your mind in the first person as "you". In a sense this is correct. But it also begs the much deeper question, who exactly is it who is in charge of your conscious cognitive structure of your mind? Who is it who is making all of these decisions in your life? In the next chapter entitled *On the Nature of Human Consciousness* I will address the questions of who you really are, how did you get here, why are you even here, and what is life all about?

CHAPTER 4

On The Nature Of Human Consciousness Or Who is Minding the Store

Introduction

Any comprehensive analysis of the human brain and mind must also address the nature of consciousness. For it is consciousness that directs the processes of the brain and mind in creating meaningful realities in life. Since we are all self aware and have volitional will we are free to create a life of meaning and happiness or a life of pain, suffering and limitation. The choice is strictly up to us. Therefore, it is imperative that we take a look at the processes of consciousness to see if we can determine why some of us make wise choices in life while others consistently make poor choices. It is the processes of consciousness that play a vital role in organizing the content of our life. As you take control over the process of consciousness you take control over how you create the content of your life. In Chapter 8 we deal with how to improve the content of your life. In this chapter we will start by describing the processes of consciousness. We will then explore the underlying energy forces of human consciousness. We will consider the early evolution of consciousness. Next we explore how the processes of consciousness organize ego personality, for when we organize ego personality we are able to create inner and outer realities in life. We also consider conscious and subconscious states. Finally we end this chapter by exploring the future evolution of human consciousness.

There is surprisingly very little in the way of systematic scientific investigation into the nature of human consciousness. Yet, all aspects of a highly evolved civilization including science, medicine, mathematics, communications, space exploration, art, literature, humanities, and all other social and cultural fields of study are a direct outcome of the creative processes of human consciousness. Without a fundamental understanding of the nature of human consciousness mankind will never be able to fully understand nor develop a comprehensive working model to describe how human consciousness creates the many realities of an advanced civilization. Of course, over the years there have been those working in such diverse fields of study as theology, philosophy, psychology, neuroscience, quantum physics, cosmology and metaphysics, to name just a few, who have, from time to time, attempted to define consciousness in some systematic way.

The major obstacle to any systematic understanding of consciousness has been that consciousness itself appears as a process function in nature and as such it does not readily lend itself to quantifiable laboratory analysis. What is meant by process function will be explained shortly. The fact that it is very difficult to quantify consciousness by the scientific method has left the field open to speculation as to the very nature of consciousness. Thus, your field of study determines, to a large extent, how you view human consciousness. For instance, if you are a neuroscientist you probably describe consciousness in terms of the neurological processes of the brain. If you happen to be a physicist you might view consciousness as an emergent property arising out of the complexity of the self organizing system of the human brain. If you are a mathematician you might view consciousness in terms of the algorithms describing the processes of the mind/brain system. If you are a theologian you probably view consciousness in terms of the human soul or spirit, while Metaphysicians will talk in terms of cosmic consciousness, universal consciousness and higher states of enlightenment. While there is probably some merit in all of these approaches none of them gives us a fundamental working model of how consciousness itself, through the operation of the human brain and mind, creates realities in life.

What then is the nature of human consciousness, and how do human beings become self aware? These questions are very relevant for the answers hold the key to the very nature of man. By understanding consciousness we can begin to take effective control over all of our inner mental states, including our thoughts, feelings and actions, and start to actually influence the direction of our own evolution and the evolution of all mankind. These questions have bedeviled mankind for hundreds of years. However, since the advent of quantum physics they have generated renewed interest. Part of the problem has to do with the way we understand the terms "consciousness" and "self awareness." Self Awareness can be defined "as the faculty by which we become aware of our own self existence." Consciousness can be defined "as the faculty by which self awareness, through the operation of the human brain and mind, organizes life realities."

Self awareness provides each of us with a sense of personal existence, an awareness of individuality and of separateness. In fact it is this very aspect of self awareness that has given rise to the human condition existing in an objective world in which everything from our skin out exists and operates separate and apart from us. Self awareness first creates for itself ego personality in order to experience this sense of individuality and separateness. Ego personality can be defined as "the interface mechanism by which self awareness is able to create for itself physical realities in a three dimensional world in time and space." Ego personality consists of the composite mental, emotional and physical characteristics and traits of an individual developed to effectively function in a physical environment. It is through ego personality that we are able to experience physical reality. If ego personality is not satisfied with its physical realities, self awareness is able to change both ego personality and those realities. How this is accomplished is explained below.

Process Functions of Consciousness

The process functions of consciousness consist of self awareness, knowledge, intelligence, and volitional will. Self awareness utilizes the process functions of knowledge, intelligence and volitional will to create ego personality. Knowledge is the process function of consciousness that organizes subconscious analog patterns so that they can be interpreted meaningfully. Intelligence is the process function of consciousness that interprets these analog patterns in some meaningful fashion. Volitional will is the process function by which self awareness is able to make its selections based on these analog patterns in the process of organizing a life. This then is what I mean when I say that consciousness represents process functions in life. Consciousness processes energy into realities which become the content of our life.

When I refer to "process functions" as opposed to "application programs" I am referring to the process functions that operate the application programs of the mind/brain. The application programs of the mind/brain form the content of ego personality and the content of our life realities. Self awareness serves a processing function at the time of conception in order to create ego personality to manifest and experience the content of our life. Self awareness itself does not represent the content of our life.[1] The content of life is represented by the characteristics of ego personality and that large body of knowledge referred to in Chapter 2. That large body of knowledge does not represent the knowledge process function of consciousness. The knowledge process function of consciousness is, however, the process function that organizes that large body of knowledge. In similar fashion the values we placed upon that body of knowledge also represent content in life and therefore do not represent the process

1 Self awareness is what mobilizes human consciousness to create an identity for itself in the form of ego personality. Self awareness is a primordial instinct of mankind already existing at the time of conception.

function of intelligence. One is process and the other is content.[2]

We normally think of consciousness as being consciously aware of ourselves as we make volitional choices in life. This is partially true, but it does not take into account the entire nature of human consciousness. We think that being self aware and having volitional choices constitutes human consciousness for we are consciously aware of these two characteristics of consciousness. However, consciousness also includes the subconscious processing functions of knowledge and intelligence of which we are not consciously aware. We normally think that the brain and mind automatically respond in some fashion to the incoming stimuli received by the system in arranging the basins of attraction and analog patterns in creating a life for ourself. However, it is these two processing functions which are the ones that actually organize environmental stimuli at the neurological level of the brain and process the arrangement of the analog patterns of the subconscious mind. Therefore, it is these two subconscious processes of knowledge and intelligence that organize the content patterns of life. It is the conscious processes of self awareness and volitional will that organize those content patterns into a meaningful life for us.

You might question whether the knowledge and intelligence processes are purely automatic functions of the brain and subconscious mind? That is, are they processes over which we have no awareness and no direct control? The answer is yes and no. Yes, we have very little control over these processes for they remain largely unconscious automatic functions of the brain and mind. However, once we understand how the system processes environmental stimuli in organizing neurological basins and analog patterns we can consciously take control over such processes and direct their organization. We can rearrange the basins and analogs, if we are not satisfied with them, and therefore create a better life for ourself. When I say "we can consciously take control of such processes" I am not talking about personally consciously experiencing the arrangement of these basins and analogs. Rather I am saying that through the intentionally of thought we can reorganize the negative basins and analogs and create new positive basins and analogs. This is precisely

2 I.Q.Tests endeavor to measure both process functions of the mind/brain as well as its content. Therefore, I.Q. Tests are difficult to interpret since they attempt to measure how you think as well as what you think about. How you think is a process. What you think about is content. While how you think determines what you think about, the I.Q. test can only measure with some degree of accuracy what you think about. That is, the test measures the content of your mind, and not the processes by which you derive the content.

what we do by applying the MetaCreation processes. In this fashion you can empower yourself!

The Energy of Consciousness

The processes of consciousness are essentially energy, a form of kinetic energy to be exact. As organized kinetic energy consciousness is forever processing realities in life. The energy of consciousness is dynamic and constantly in motion. Consciousness is not static, fixed or characterized as being a thing, such as a physical or material presence. Consciousness does not have an identify such as ego identity. Consciousness does not exhibit specific characteristics of the human condition. Consciousness does not determine the color of your eyes, the color of your skin, what sex you are, or any of the other hundreds of characteristics making up your composite physical appearance. Consciousness, however, is what operates the application programs of the brain and mind which, in turn, facilitates the manifestation of these human characteristics as well as the content of your life. Consciousness, through one's intentional thoughts, can be expanded to the point where we are able to attain higher states of consciousness and expanded states of awareness. In this respect consciousness exhibits morphological processing capacity. Such expansions in consciousness are commonly referred to as stages of self actualization and enlightenment. For a conceptualization of human consciousness see Chart 14 at page 127, which correlates the organized field of pure kinetic energy with human consciousness.[3]

We have seen human consciousness is basically formed out of organized kinetic energy as is the energy at the quantum level in the formation of all of nature. Energy throughout all of nature takes many forms. For instance, the normal range of the energy frequencies of the electromagnetic fields forming energy patterns of the human body are in the range of 62 to 68 hertz.[4] When these frequencies drop below 58 hertz the body is susceptible to disease.[5] However, the frequencies of the entire electromagnetic energy spectrum itself flowing throughout all of nature ranges far above and well below the range of these electromagnetic frequencies of the human body. Chart 15 at page 128, *The Electromagnetic Spectrum,* illustrates how narrow the energy spectrum of human vision is in com-

---

3 The reference to the "organized field of pure kinetic energy" refers to a theoretical field. However, there is a correlation between it and the electromagnetic spectrum appearing throughout nature. The energy of human consciousness and the energy of the electromagnetic spectrum at the quantum level consist of organized kinetic energy.

4 Hertz is a unit of frequency measured at the rate of one cycle per second.

5 D. Gary Young N.D., *Aromatherapy The Essential Beginning*, p. 38.

CHART 14
Correlation Between the Organized Field of Pure Kinetic Energy and Human Consciousness

1. The Components of the Organized Energy of the Field

   Nonlocal Motion (organizing component) +
   Nonlocal Configuration (organizing component) +
   Evolution of the Field (unity, order, coherence, directional flow) =
   Organized Field of Pure Kinetic Energy

2. The Components of Localized Kinetic Energy Field

   Locality of Motion (organizing component) +
   Locality of Configuration (organizing component) +
   Evolution of the Field (unity, order, coherence, directional flow) =
   Local Field of Pure Kinetic Energy

3. The Components of Human Consciousness

   Locality of the Field of Pure Kinetic Energy +
   Life Energy & Mental Image of Ego Self +
   Space and Time +
   Human Embodiment =
   Human Consciousness

4. The Components of Expanded Consciousness (Self Actualization)

   Human Consciousness +
   Control over Emotional Kinetic Energy (Basins & Analogs) +
   Control over Mental Kinetic Imagery (Basins & Analogs) +
   Control over Space and Time +
   Control over Creation of Inner and Outer Realities =
   Expanded Consciousness (Self Actualization)

5. Experiencing the Field of Consciousness (Enlightenment)

   Expanded Consciousness +
   Experience Self as Source of Energy +
   Experience Self as Source of Imagery +
   Experience Self as Source of Space & Time & Realities =
   Experience Self as the Field of Consciousness (Enlightenment)

Note: This Chart illustrates how the Field of Pure Kinetic Energy holds the potential for a much higher order of organizing complexity than presently experienced by Human Consciousness.

CHART 15

Illustrating how limited Human Vision, "V" is when compared to the entire range of energy frequencies of the Electromagnetic spectrum.

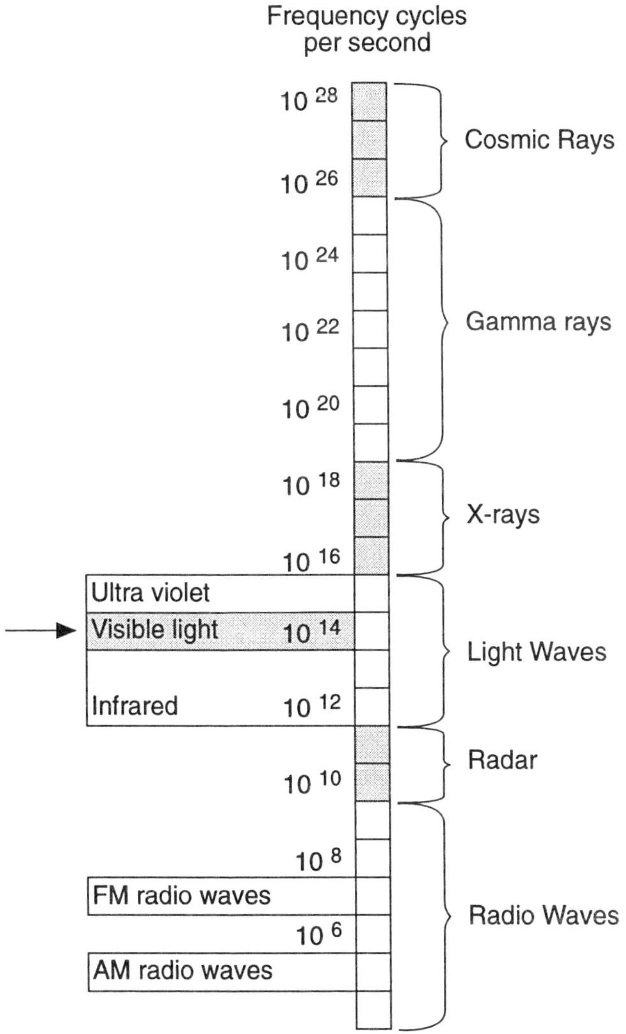

This chart illustrates graphically how limited mankind's sensory receptors, primarily K/V/A, are in experiencing directly the submodalities (the basic building blocks) making up mankind's inner and outer realities. Accordingly mankind is, for the most part, unaware that whenever it rearranges its K/V/A imagery of the mind it is also having an effect of rearranging its inner and outer realities in life.

parison to the entire electromagnetic spectrum existing throughout all of nature.[6]

The process components of consciousness, namely, self awareness, intelligence, knowledge, and volitional will are formed out of kinetic energy. Science has shown that the energy existing throughout all of nature, like consciousness itself, is following an orderly evolutionary path of unfoldment. The appearance of disorder and chaos are surface features only in the path of evolution. However, order and unity appear to exist at the deeper levels of nature as well as at the deeper structures of human consciousness. The inherent flow of evolution is intrinsically an orderly one. The surface features of chaos and the deeper features of order and unity are characterized by theoretical physics as "chaos theory". In many respects the human condition on the surface exhibits the chaos of lack, limitation, suffering and pain. Yet, the deeper structures of human consciousness exhibit order and unity similar to the characterization of chaos theory.

The kinetic energy of human consciousness is not chaotic energy but organized energy. It exhibits a type of unity, order and cohesiveness as it processes mental and physical realities in life. Human consciousness also appears to involve, among other things, higher levels of complexity within its self organizing systems. The evolutionary path of human consciousness appears to be moving from simpler levels of organization to higher levels of organization. This higher level of organization embodies greater capacity for creating levels of complexity leading to higher states of consciousness and expanded states of awareness.

There are those who would argue that the evolutionary path of human consciousness is the result of random chance alone. I do not believe that human consciousness, like the expansion of the universe itself, is the result of random chance. However, even if it were arguably the result of random chance, the fact of the matter is that we are right here now. Since we are right here now we might as well figure out what it is about the processes of human consciousness that create realities in the spacetime of right here now. For then we might be able to create for ourselves more prosperous realities right here now. Chart 16 at page 130 illustrates the directional flow of human consciousness from lower levels of organization to higher levels of complexity in its evolutionary path.

I use the metaphor expressing the unity, order and evolutionary path of consciousness as the "wholeness, perfection and completion" of the human condition. This metaphor also expresses

---

6 Fritjof Capra, *The Tao of Physics*, page 60.

CHART 16

Relationships Between the Organized Field of Pure Kinetic Energy and Human Consciousness

The Source	The Source (The Unknown - The Unmanifest)	The Source

Field of Pure Kinetic Energy - Its Characteristics

Process Components	Infinite Energy	The Flow - Evolutionary
Motion	Organizing	Totality
Configuration	Coherence	Fulfillment
Perfect Order	Unifying	Completion
Potential	Synthesizing	Integrating
Wholeness	Maximizing	Perfecting
Possibilities	Correlating	Sufficiency
Originating	Flowing	Developing
Potentiality	Energizing	Achieving
Differentiation	Existing	Infinite Field

Human Consciousness - Its Characteristics

Process Components	Life Energy	Life Flow - Direction
Consciousness	Pure Awareness	Evolutionary
Self Aware	Individual Self	Self Evolving
Ego Self	Self Experience	Existence (Space/Time)
Volitional Will	Individual Potential	Directional
The Mind	Unlimited Potential	Unlimited Possibilities
Mental Capacity	Formative	Potential for Manifestation
Cognitive Structures	Capacity for Self Experience/Expression	Potential for Creation of Realities
Thoughts	Gives Definition to Mental Energy	Created Reality (Physical Manifestation)
Belief Systems	Organizes Energy	Directed Realities
Self Actualization	Directs Energy Flow	Expanded Realities
Self Realization	Control over Both Energy and Thoughts	Expansion of Awareness
Enlightenment	Experiences Self as Source of all Energy	Fulfillment - Completion

Enlightenment - Its Characteristics (Field of Kinetic Energy)

Process Components	Infinite Energy	Evolution - Fulfillment - Completion

Experience the Source	Experience the Source	Experience the Source

the deeper order of all physical manifestation from the quantum particle to the macro level of the universe. We do not fully experience this wholeness, perfection and completion in our life because of our childhood programming and our adult enculturation. Our programming and enculturation have, in so many ways, limited us from a deeper, more direct perception and awareness of these fundamental characteristic of our very own human nature. Refer again to Chart 16 to get a sense of the relationship between the field of pure kinetic energy, as represented by the metaphor of wholeness, perfection, and completion, and the directional flow of human consciousness toward self actualization and enlightenment. As we have seen from relativity and quantum theory, time, space, mass, matter, and energy are all interconnected and relative to one another. This quantum model accurately describes at the quantum level of our own existence the organizational processes of the mind of man in creating realities. Chart 17 at page 132 entitled,*Conceptual Structures of the Field of Consciousness*, gives a little different perspective in the relationship between the field of pure kinetic energy and the organization of human consciousness.

This short description on the energy of consciousness illustrates that by changing the energy and images of our mental and emotional constructs of reality we change those realities themselves and in so doing we change our level of consciousness. For instance, Chart 18-1 at page 133 shows how changes in the mental energy patterns of the subconscious mind change the images in the mental field of the mind. Chart 18-2 shows how the subconscious mind organizes the mental imagery arising in the mental field of the subconscious mind into patterns of conscious and subconscious inner realities. Chart 18-3, building on these two preceding charts, illustrates how the subconscious mind is then able to organize the mental imagery of the mind's inner realities into almost endless patterns of outer physical reality. These charts illustrate the processes by which the subconscious mind brings to conscious awareness patterns of outer physical reality. If we have no conscious awareness of our outer reality it holds no meaning for us. It is as though it does not personally exist for us. It simply is not a part of our personal frame of reference within our three dimensional world. In this sense we manifest reality and reality reflects us! These three charts, together with the preceding charts on the directional flow of consciousness, probably best illustrate how the subconscious mind brings to conscious awareness experiences of outer physical realities in time and space.

In summary, any threshold definition of human consciousness must at least include some degree of self awareness, know-

CHART 17

Conceptual Structures of the Field of Consciousness

The structure of consciousness can be conceptually visualized as nesting boxes, within nesting boxes within nesting boxes to infinity. Or it can be described as a Mobius ring which continuously folds back on itself to infinity. The hierarchy of consciousness looks something like this.

1. The Source which consists of the entire field of pure kinetic energy, including the known and the unknown (the unknown being defined as everything else outside the known), contains within itself the local energy fields, individual human consciousness, life, mind including its cognitive structures, belief systems, thoughts, time, space, potential for matter and energy.

2. The Entire Kinetic Energy Field, consists of, and contains within itself, local energy fields, individual consciousness, life, mind, including its cognitive structures, belief systems, thoughts, time, space, the potential for matter and energy.

4. Local Energy Fields consists of and contain within themselves, individual consciousness, life, mind including its cognitive structures, belief systems, thoughts, time, space, the potential for matter and energy.

3. Individual Consciousness consists of and contains life, mind including its cognitive structures, belief systems, thoughts, time, space, the potential for matter and energy.

4. Life consists of and contains mind, including its cognitive structures, belief systems, thoughts, time, space, the potential for matter and energy.

5. Mind consists of and contains, its cognitive structures, belief systems, thoughts, time, space, the potential for matter and energy.

6. Belief systems consist of and contain, thoughts, time, space, the potential for matter and energy.

7. Thoughts are contained within time and space, and hold the potential for the creation of matter, and energy.

8. Time and Space consist of energy and contain matter.

9. Matter consists of energy and resides in time and space.

10. Energy is the power which gives rise to, forms, and sustains all.

11. The Source of all consists of the entire field of pure kinetic energy, the known and the unknown, and contains within itself the fields of kinetic energy, human consciousness and - - - - - - - - - -

ledge, intelligence and volitional will. This is necessary in order for there to be something definite, even if intangible, which individual awareness can act upon in order to organize and maintain a separate sense of self identity and to operate effectively in a three dimensiional world. Without at least some degree of knowledge and intelligence we probably would not be much better off then the apes and monkeys swinging from the trees in the jungle. Accordingly, it is these process functions of human consciousness that activate the mind and brain to start organizing analog patterns of our subconscious mind into the imagery of our inner and outer realities in life.

The Early Evolution of Human Consciousness

In the early evolution of human consciousness our ancestors perceived that their world environment was hostile to their well being. They knew that for their personal survival they better learn how to meet the harsh contingencies of life which they encountered on a daily basis. For instance, our ancient tribesman quickly learned at an early stage of his life that saber tooth tigers and woolly mammoths were detrimental to his health. He soon figured out (that is, he adapted his primitive internal reality programs of his cognitive structure) to take effective evasive action, and if he was lucky enough with his club that day he might still be alive by nightfall and, in addition, have plenty to eat. If he wasn't quite so lucky he was the saber tooth tiger's next meal instead of the other way around. It never occurred to him (and for that matter it has never occurred to us either) that by so adapting to his environment, in this survival of the fittest mode, his experiences of reality were perceived to exist outside of himself. He therefore concluded that physical reality was largely outside of his control. He quickly learned about how physical laws affect time and space. He didn't know anything about spacetime of the fourth dimension. From his point of view he had to judge the distance between himself and the saber tooth tiger and determine how fast he would have to run in order to reach the safety of his cave. If he miscalculated he was in big trouble. Remember during these prehistoric times he didn't know about Einstein's theory of relativity nor quantum physics. Frankly, he didn't much care about the theoretical basis of time and space nor the theoretical relationship between energy, mass and matter, particularly when he was trying to outrun a tiger.

Early evolution could have taken a much different turn of events and our tribesman might have felt that perhaps, rather than adapting his thoughts patterns to the world events around him, he would be better served by attempting to adapt the world and its events to align with the energy and images of his own thought processes. Perhaps if he had done so he might have discovered that

CHART 18 - 1

How Changes in the Mental Energy Patterns of the Mind Changes the Images in the Mental Field of the Mind

Changes in Your Mental Energy Patterns	Changes in the Images in Your Mental Field
1. Change Energy of the State- - - - (Kinesthetic State Changes)	Changes Image's Configuration Between Positive and Negative
2. Change Energy Intensity- - - - - - -	Intensifies Images of Reality (The Image becomes more Vivid)
3. Change to Positive Frequency- -	Results in more positive, uplifting and enjoyable Imagery
4. Change to Negative Frequency- -	Results in more negative and more painful Imagery
5. Change Direction of Energy Flow -	Changes from Inner to Outer Images
6. Expand the Energy Field- - - - - -	Expands the Outer Limits of Your Field of Realities
7. Reduce the Focus of Attention- -	Reduces Size and Intensity of Images
8. Focus on the Field's Background -	Increases Background and diminishes foreground patterns
9. Discordant and Confused- - - - - - Energy Patterns	Creates Confusing Imagery, Creates Confusing Realities
10. Accelerate Energy Pacing- - - - -	Increases Pacing of Mental Imagery
11. Slow Down the Pacing- - - - - - - -	Reduces Pacing of Mental Imagery
12. Increase the Rhythm and- - - - - - Harmony within Energy Field	Soothing more coherent and Harmonious background Field
13. Intensify Distortions in- - - - - - - the Energy Field	Creates Distortions in mental imagery
14. Intensify the Energy- - - - - - - - - - of the Object Pattern	Intensifies the Definition of the Imagery
15. Intensify the Energy of the - - - - - Object Identification	Intensifies the Boundaries and the Imagery's Substance or mass
16. Shift the Energy from one- - - - - - pattern to another pattern	Shifts attention from one Image to another Image
17. Shift Energy to- - - - - - - - - - - - - - an out-of-focus state	Dissolves the Image's Distinctiveness against the Field's background
18. Remove all Energy- - - - - - - - - - - from the Image Pattern	Dissolves the energy of the Object and diminishes clarity of the Object
19. Withdraw all Energy- - - - - - - - - - from the Mental Field	Reduces both the Imagery and the Background Field
20. Stop all energy flow- - - - - - - - - - to Patterns in the Field	Removes all Imagery and Collapses the Field, Collapses Realities

How the Processes of the Mind Transforms Energy in the Form of K,V,A, into Images in the Mental Field of the Mind.

CHART 18 - 2

How The Subconscious Mind Organizes the Mental Imagery of the Subconscious Mind into Patterns of Conscious and Subconscious Inner Realities

Conscious Mind Creations of Mental Patterns in the Form of Thought	Subconscious Mind Formation of the Mental Patterns into Inner Realities
1. Mind prior to a creative thought- -	-The undifferentiated field of the subconscious mind
2. Belief in the creative process- - -	-Affirming the mind's ability to create
3. Focusing attention on - - - - - - - - - the initial desire	Impressing upon the subconscious mind the energy of that desire
4. Energy and image of our- - - - - - creative thought	Attracting and coalescing the energy of the image in the subconscious mental field
5. Believing in our reality- - - - - - - - -	Acknowledging the subconscious mind's ability to create
6. Experiencing our thought- - - - - as real	Allowing the subconscious to organize our thought as inner reality
7. Removing all mental and - - - - - - - emotional blocks to the realization of our thought	Removing all negative energy which might impede the reality of our thought
8. Accepting our thought- - - - - - - - as representing our reality	Facilitating the subconscious mind's formation of our thought as inner reality
9.Cooperating with the sub-- - - - - - - conscious mind in manifesting our thought as real	Allowing the subconscious mind to organize our thought as reality
10.Experiencing the manifestation- of our thought as inner reality	-Experiencing our thought as separate from the energy of the entire field of our subconscious mind

The involvement of our conscious and subconscious mind in the process of interpreting the energy of our conscious thought into patterns of our inner reality

Chart 18 - 3

How the Subconscious Mind is Able to Organize Mental Imagery of the Mind's Inner Realities into Almost Endless Patterns of Outer Physical Reality

Thoughts arising within the Energy Field of the Subconscious Mind	The Effects the Energy Field of the Subconscious Mind has in Organizing Outer Reality
1.The subconscious mind prior to a creative thought	The undifferentiated field of the subconscious mind
2. A thought representing our belief in the creative process	Affirming the subconscious mind's ability to create
3. Focusing attention on our initial desire	Impressing upon the subconscious mind the energy of our desire
4. Energy and image of our creative thought	Attracting and coalescing the energy of the image in the subconscious mind
5. Cooperating with the subconscious in our ability to manifest outer reality	Allowing the subconscious to bring the object of outer physical reality into our conscious awareness
6. Believing in our outer physical reality	Allowing unlimited correlation to operate in the subconscious mind to bring awareness to our outer reality
7. Removing all mental and emotional blocks to its creation as our reality	Removing any negative energy which might impede awareness of our outer physical manifestation
8. Experiencing the object of creation as real to us	Acknowledging the subconscious mind's ability to organize awareness of our outer reality
9. Releasing the creation knowing its already exists as our outer reality	Facilitating an awareness of our manifestation in the time frames of our outer spatial field
10.Experiencing our manifestation as our outer physical reality	Experiencing our manifestation as our outer physical reality

---

This chart illustrates the process by which the subconscious mind brings to conscious attention patterns of outer physical reality. It is this process by which we become aware of our outer physical realities. It is what allows consciousness to manifest reality. If we have no conscious awareness of reality it holds no meaning for us. It is as though it does not personally exist for us.

the world was not quite so hostile after all. He might have found that it was simply his perception of his world and his defensive behavior and fearful thought processes which made it so. Remember, he didn't know anything about analog thinking in the spacetime of the fourth dimension. However, if he had he might have stopped and thought for a moment about other creative possibilities of effectively adapting his environment to his thought processes in order to live a more stress free life style, (maybe we should do the same). For starters, he might have adopted a more healthy life style and perhaps even thought of becoming a vegetarian and stop chasing tigers and mammoths for his next meal. Who knows, maybe if he had just stopped fearing tigers and mammoths and left them alone, they might have, in turn, just ignored him. The point I am making here is that he had other options available to him if he just expanded his awareness of other possibilities. If he had just entertained analog thinking for a minute and contemplated his options other possibilities would have opened up for him. He could have empowered himself and taken more control over his world of saber tooth tigers and woolly mammoths.

Unfortunately, these alternatives never even occurred to him and so today we have inherited, and are stuck with, a very self limiting cognitive structure. A cognitive structure we inherited and we have been struggling with from time immemorial to adapt as best we can, one way or another, to our perceived hostile and self limiting world environment. It is unfortunate that we never realized from the start that the energy behind our thoughts and the field from which they arose, our brain, subconscious mind, and the cognitive structure of self awareness, all exist outside the space and time framework of our lives. This realization alone would have changed the course of human events. Right from the start we would have had the ability and power, drawn from our own expanded self awareness and our experiences of world events around us, to consciously create a more hospitable living environment. Instead, we have been continually adapting, existing, and struggling with life and its many problems since the beginning of time.

The thing that we were never aware of was that our very own creations of life realities, both negative and positive, are what form our experience of who and what we are. It is what gives definition and substance to our perceptions of ourselves and of our world. We didn't fully realize that we, ourselves, are creating all aspects of our life realities, moment by moment, each and every day of our lives.

It is the effect of our limited awareness that has inhibited a more accelerated growth in the evolutionary development of expanded states of consciousness for all mankind. While mankind thinks that it represents a very advanced state of intelligence it remains, as to the understanding and full utilization of the internal dynamics of the human brain and mind, not much better off than the awareness of these early tribal ancestors. Even today mankind still acts from this primitive style of thinking rising out of a vastly restricted cognitive structure. The Darwinian theory of evolution based on the survival of the fittest holds true today as it did hundreds of thousands of years ago. *Mankind has never been able to expand his cognitive structure to the point that he is able to fully realize and experience himself as the creator of his own realities! We never quite realized, nor personally experienced, that we are all one and interconnected within the creative force which gives rise and sustains the entire universe and all of mankind within it. We never understood in any conscious cognitive way that we collectively have the capacity and ability to create whatever type of world we envision for ourselves and our follow men!*

Consciousness as an Emergent Property of the Brain and Mind

A major point of contention in the scientific community has been whether or not human consciousness is an emergent property or an inherent property of mankind. As an inherent property consciousness would be considered to have arisen out of the anatomical development of the billions of neuron cells of the brain of prehistoric man. As an emergent property consciousness would be considered to arise out of the complexity of the self organizing systems of the brain and mind of each newborn child. The answer would appear to be that consciousness of man, that is, his capacity for knowing and his sense of self awareness, arose historically and therefore represents inherent properties of the brain and mind of modern man. The brain and mind of our early ancestors evolved to such a complex level of self organization that consciousness itself arose from within the mind/brain structure itself. Consciousness then is not an emergent property of the brain and mind of each newborn child.

In similar fashion the form of the primitive physical body of man emerged at even a much earlier stage in time than the time human consciousness first appeared in the evolutionary chain of development. Today each time a newborn baby comes into this world it does not recreate consciousness and a sense of self awareness. The newborn child does develop for himself a personal identity and operational framework for life. However, this personal identity and these operational programs are far different from consciousness and self awareness. It is the personal identity and operational programs that could be described as emergent properties of each newborn child. Self identity and these operational programs start to develop during the time the child is in the mother's womb. On the other hand, self awareness and the faculty of cognition arose from the process functions of knowledge and intelligence and represent inherent properties of the human condition. They are already in place when the child is born. It is from the environmental stimuli the child receives, and from its inherited genes represented by its DNA mental precursors, that consciousness starts to organize the basic parameters of the primary personality and operational systems of the mind. In simple terms it is consciousness and a sense of personal self awareness that are present at birth. The child then uses his sense of self awareness and his consciousness to create for himself an ego personality. He then uses his ego personality to perceive, respond to, and create a life for himself.

If self awareness and the processing functions of human consciousness were not in place at the time of birth we would be mentally brain dead on arrival. For we need the processing functions of consciousness in order to create a life for ourself. The newborn baby has an undefined sense of self, all of the essential neurological cells it will ever need in developing a life, and the inherent processing capacity that is required to operate effectively during its entire lifetime. However, self awareness, in order to identify with its human body, needs to develop an ego personality. It also needs to learn from the time of birth how to acquire, through the process functions of consciousness, that large body of knowledge and an attitude we referred to back in Chapter 2. To simply state this another way, the child immediately starts to program his personality and operational systems by using inherent processing functions of human consciousness in order to create a life for himself.

Consciousness of self awareness first appeared hundreds of thousands of years ago when prehistoric man evolved to the point when he finally climbed out of the trees and said, "I Igor, you Jane." At that point in time mankind became self aware for he experienced and understood his separateness from others and his environment. Since that time mankind has evolved to the level of consciousness he experiences today. Each child therefore enters his world with a primordial sense of self awareness and uses this sense of self awareness to develop for himself an ego personality for both responding to, and interacting with, his environment. Consequently, each child develops differently in response to his environment. No two people are exactly the same, not even identical twins, nor are their brains hard wired in the exact same manner. Consequently, as mankind became self aware hundreds of thousands of years ago mankind started to develop a rather elaborate ego personality in order to more fully experience his environment. How we each develop that elaborate ego personality is addressed in the next section.

Present Day Consciousness and the Formation of Ego Personality

At the time of conception through the gestation period the anatomical growth of the child takes place inside the womb at an unprecedented rate. By the time the child is born he weighs on average five to over seven pounds and his internal organs and exterior limbs are fully developed. During this period self awareness, by utilizing the process functions of consciousness, organizes the basic application programs of ego personality. These application programs create the content of ego personality and its operational style for effectively living in a physical body in a three dimensional world. Only by creating ego personality is self awareness able to fully function in the physical world. By creating ego personality self awareness is able to experience mentally, emotionally, and physically the infinite number of realities it creates which form the many aspects of its life. Without a physical body in which to exist in human form in the world and without an ego personality in which to experience that world, self awareness, without identification, would be unable to experience its physical environment. Unfortunately, we have identified with our physical body and ego personality to such an extent that we find it extremely difficult to detach our self awareness from them for even a brief moment. Detachment is needed for it allows us to obtain a broader perspective on life in order to more effectively resolve conflicts in life and create expanded realities. Analog thinking and MetaCreation help in separating your sense of self from your ego personality and physical possessions in life. More often than not because of our total absorption and identification with ego personality we are almost totally unable to rise to a higher level of mental organizing synthesis, a level needed to resolve life style issues and create a better life for ourself. How to momentarily detach awareness from ego personality and gain a greater perspective on life in order to live life more effectively is dealt with in detail in Chapters 9 and 10.

Chart 19, at page 142 entitled, *Creation of Consciousness of Self and Ego Personality,* sets forth how self awareness accomplishes the task of creating ego personality. Briefly, self awareness translates the process functions of consciousness without physical identification into process functions of ego personality in order to create realities for itself. The components of consciousness, namely, self awareness, knowledge, intelligence, and volitional will all serve as the process functions of ego personality. The process functions of ego personality become its operating drive systems for the formation of a personality and its operational style for living life effectively in a three dimensional world.[7] You will note that the process functions of consciousness line up with the process functions of ego personality and with the four archetype prime directives of ego personality. The reason for this is that it is the components of consciousness which constitute the drive system creating a life for us, not ego personality, not the content of our character nor our possessions and status in life. Unfortunately, far too often we believe that it is ego personality, the content of our character and our possessions, that determine who and what we are in life.

These process functions of ego personality are used by self awareness to created the primary programs which are used to organize the content of ego personality and its operational style. The first four primary content programs of ego personality are referred to as the four Archetype Prime Directives™ in life. The Archetype Prime Directives™ constitute the foundation upon which all of our primary personality and operational programs are organized by the mind and brain. All conscious secondary thoughts arise within these primary programs. These four prime directives start organizing and processing for us experiences of our inner and outer realities in life from the time we are in the womb and continue after birth.

The process functions of consciousness and their corresponding process functions of ego personality used to create the first four Archetype Prime Directives™ in life are described as follows:

Process functions of <u>Consciousness</u>	Process Functions of <u>Ego Personality</u>	Four Archetype Prime Directives <u>of Ego Personality</u>
1. Self Awareness	"I am Me,	"I like who I am" This is me"
2. Knowledge	"I know, I organize information"	"I have a right to be here,"I have a right to a life"
3. Intelligence	"I think, I interpret information"	"I express myself fully"
4. Volitional Will	"I decide, I make my own decisions"	"I am capable, I can succeed in life"

---

7 The process functions of consciousness constitute elements of the child's DNA. Normally we think of DNA as the precursors for the physical attributes of the child. The DNA also embodies the process functions of consciousness as well as the application functions making up the characteristics forming the physical identity of each child. Without process functions there would be nothing to activate and mediate the application functions of the mind in order to create the characteristics of childhood individuality.

CHART 19

CREATION OF CONSCIOUSNESS OF SELF AND EGO PERSONALITY

Conceptual relationship among the components of Consciousness, the process functions of Ego Personality, the organization of the Subconscious Mind, and the organization of the neuro networks of the Brain, in creating a functional Ego Personality operating in a physical environment

Components of Consciousness	Process Functions of Ego Personality	Organization by the Mind	Organization by the Brain
1. Self Awareness	"I am Me, this is Me"	Organizes analogs to differentiate sense of self from others	Organizes neuro networks to recognize the concept of a self who is self aware
2. Knowledge	"I know, I organize information"	Organizes analogs into meaningful imagery	Organizes neuro networks to process stimuli into meaningful imagery
3. Intelligence	"I think, I interpret information"	Organizes meaningful imagery into hierarchy of cognitive structures	Organizes neuro networks to process hierarchy of cognitive structures
4. Volitional Will	"I decide, I make my own decisions"	Selects imagery based on analogs reflecting volitional decisions for a meaningful life	Organizes neuro networks to process conscious decisions to be reflected in the subconscious analogs on the mind

NOTE: The components of consciousness, namely self awareness, knowledge, intelligence, and volitional will, are all process functions of the mind and brain. That is self awareness utilizes its intelligence, knowledge and volitional will to formulate for itself the basic functional processes of an ego personality. The four Archetype Prime Directives are then created by the functional processes of ego personality using self organizing processes in order to create for ego the content of its ego personality and its operational programs for the creation of its inner and outer realities in life.

The four processing functions of consciousness are subconscious processing functions of the mind/brain. The four processing functions of ego personality are also largely unconscious process functions of the mind/brain because we seldom consciously reflect upon them as such, and consequently, they remain largely unconscious functions. The four Archetype Prime Directives™ in life are also largely unconscious content components of ego personality, for again, we seldom consciously reflect upon them. The above descriptions of the process functions of consciousness and ego personality describe the relationship between the process functions of consciousness and the process functions of ego personality and how they relate in organizing the four primary content characteristics of ego personality. The process functions of ego personality namely "I am me", "I know", "I think" and "I decide" are stated in neutral terms for they represent process functions only and not content functions. As process functions they organize the content of ego personality.

As shown on the chart above these process functions of ego personality are then translated into the content of the four archetype prime directive in life represented as "I like who I am", "I have a right to be here", "I express myself fully", and "I am capable". These prime directives are stated in positive terms for they are intended to represent a positive personality and operational program for effective living. However, far too often, as was the case in the life of Tommy and Ricky, these prime directives are negatively organized as a result of an abusive or neglectful childhood. In such cases the content of such prime directives would appear as "I don't like who I am", "I don't have a right to be here", "I can't express myself fully", and "I am not capable". It is the negative energy and matrix configuration of these negative prime directives in life which set the tone for all of the pain, lack and limitation a person experiences throughout his entire adult years. In order to change the negative energy of these prime directives to positive energy you need to address the analog patterns of the subconscious mind and not just address the problem at the level of conscious cognitive thought. The four archetype prime directives are dealt with in detail in Chapter 8. Chart 19 describes the structure and functions of the mind and brain involved in organizing these process functions of ego personality.

How The Process Functions of Ego Personality Creates A Life For Us

Self awareness organizes the process functions conceptually describing the sense of ego personality as "I am me" for purposes of self identification with a physical body. Self awareness moves from a sense of self without identification to ego identification through the instrumentality of the "I am me" process function of ego personality.

Thereafter interaction with his environment causes the child to develop in detail what the concept of "I am me" ultimately means. When the child is first born it does not have a clear sense of identification or boundaries between himself and his environment. He does not have a well defined sense of ego self. Such an identification is rapidly developed from the sensory stimuli coming from his environment. Initially such sensory stimuli is provided by the primary caregiver. If the primary caregiver fosters a nurturing, loving, supportive environment the child will develop a positive ego personality and a positive outlook on life. On the other hand if the child is raised in a neglectful or abusive environment much like Ricky and Tommy the child will develop a negative ego personality with a limited or negative outlook on life.

Recent scientific studies by neuroscientists have shown that in the argument on which is most important, genes or environment in the development of the child, environment is winning hands down.[8] The reason for this outcome is that, all things being equal, a normal child starts out in life with more than sufficient gray matter, over two hundred billion neuron cell connections to be exact, in order to develop properly. Probably just half of these would be sufficient to get the job done. However, whether or not he will actually develop properly depends to a large extent on the environment. Just how we reverse the adverse effects on the child of a neglectful or abusive environment later on when he is an adult is addressed in Chapter 8.

The second component of consciousness is knowledge. Knowledge is translated conceptually by ego personality into "I know, I organize information". This process function allows consciousness of ego personality to organize incoming stimuli into meaningful analog patterns which can then be acted upon by ego personality to create a life for itself. The third component of consciousness is intelligence. Intelligence is translated by ego personality into the process function conceptually described as, "I think, I interpret information". This process function of ego personality, "I think", activates the organization of the mind and brain to start processing stimuli into meaningful imagery forming thoughts and beliefs that can be inter-

8 See, *Rethinking the Brain,* by Terry Shore.

preted by ego personality into experiences of reality.[9] As indicated earlier, intelligence in this respect is simply a process function of the mind and is not intended as a measurement of the I.Q. of the child

The final component of consciousness is volitional will. Volitional will is translated conceptually by ego personality into "I decide". It is this process function of ego personality which activates the mind and brain to determine which experiences out of literally hundreds of sensory stimuli that it receives at any given moment it will integrate into its cognitive structures of thought for the purpose of creating its realities. The concept of "I decide" is usually given full expression by the time child reaches the stage of the "terrible twos" when he finds out the power in the word "no". He finds that by simply using the word "no" as in "no, I won't go"or "no, I won't eat my vegetables" or "no, I won't go to bed" he is able to take much greater control over his environment than ever before. Thank goodness this period doesn't last very long before the child finds out the outer limits to the command "No" and that there are other, more appropriate, ways of resolving conflict than a direct frontal attack on his parents' authority.

One important note: these translations of the components of consciousness into process functions of ego personality are abstract conceptualizations stated in linguistic terms, which is the processing medium of the mind. However, this conversion process takes place on the subconscious level of the mind of the child. Neither children nor for that matter adults are consciously aware of the internal processes of the mind in translating external stimuli into thoughts which can then be experienced by the ego personality. The mind and brain created the system of conscious and subconscious processing as the most efficient manner of operating the mind/brain system. If your conscious self was aware of all of the many thousands of internal processes going on each moment in creating reality you would be totally overwhelmed and not be able to function effectively in making conscious day to day decisions for yourself.

---

9 One interesting historical observation which comes to mind when you hear the process function "I think" is the statement by the famous sixteenth century philosopher Descartes, "I think, therefore I am". Descartes unwittingly combined the process function of self awareness described as "I am me" with the process function for intelligence as described as "I think". He then concluded that since he was a thinking being he must be a self aware being. And you know what, Descartes was absolutely correct. Unfortunately, he took his theories and developed what has become known as the Cartesian way of thinking which as we have seen dominated Western thinking for over three hundred years.

You may be wondering why consciousness forms a sense of ego personality by identifying with the process functions of "I am me", "I know", "I think"; and "I decide". The reason for the translation of these abstract concepts to personal functional subconscious processes of ego personality is that the mind and brain evolved in a way that processes thought into an awareness of reality. That is, the ego personality evolved its neuro networks of the brain and the cognitive structures of the mind in order to process data to form thoughts. The brain processes data in the form of sensory stimuli into neurological patterns and the mind interprets these patterns into meaningful thoughts and beliefs which in turn create our experiences of inner and outer realities in life.

The mind and brain could have evolved to only process numerical computations, or musical computations, or even developed to only process sensory stimuli into symbols and pictures. In this case self awareness would have translated the components of consciousness into numerical processes in the case of a mathematical mind; musical notations in the case of a musical mind; art symbols in the case of an artful mind; and symbolic processes in the case of a symbolic mind. As it turned out we developed linguistic patterns to articulate the expressions of reality in the form of words, phrases and sentences which hold meaning for us. Consequently, the brain developed its neuro networks to translate sensory stimuli into mental imagery which the mind was then able to translate into linguistic patterns of thought in order to give us a conscious experience of reality.[10]

In summary, self awareness from the time the child develops in the womb and throughout its entire adult life, through the process components of its consciousness, creates for himself an ego personality that represents the initial point of contact for self awareness to experience its physical world. Ego personality, through the components of consciousness, namely, knowledge, intelligence, and voli-

---

10 In the past the mind did in fact develop methods for communicating reality in forms other than linguistic patterns. For instance, the early Egyptians extensively used pictographs for communication. The early North American Indians used symbols in the form of sand writing and other totems representing spirituality, their early ancestors and seasonal phases to construct the realities of their life. Today modern theoretical mathematicians use mathematical computations to express the reality of nature. The artist uses pictures, sculpture and other symbols to create experiences of reality often quite different from what we experience in everyday life. Just look at the art of Pueblo Picasso and the music of Mozart. All of these systems provide different expressions of reality other than the expressions of reality described in linguistic terms alone.

tional will, creates for itself, piece by piece, primarily using linguistic symbols, in the form of words and phrases, the many aspects comprising the complex structure of its personality. Ego personality continues this process of assimilation, through the medium of linguistic symbols, to experience physical reality. Self awareness in this way, through ego personality, by continuing this creation process creates for itself a system of hierarchy of cognitive structures of thought which provide ego personality with the means to create and experience a greater variety of complex realities. This system of cognitive structures is addressed in Chapter 9. The archetype prime directives are described in detail in Chapter 8. One last observation before moving on is that these four basic processes of the human mind set the foundation for the organization of all of the other subsidiary interconnected and interrelated process functions of the mind. These subsidiary process functions are referred to as the "self organizing processes" of the cognitive structure of the human mind. These other self organizing processes are set forth in Appendix Two.

Degrees of Awareness of the Human Mind

Previously we explored the process functions of consciousness. Here we explore states of personal conscious awareness and their effect in our life. These states include full conscious awareness, nonconscious, unconscious and subconscious mental states in time and space. The full conscious awake state is the normal awake state of everyday life. The subconscious state is the state in which we have little to no conscious awareness of internal events taking place. It is the state where all internal mental processing take place.The unconscious state is the state where we have no conscious awareness of either our inner or outer reality and is often referred to as the state of fainting, passing out or being knocked unconscious. The term "transitional state" shown on Chart 20-1 at page 149 needs a little explaining. The term "transitional state" simply describes the state between full conscious awareness and the subconscious state. This transitional state is experienced when using the MetaCreation processes. While in the subconscious state when using the MetaCreation processes we are still consciously aware. However, conscious awareness is not actively involved in the process. Our conscious awareness simply acts as a passive observer. If conscious awareness were involved in the process, conscious logical thought would inhibit our ability to process realities by the subconscious analog system. I use one additional term in Chart 20-3, at page 151, referred to as the "active awake state." The active awake state is obviously a part of the full conscious awake state. However we added the active awake state for it almost always takes place in linear time and fixed space, whereas the conscious mind can distort and displace time and space. One additional point, and that is the process functions of human consciousness discussed in the preceding section are entirely different than these various states of personal conscious awareness. One deals with the process functions of the mind/brain and the other deals with how awake and aware we are concerning our inner and outer world.

Chart 20-1, at page 149, entitled, *Overlapping States of Consciousness*, illustrates the directional flow of the conscious state moving from the outside to the inside and from conscious awareness to the subconscious state. Chart 20-2 *States and Degrees of Conscious Awareness of the Human Mind* and Chart 20-3 *Mental States of the Mind in Time and Space* both illustrate that we are almost totally unaware of the mental programs organizing our internal and external realities. If we are not aware of the effects of the cybernetic feedback loops in life we have few possibilities for altering or changing the course of events in life. As you look at Charts 20-2 and 20-3, it becomes clear why we are unaware of much of the internal mental processes creating realities in life. Specifically, you will note from Chart 20-2 that there is no awareness of the field of pure consciousness. There is no awareness at the subconscious level of the primordial instincts of mankind, dream states, and disassociated states of consciousness. There is very little to no actual awareness at the preconscious level of the primary belief systems from which we derive our self identity and performance models for living. You will note that it is not until we reach the level of our secondary beliefs of our personality and life styles that we experience full conscious awareness! Finally, there is very little awareness at the post conscious level of the cybernetic relationship between our internal mental programs and the creation of our outer physical realities in life. We have little awareness of how the cybernetic loops feed data back into the system. The purpose of this chart is to make the reader aware of how really unaware we are concerning how we form the patterns of our life!

Another way of illustrating how unaware we are is that mankind assimilates about only 10% of the hundreds of bits of information available to our sensory receptors at any given moment of the day. Of this 10% of the data that is received by the neurological structure of the brain, we are unaware of the assimilation of approximately 90% of such data. Stop and think for a minute, if we are only aware of receiving 10% of the incoming data, and if we are only consciously aware of processing 10% of the data that is received, then mathematically this amounts to only about 1% to 2% of the total of all such data potentially available to our sensory receptors and consciously pro-

CHART 20-1

OVERLAPPING STATES OF CONSCIOUSNESS

Directional Flow of Conscious Attention - Moves from the Outside to the Inside and from Conscious Awareness to Subconscious States

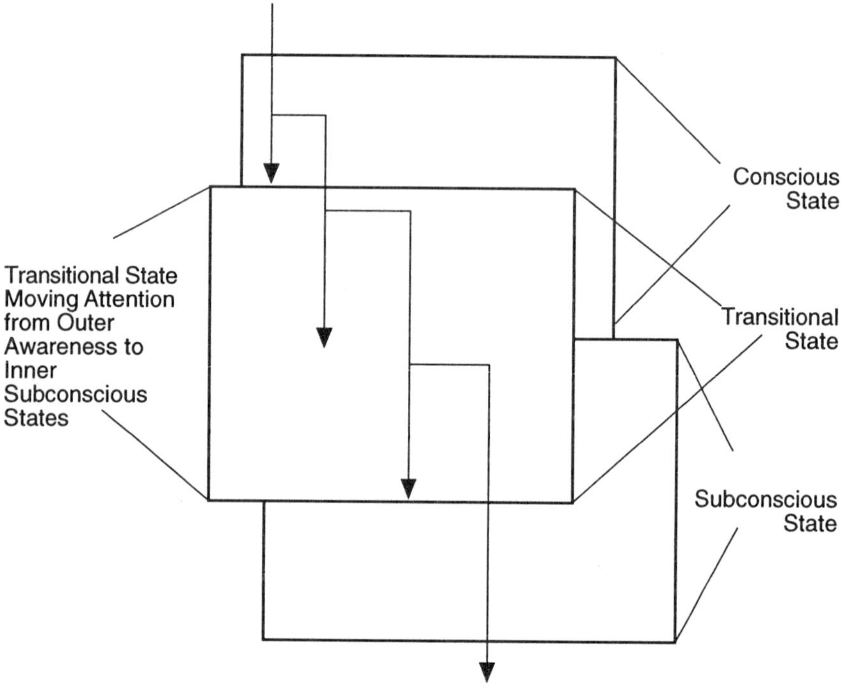

The MetaCreation Exercises move the conscious mind to the transitional state where the mind can access the subconscious mind to create alternative expanded realities outside of conscious awareness, outside of cognitive thought, and outside of the sequential time and space of a three dimential world.

CHART 20 - 2

STATES AND DEGREES OF CONSCIOUS AWARENESS OF THE HUMAN MIND

THE KINETIC ENERGY FIELDS OF THE MIND
FIELD OF CONSCIOUSNESS (No Awareness)
Pure Awareness without Imagery within
time and space of a three dimensional world.

|

SUBCONSCIOUS STATES (No Awareness)
Primordial Instincts of Mankind, Dream States, and Disassociated States
Represented as Symbols and Images without conscious cognitive thought
or content.

|

PRECONSCIOUS STATES (Limited Awareness)

Self Identification Programs for the formation of Personality	Operational Models for functioning in society
|	|

FULL CONSCIOUSNESS - PRESENT AWARENESS

Secondary Beliefs as to manner, style & mode of self identity	Secondary Beliefs as to manner, style & mode of performance
|	|

FULL CONSCIOUSNESS - PRESENT AWARENESS

Personal Actions & Behavioral Styles	Interactive & Behavioral Styles
|	|

POST CONSCIOUSNESS (Limited Awareness)
Creation of Personal Realities

|

POST CONSCIOUSNESS (Limited Awareness)
Data Feed Back Loop to Primary Belief System Structures

|

POST CONSCIOUSNESS (Limited Awareness)
Modifications in Secondary Beliefs
Modifications in personal appearance, behavior and actions

|

POST CONSCIOUSNESS (Limited Awareness)

|

Modifications in Inner and Outer Realities

CHART 20-3

MENTAL STATES OF THE MIND IN TIME AND SPACE

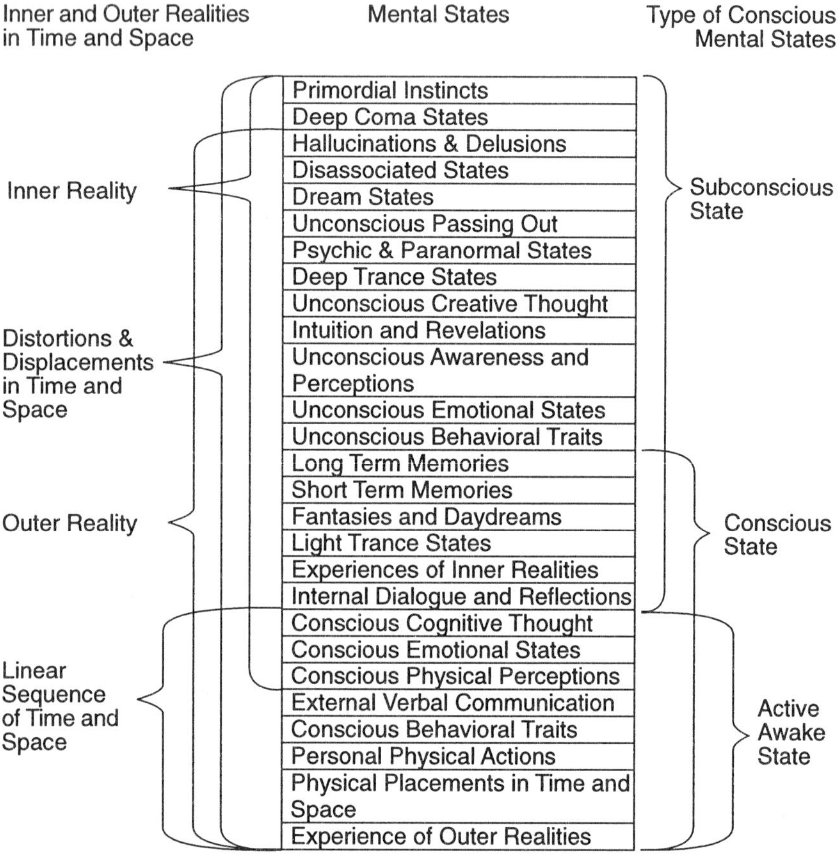

This Chart show how the involvement of the mental field of energy and images with the Subconscious, Conscious, and Active Awake States of mind and with our physical placements of inner and outer realities within time and space all have an effect on the creation of our inner and outer realities in life.

cessed by the cognitive structure of the mind.[11] Chart 20-2 graphically illustrates the extent that creation of realities in life takes place outside of our awake state of awareness. It is because of our unawareness of how we process inner and outer realities that mankind has not evolved to higher states of consciousness. Regardless of how advanced mankind has become by reason of the advances we have made in science and technology, mankind still has not fully realized in any systemic way, how the mind and brain actually work in creating realities in life. In this respect mankind has not progressed in human consciousness much beyond that of our early ancestors.

Few people recognize the extent of our unawareness in how we process realities and its consequences in our life. The impact on mankind will have earth shattering implications once people realize to what extent they are limiting themselves in creating a life. It is amazing that all of our actions are based, for the most part, on our conscious awake state. Yet almost the entire transaction of creation of our realities take place on the subconscious levels of our mind outside conscious awareness. Yet seldom, if ever, do people make a conscious effort to access the mental processes by which they create a life. As we have seen we are actually only consciously aware of maybe 1% or 2 % of all of the processes of how we actually create a life. That means about 98% to 99% of our reality is created at a level outside of our conscious awareness.

People need to become more consciously aware of themselves and their surroundings and the effect one has on the other. The fact of the matter is that they are creating realities randomly without clearly knowing what they are actually doing. What they are actually doing is physically trying to force things into manifestation solely though their conscious mental effort and through their physical labor in making a living. Physical effort in bringing into manifestation physical reality is actually an outdated way of creating things. This is the way our early ancestors went about ordering their life. Modern man is far more capable of achieving a vastly more fulfilling life for himself than merely creating a life through physical labor alone. The Meta-Creation theory and processes hold the potential for quantum leaps

---

11 This analysis is purely hypothetical in order to make a point. The automatic systems of the mind and body of which we are unaware process considerable data which is not really relevant to these comparisons. For instance, reflex reactions, unconscious learned action, intuition, primordial instincts, mechanisms for survival, such as our automatic immune system, our unconscious sense of time, space, direction and basic, but hidden, values and purposes in life, dream processing, and just simply gut feelings and awareness on other levels of assimilation lying just beyond the surface of conscious awareness are also operating all the time in our everyday life.

in the evolutionary development of all mankind.

Brain Wave Frequencies Of The Human Brain and Mind

In this section we address brain wave frequencies which measure the internal activity of the brain and mind in organizing patterns of reality. Chart 21 at page 154 entitled *Brain Wave Frequencies* sets forth the cycles of brain wave frequencies during various states of consciousness. Beta frequencies of the brain cycle between 14 to 33 cycles per second, with an average of about 22 cycles per second. Beta levels of the brain are characterized by the awake state of the mind when an individual is involved in his active daily activities. The more intense the mental activities the higher the brain wave frequencies. Because of the high brain wave activity in the Beta range the greater the mental chatter of the mind. High Beta frequency causes an increase in mental chatter. In turn, mental chatter causes an increase in brain wave activity. The mind is trained to think in terms of language patterns during our awake state to formulate our thoughts and to communicate them to others. This entrainment automatically continues the mental chatter of the mind when we are not consciously or deliberately processing language into thought patterns in our awake state.

Contrasted to the rapid brain wave activity of the Beta state is the Alpha level with frequencies between 7 to 14 cycles per second. Alpha states of mind are characterized by relaxation states and dream states. The dream state serves the purpose of balancing and releasing tensions and stresses arising in the mental energy field of the mind during times of stress in daily living. Theta and Delta levels cycle at even slower frequencies. The frequency of the Theta state is about four cycles per second. The Theta state represents deep sleep beyond the dream level where the body does a considerable amount of healing, replenishing and nourishing of the systems of the physical body. The Delta state cycles at about one half to one cycle per second and represents a very deep level of brain wave activity similar to a deep coma or trance-like state where almost all of the mind, brain, and bodily functions cease to operate. This state could be characterized as a deep hibernation of all of the human systems.[12]

You may be wondering what is the importance of brain wave frequencies on the human brain and mind? Simply this. We have seen that most of the functions of the creative process takes place on the subconscious level of the mind. Brain wave cycles slow down

12 Some eastern Yogis have trained themselves to sustain this level of deep slowing down of their mental and bodily functions for considerable periods of time. For an excellent discussion on yoga theory and practices see *Yoga and Psychotherapy, The Evolution of Consciousness*, by Swami Rama, Rudolph Ballentine, M.D. and Swami Ajaya, PhD.

CHART 21

BRAIN WAVE FREQUENCIES

Number of Cycles per Second	Frequencies	Cycles	Conscious State
33			
22 14	BETA		Active Conscious Activity
7	ALPHA		Relaxed States and Sleep States
4	THETA		Deep Sleep
1/2	DELTA		Deep Coma Like State

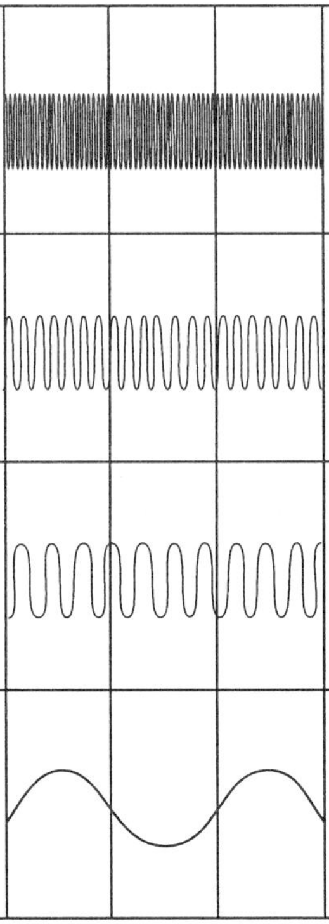

1 Sec. 2 Sec. 3 Sec.

Cycles Per Second

in the relaxation Alpha state where, for the most part, the mental chatter of the mind is eliminated. Remember, it is at the level of the subconscious mind that the conscious mind accesses the energy and images of the subconscious analog system for the purpose of creating realities. It is also the subconscious level of the mind that space-time exists outside of the restraints of the physical time and space of our daily lives. We orient ourselves to physical time and physical space in our awake state in order to function effectively in our daily affairs. However, at the subconscious level of the right side of the brain we are involved in creative thought outside the constraints of linear time and fixed three dimensional space. Therefore, relaxation states are conducive to creative thought in altering reality. The more we access the subconscious mind through deep relaxation the more we are able to access the creative powers of the mind in creating the type of life we desire. This is not to say that no creative processing takes place on the Beta level of awake consciousness. However, it is limited in its scope and effect whereas creative processing at the Alpha level of deep relaxation is much more conducive to altering realities in time and space.

You are familiar with the distinction that it is the right side of the brain that is used for creative holistic intuitive thought and it is the left side of the brain that is used for logical linear rational thought. The reason for this is that on the Beta level of awake consciousness we primarily use the left side of the brain and think in terms of language patterns primarily to communicate thoughts and ideas and only secondarily to process creative thought into reality. We spend most of our awake state in communication with others and often not enough time in inner contemplation and creative thought. Some people spend endless hours in meaningless talk and chatter, and when no one is listening to them, they chatter silently to themselves. It is precisely because we have effectively trained ourselves to use language to express ourselves during our awake state that when we are not communicating, the mind automatically resorts to a stream of mental chatter to express internal mental states and realities.

Whereas on the subconscious level of the mind we use imagery, symbols, patterns, and metaphors, as well as nonlogical thinking, intuition and language patterns, to process data into meaningful and useful patterns of creative thought. It is this creative thought which, in turn, creates experiences of realities. All of this takes place outside of the constraints of physical time and space. We seldom use rational linear language to process data on the subconscious level of the mind. Also on the subconscious level we create thought in multidimensional fashion rather than in linear spatial dimensions. For these reasons our greatest creative ideas flow from the subconscious unrestricted by language, causal and logical constraints, and time/space barriers.

Conscious and Subconscious States in Time and Space

With this background on brain wave frequencies we will now take a closer look at the considerable overlapping effect between conscious and subconscious states and distortions and displacements in time and space as set forth in Chart 20-3. The reason for this overlapping and distortion effect of various states of mind is that the human brain and mind work in combination in an overlapping and stacking fashion. Data in the form of stimuli upon entering the brain is acted upon by the brain and is arranged into electrochemical stimuli of the brain's basins of attraction which are converted into images by the subconscious mind. This process is a very elaborate and complex process that involves millions upon millions of neuro synaptic connections whereby previous information stored in memory is used to interpret new information. The result of this process is that the primary personality and operational programs of the subconscious mind form secondary thoughts and beliefs about how this newly acquired information can best be assimilated and used in the life of the individual.

This process goes on continuously and cycles back and forth through the conscious and subconscious centers of the mind/brain in cycles of stacking, overlapping and processing. This mental process continues until the new information is finally stored in memory for future use or is currently acted upon in creating thoughts, perceptions and behaviors. These processes go on throughout our life and appears during sleep time as images in our dreams. The mind, like our heart, never stops and continues this elaborate and complex stacking, overlapping and processing of stimuli on the conscious and subconscious levels of the mind in the creation of realities in our life up until the moment we die. For a schematic diagram of how energy and imagery patterns cycle back and forth from the conscious to the subconscious mind in forming thoughts and realities in life refer to Chart 22 at page 157.

Often, however, where the states of inner and outer realities, states of time and space, and conscious and subconscious states of mind all overlap with one another there is the potential for cognitive dissonance, mental confusion, disorientation, disequilibrium and destabilization in the operating systems of the mind. For instance, in referring back to Chart 20-3, the state of an individual's outer reality and the state of his linear sequence of time and space, and long and short term memories, all overlap with the state of inner reality, and

CHART 22

ENERGY CYCLES IN THE FEED BACK LOOP OF CONSCIOUSNESS

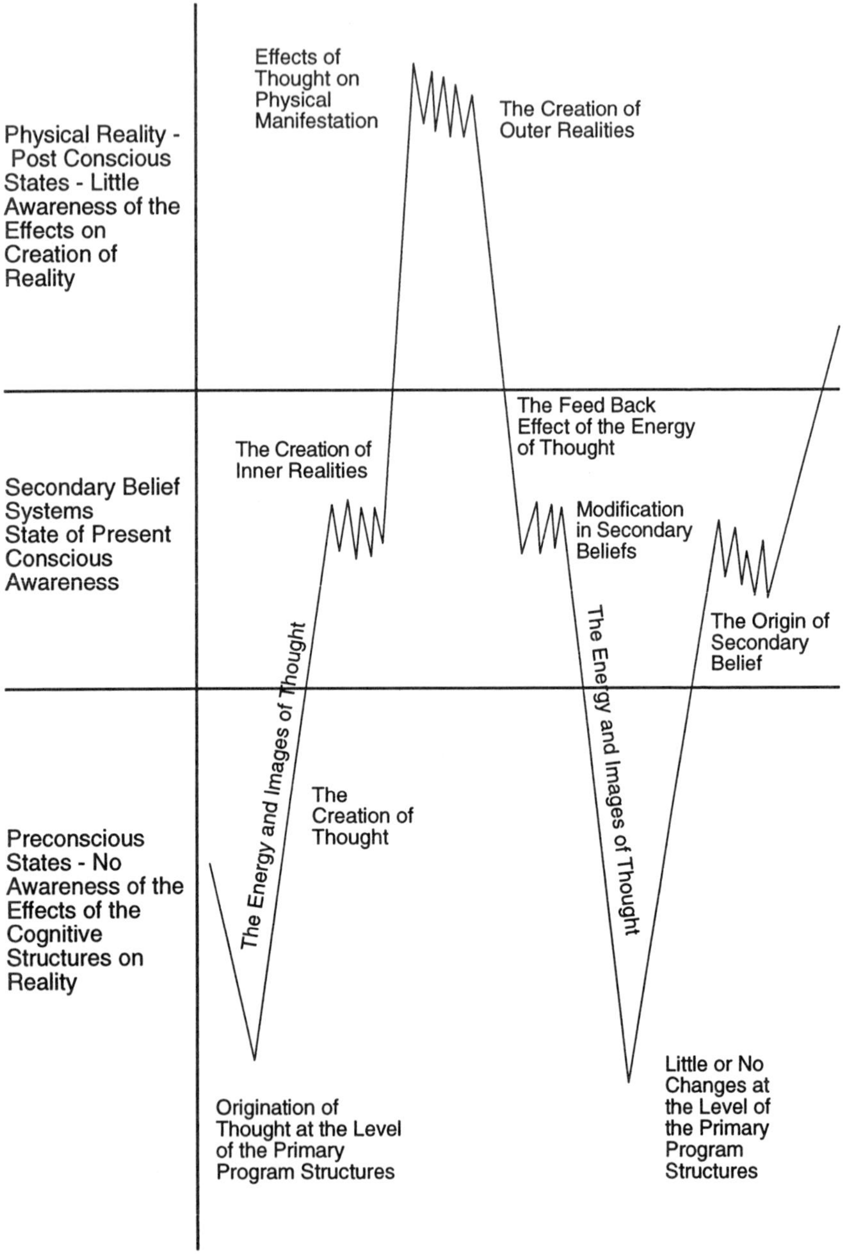

therefore hold the potential for distortions and displacements in the creation and placement of either inner or outer realities in time and space. When you look at Chart 20-3 one feature of the chart that becomes very apparent is that a considerable amount of our time is spent each day in subconscious states existing as our inner reality. This inner reality appears as our internal dialogue, unconscious creative thought, introspection, daydreams, fantasies and other inner states. These internal states often create distortions and displacements in the timing and spacing of our inner and outer realities. This is not to imply that this is necessarily a negative or bad thing. In fact, this often serves a very beneficial purpose. It allows for the development of innovative and creative, thoughts, concepts and new ideas in the way we see and do things unrestricted by cognitive thought and the linear sequencing of time and space of our outer three dimensional world.

The productive use of time during subconscious mental processing with deliberate focus and intention as to how we can best use these internal states have a beneficial effect in creating a better life for ourself. This is a basic premise of analog thinking. However, we must first take control of these inner states and start to program them for more effective ways in creating that better life. We put far too much attention each day in making a living in our physical three dimensional world in which we exist in order to maximize pleasure and minimize pain. This orientation to life places a significant restriction upon our capacity for creating a vastly superior state of awareness for the enhancement of life in physical space. The reason for this is that the very creation of such expanded states of awareness, for the most part, lie beyond our awareness of physical reality. Chart 20-1, *Overlapping States of Consciousness*, illustrates the directional flow of consciousness from the conscious awake state to the subconscious state. As the conscious mind accesses the subconscious mind we can create alternative expanded realities outside of conscious awareness, outside of cognitive thought, and outside of the sequential time and space of a fixed three dimensional world.

Another example of distortions and displacements in time and place is when a person recalls from memory, people, places, or things in an out-of-sequence order in which they took place. Often they are remembered out-of-sequence, or in the wrong location in space. They can even create distortions and inaccuracies in the specific details concerning the recalled people, places or things themselves. This model of conscious and subconscious states in time and space, as illustrated by Chart 20-3, indicates that recall from short or long term memories are not always as reliable or accurate as one would like them to be. This model, however, does not address why these distortions and displacements take place. The internal primary ego personality and operational programs of the mind represented by our internal somatic system do not always line up our sense of what our reality should look like with the actual external sequencing of that reality in time and space. That is, the mind often doesn't recall the precise manner in which events actually took place like an exact photograph of the event itself. Our memories are subject to the internal somatic system and therefore are subject to being created, formed, molded, altered, expanded, contracted and eliminated entirely by the subconscious mind. Such distortions are the result of the organizational patterns of the ego personality. In effect we distort time, space and reality in order to meet the expectations of our subconscious image of ourself. We create an image which we portray to the world in order that we may be seen and understood in the manner we believe ourselves to be. That is, we endeavor to portray ourselves to the world in the most favorable manner possible for we always want to appear at our best. However, this is not always a conscious decision. If the distortions were not in our perceived best interest we would not subconsciously distort events in the space and time of our life. There is a saying for this somatic effect: "every conscious or subconscious thought or action, however negative, has a positive intent behind it whether we are aware of it or not".

Various trance states of hypnosis overlap with the subconscious and conscious states as well as with our states of inner and outer reality. The reason for this overlapping effect is that trance states of hypnosis can exist in states of either inner or outer reality or both simultaneously as well as exist in both the subconscious and conscious states of mind. Remember, the subconscious state is the state of mind conceptually lying below the cognitive thought processes of conscious awareness and is the state lacking in the capacity for conscious external sensory perception. A person can be in deep hypnosis and be in a subconscious state where suggestion by hypnotic induction is extremely effective since cognitive processes are almost entirely suspended. On the other hand, the person can be in a light trance state where cognitive processes are operative and the subject can cognitively process and integrate inductive suggestions on his own. Hypnotherapists are fully aware of this distinction and take advantage of it in their practice.

Another example of the interrelationship of states of mind illustrated by Chart 20-3 is how creative thought arises within the subconscious mind. When arising from the subconscious state, such as a deep dream state, if remembered upon awaking, both subconscious and conscious states can potentially come into play. For instance, creative thought arising from the subconscious dream state, upon awaking, starts to filter through our conscious mind where cognitive processing can take place. We may then consciously, through the processes of inductive and deductive reasoning, rationally and logically analyze the creative thoughts and organize them into meaningful patterns of interpretation, patterns which can be understood and explained by the conscious cognitive mind. This is the process I used to write, or should I say, create this book. Much of this book was revealed to me from the preprogrammed analogs of my subconscious mind during my early morning walks along the Chattahoochee River.

Once you begin to understand the internal dynamics of how these overlapping and stacking states of mind affect one another in the creation of patterns of reality you can start to use this knowledge to your advantage by intentionally using your ability to alter states in order to achieve the desired results in life. As mentioned above, most of this book was written by the process of using the subconscious mind in deep relaxed states, including sleep states, to process information into meaningful patterns and relationships. Since the subconscious mind is not restricted by the dimensions of a physical world and the logical processes of the conscious rational mind it can freely draw on its storehouse of information and our many past experiences. With this information the subconscious mind can conceive of new and unusual pattern interpretations which the conscious analytical mind can then logically and rationally set forth in a direct simple fashion that anyone can understand.

A simple way to use creative thought in this fashion is to write out an incomplete thought which is left dangling as to the information needed to complete the thought. For instance, if you did not know in what direction your next move should be in pursuing a particular task, career or relationship you might write out on a sheet of paper something like the following. "With respect to the job that I am seeking I have sent out my resumes, networked with my friends, and talked to several professional headhunters. In pursuing the perfect job the next thing I need to know or do is. . . . ." You might write out your incomplete thought on a sheet of paper just before going to bed at night. It helps immensely if you do this exercise with the expectation of positive results. If you just go through the motions believing that it will never work, guess what, it won't work. Remember, the mind/brain doesn't like unresolved tensions in its neuro networks consisting of basins of attractions and attractor pathways. The mind/brain seeks stability, equilibrium or mental homeostasis in its operating systems.

Accordingly, the subconscious mind will continue to process the question or problem posed by the unresolved thought presented to it while your conscious mind is preoccupied with something else or is fast asleep. Quite often in the middle of the night, or first thing upon awaking in the morning, you will have your answer. You should have a pencil and paper handy because quite often you will very quickly forget the answer due to other matters occupying you conscious awake state of mind upon rising in the morning. The desired information will usually come in a flash when you are not thinking about the problem at all. This usually occurs during relaxed states when you are exercising, or you are not thinking about anything in particular, and your mind is parked in neutral. That is you are not processing anything in particular, and certainly not the problem set forth on the piece of paper.

Often after typing difficult questions into my computer or setting forth a question concerning how I might visualize or conceptualize a very difficult abstract concept, I would go for a long walk along the Chattahoochee River. I would not be thinking about anything in particular and in a flash the entire diagram or information I was seeking would be immediately visually present in great detail. I was able to create most of the charts in this book in this fashion. I would immediately write down the information or design and then start to refine and elaborate upon the information I received. The subconscious mind will continue to process the question or problem posed since it seeks resolution to any dynamic tension set up in the mind/brain system. The subconscious mind can draw on an almost infinitely large storehouse of information in which to find the most appropriate solution. After awhile you don't even need to write out the question. You only need to pose the question in your mind and let it sit there. For a comparison summary of how thoughts are processed by the conscious and subconscious mind refer to Chart 23-1 at page 162. For a comparison summary of the significant characteristics of the conscious and subconscious mind in organizing thoughts in life refer to Chart 23-2 at page 163.

<u>The Future Evolution of Consciousness</u>

As we approach the next millennium it is becoming increasing apparent that mankind should take a more active role in his own evolution. The next evolutionary stage for mankind is to develop self awareness beyond simple ego identification, where mankind is able to experience firsthand the state of self awareness apart from its many creations in life including its creation of ego personality. Mankind now has available both the theoretical basis and the emerging technology, even though in rudimentary form, to start to explore

CHART 23 - 1

Comparison of the thought processes of the Conscious and the Subconscious Mind

Processes of Conscious Mind	Process of Subconscious Mind
1. Only experiences reality doesn't create reality	Only creates patterns of reality, doesn't experience reality
2. Doesn't create emotions only experiences emotions	Only creates the components of emotionality, doesn't experience emotions
3. Limited possibilities in arranging thought	Wide range of possibilities in creating analogs forming realities
4. Can live with conflict	Always seek resolution to tension
5. Limited processing potential	Almost unlimited processing potential
6. Primarily thinks analytically	Wide range of possibilities in processing analogs into thoughts and feelings
7. Generalized thought	Specific representations of thought
8. Negative thinking	Makes positive representations of a negative thought
9. Makes value judgments	Non judgmental in processing thought
10. Belief- conscious acceptance of the validity of a thought	Acceptance of a thought without critical evaluation
11. Consciously determines mission or goal in life	No discernment as to life's goals
12. Uses linguistic patterns to express self	Uses symbols, metaphors and images
13. Thinks in linear sequential patterns	Organizes multi dimensionally
14. Processes data logically	Processes data in analog patterns
15. Creates cognitive thoughts	Only acts on cognitive thought, doesn't create cognitive thought
16. Thinks in linguistic models	Creates analog patterns of reality
17. Processes thought in time and space	Creates analogs patterns outside of time and space
18. Conceptualizes in graphic three dimensional images	Organizes in multidimensional metaphor and symbols
19. Usually has a goal or purpose in mind when thinking	Responds to thought without a purpose or goal in mind
20. Usually creates within physical reality	Ability to create outside physical reality
21. Consciously aware of spatial relationships	Can create realities outside of spatial relationships
22. Thinks in time sequences	Creates thought outside time sequences
23. Limited number of variables	Unlimited number of variables
24. Short and long term memory fades with nonuse	Short & long term memory permanently stored but not always easily recalled

CHART 23 - 2

Summary Of The Significant Characteristics Of The Conscious And Subconscious Mind In Organizing Thoughts

Characteristics of Conscious Mind	Characteristics of Subconscious Mind
Belief in the validity of a thought	Only manifests valid thoughts held in conscious belief
Creates beliefs within time and space	Creates realities unrestricted by time and space
Forms thought linearly within time sequenced units	Manifests thought multi-dimensionally unrestricted by time sequencing
Limited in the number of variables it can handle at any one time	Capable of sorting and organizing an unlimited number of variables anytime
Experiences, but doesn't create inner and outer realities	Manifests, but doesn't experience, realties in accordance with valid thought
Experiences, but doesn't create inner and outer emotions	Manifest, but doesn't experience expressions of emotions
Views world reality objectively as existing separate from self	Capacity to internalize objective world to create alternative world realities
Limited ability to correlate variables of thought at any given time	Almost unlimited capacity to correlate thoughts to manifest as realities
Limited ability to select the most appropriate thought to complete a task	Ability to sort for the most appropriate people, places and things in order to manifest the thought as reality
Seldom accepts the validity of a thought not rationally understood	Will only express reality based on a valid thought
Primarily limited to accepting and expressing thought only held to be true	Holds unlimited potential for creating realities based on thoughts held to be true
Thought that is not considered to be true will usually be rejected by conscious mind	Will only act on expressing thoughts as reality if considered to be true by the conscious mind
Conscious mind usually unaware of the deeper cognitive structures governing thought	Subconscious mind has full access to the deeper cognitive structures of the mind but unable to convey these structures directly to the conscious mind
Conscious mind usually attempts to resolve conflict by reorganizing secondary beliefs bearing upon the conflict or problem	Subconscious mind will only act upon the secondary thoughts of conscious mind without attempting to reorganizing the deeper structures giving rise to the conflict or problem

dimensions of consciousness beyond simple conscious awareness of its daily existence. Man can, and should, direct more of his energies at taking much greater control over the processes of his mind. The reason for this, as we will see from the remainder of this book, is that the mind/brain self organizing system is a morphological system. The human mind has the capacity to bootstrap itself to much higher levels of organizational synthesis and in fact has been bootstrapping itself for over several hundred years or more. The human mind has the capacity to self program and self experience higher states of consciousness and expanded states of awareness. Many in the field of applied mind science are now beginning to realize that the basic programs of the mind are in fact morphological in nature. They can reprogram themselves for greater awareness. In turn, these newly expanded programs can themselves enter into the organizing and processing arena. There they can start to create even greater more complex programs in order to attain higher states of consciousness and expanded states of awareness for the purpose of creating more expanded realities in life both individually and collectively.

The human mind has the latent capacity, if harnessed, to allow mankind to not only create greater realities in life but to experience the processes by which these creations are brought into manifestation. In effect, self awareness not only has the capacity to become fully actualized but also holds the capacity to attain states of consciousness referred to as enlightenment whereby self awareness is able to experience the source of its creations. This may sound very farfetched and preposterous to many readers. However, mankind needs to acknowledge its own potential and take control over its own destiny. If we fail to do so mankind is doomed to continue making the same mistakes its has been making for the last several hundred years or more. For a conceptualization of the evolutionary flow of the universe and the evolution of human consciousness leading to states of actualization and self realization refer back to Chart 16 at page 130 entitled, *"Relationships Between The Organized Field of Pure Kinetic Energy and Human Consciousness"*.

The simple fact of the matter is that we still do not know how to stop regional wars, global pollution, poverty, crime, disease, mental illness, child abuse, racial prejudices, premature deaths, broken homes, social violence, excesses in industrial, economic and social relationships, and the list goes on and on. The human mind holds the capacity to effectively deal with these issues and much, much more. The theory to create such a reality, even in its very basic and preliminary form and the emerging technology is increasingly being developed, though not widely known, nor refined in all respects. It is still in its infancy and many breakthroughs are still needed. However, major advancements are being made in the development of this technology in such diverse fields as robotics, artificial intelligence, and neuroscience with its sophisticated brain mapping technology. In neuro and biochemistry research it is now possible to alter our neurotransmitters and thereby directly influence our feelings, thoughts and behavior. Scientists have developed the use of computer models for testing out theories on human cognition, including classical cognitive models and the newer connectionism models. Psychology, through such sophisticated methodology as KV disassociation, modality and submodality work, time and space exercises, and hypnosis, are all effective in accessing and altering the energy and images of the subconscious mind. Quantum physics with its breakthrough research in particle physics and relativity theory has contributed to our understanding of the nature of the universe and the nature of complex dynamical systems. Cosmology has contributed in our understanding of the relationship of man and his world in its exploration into the nature and origin of the material universe itself. These widely diverse sciences have in recent years all converged to a point that mankind is on the very brink of paradigm breakthroughs in the understanding and application of how the human mind and brain work to create the many realities we experience each and every day of our lives. Refer to Chart 24 at page 166 for a list of the various disciplines involved in the understanding of the human mind and brain in creating reality.

However, even today the existing theory and technology, in its present form, is sufficiently developed to allow mankind to act upon it, both individually and collectively, in order to create a better world for ourselves and for future generations. Frankly, mankind has little other choice. Either continue to slowly stagnate on its evolutionary path or even worse yet, possibility destroy itself, or move on to a much higher cognitive level of mental organizational synthesis where it views itself, others and the world from an entirely new and expanded vantage point. From this higher vantage point it experiences and acts on the basic premises: (1) that all mankind is interrelated and interconnected, (2) that we are a smaller part of the greater whole making up the entire universe, (3) that we ultimately have the ability within ourselves to take control over the creation of our own realities, (4) that we have the capacity to collectively take control over our world reality, (5) that we also have the capacity to take control over the time and space of those realities, and (6) that ultimately we have the capacity to take control over our own destiny and the destiny of future generations to come.

This is not just some idle or rarefied idealism. There exists

CHART 24

Interdisciplinary Sciences Involved In the Study of the Structure, Functions, and Processes of the Human Brain and Mind

Interdisciplinary Sciences

Physics

- Astrophysics
- Quantum Physics
- Neuro Physics
- Mathematics
- General Systems Science
- Computer Science
- Molecular Physics
- Theoretical Physics
- Environmental Science

Medical Science

- Medicine
- Genetic Science
- Neuroscience
- Immunology
- Psychoparmacology
- Biochemistry
- Neuro diagnostics
- Neuropathology
- Biology
- Bioenergetics

Psychology

- Clinical Psychology
- Cognitive Psychology
- Developmental Psychology
- Social Psychology
- Humanistic Psychology
- Transpersonal Psychology
- Neuro Linguistic Programming
- Psychometrics
- Clinical Hypnosis

Philosophy

- Social Science
- Logic
- Linguistics
- Psycholinguistics
- Epistemology
- Classical Cognitive Science
- Artificial Intelligence
- Robotics
- Connectionism
- Anthropology
- History

today a real and practical way for developing our mental capacities in order to achieve paradigm shifts in how we view ourselves and how we create our world realities, paradigm shifts that will allow us to create a far better world then we can ever imagine. Each one of us can start this process right here, right now, by developing such a sense of self awareness by experiencing self as existing separate and apart from ego personality. A step in this general direction would be to get inside the enormously complex cognitive structures and analog system of the mind and experience first hand both their process and application functions. In order for self awareness to evolve to higher states of consciousness you need to start experiencing from the inside of your own cognitive structures the processes of your own mind, processes which organize your self identity, your personality and your many inner and outer realities in life. In doing so you will begin to separate yourself, that is, your sense of self awareness, from your many creations including the creation of your own ego personality. Simply stated, separate your sense of being from your sense of self identity and personally experience the source of your creations and not just the creation themselves. The MetaCreation processes are intended to increase your awareness and give you an experiential sense of such an expanded awareness of self and of your world. However, a word of caution. No exercise, no technology, and no theoretical principle or theory can or will ever do for you what you can only do for yourself. In this respect life itself is truly a solo trip.

Even today most people including many physicists and other scientists continue to view their personal world from the old Cartesian mechanical viewpoint of the zone of the middle dimension. At the same time they acknowledging that at the quantum level of existence and at the macro level of the universe relativity theory and quantum physics prevail. These different perceptions of reality at all three levels of existence are counter intuitive. This just goes to show how resistant to change mankind has become even in the face of overwhelming evidence to the contrary. However, we are beginning to see that the implications of how quantum physics affects the organization of reality in the physical world is slowly making itself known, first within the scientific community, and then throughout the general population.

Suffice it to say at this juncture that our resistance is not a reflection of our understanding of what the Cartesian mechanical method verses quantum theory really means. It is a resistance arising within the mental processes of the mind itself from which we derive a sense of self identity and a perception of how we relate to one another and create a life for ourselves. What is happening is not a revolution over scientific principles, for the war between the old Cartesian way of thinking and quantum theory is largely fought and won. Rather it is a revolution of how mankind views himself in his world based on scientific principles. A paradigm shift is taking place in how we view ego personality and the world in which we live and a paradigm shift in how we view the realities of that world. Our present realty is rapidly being swept away in favor of entirely new and expanded realities of ourself and of our world!

CHAPTER - 5

Neurological Structures Of The Brain

Introduction

Your brain contains over 200 billion neuron cells which have the capacity to interconnect with over 10,000 to 100,000 other neuron cells which provide us with a vast network of interrelated programmable neural networks. These networks have an almost unlimited capacity for encoding electrochemical impulses used for the creation and transmission of millions of data bits of information. Accordingly the human brain has the capacity for creating realities which can easily be characterized as being almost infinite in nature.[1] The human brain is often characterized as a multidimensional self organizing dynamic system which creates a huge morphological field. This morphological field has the capacity to recreate itself in many different types, forms, shapes, dimensions and states by sorting, organizing, arranging and rearranging stimuli into mental imagery.

Just to give you an idea of how complex this morphological field of the self organizing system of the human brain is imagine for just a moment that you are standing at the south entrance to a very large maze. The maze is approximately one mile long by one mile wide. The walls of the maze consist of a very dense impenetrable hedge over 10 feet tall. The exit is on the far north side of the maze. You think to yourself that before you are going to enter such a complex maze and get totally lost within its many twisting and turning pathways you will outsmart it by renting an airplane, fly over it, and take a photograph of the entire maze. In this way you will be able to map out the entire structure of the maze before you enter it.

Armed with your detailed map you feel confident you will be able to easily navigate its many pathways and within a half hour or so you will be able to emerge from the north exit. You confidently proceed into the maze and after the very first few steps the entire structure of the maze morphs into an entirely new structure right before your very own eyes! In a panic you turn to leave and to your horror there is now a solid wall where the entrance used to be. You then try to proceed and run headlong into a dead end where a pathway leading to the right used to be. You panic for you are at a total loss as to how or why the maze changed. As you run forward in desperation the maze morphs and changes, again and again, with every

1 An excellent description of the dynamic structure and processes of the brain is found in Mark E. Furman's book entitled, *Minds in Motion, The Human Performance Technology for the next Millennium.*

step you take in any direction. You finally realize that the mere act of walking causes the maze to change. But you haven't a clue as to what causes the maze to change every time you take a step. You soon realize that if you are ever going to get out of there alive you must figure out the underlying principles which cause the maze to change as you proceed to walk through it. Welcome to the morphological world of the dynamical nonlinear self organizing system of the human brain![2]

Remember, it was indicated earlier that the primary purpose of this morphological field of the brain is to assist the mind in creating realities in life. Thousands upon thousands of realities are created each and every day of our lives, realities which combine together to form a continuous stream of consciousness making up the physical content of our world. In order to achieve its goal of creating for us more effective inner and outer realities the human brain in combination with the mind performs many thousands of processes and functions for the orderly maintenance of the body and in organizing for us a meaningful life. You can now begin to see why we say that the human brain is by far the most complex and efficient self organizing system in existence in the entire universe. In a word, the human brain is capable of metaprogramming itself for the creation of entirely new and expanded realities in life! Yet the human brain's potential for creating expanded realities has hardly been tapped into and used in any significant way to better mankind's existence here on earth. In spite of mankind's enormous strides in all other areas of his life, including science and technology and many other achievements of an advanced civilization he has not been able to take effective control and direct the process of the brain in creating expanded realities for himself. How we can access and use more of this potential is addressed in this Chapter.

The Anatomical Structures and Functions of the Human Brain

The human brain consists of a forebrain, midbrain and hindbrain. The forebrain consists of the cerebral cortex which is often just referred to as the "cortex" or "neocortex". It is the cortex which occupies approximately 80% of the brain's weight. The brain weighs approximately 3 pounds yet consumes over 20% of all of our body's nutrients in order to maintain an efficient level of performance. The cortex regulates higher cognitive, emotional, and motor functions. The cortex consists of a left hemisphere which operates along the lines of logical, rational, linear, sequencing of thought processes. The right hemisphere operates in a creative holistic, nonlinear or multidimensional fashion. It processes information in the form of sym-

2 Ibid. p. 40.

bols, metaphors, pictures and images. The right hemisphere can perceive, think, and form creative thoughts and ideas outside of linear time and fixed three dimensional space and outside the normal constraints of rational, logical thought. This aspect of the right hemisphere is important to note since its characteristics are the ones used in the MetaCreation processes in creating our physical realities within time and space. The left and right hemispheres are connected by the corpus callosum.

The cortex includes the premotor cortex which deals with physical movement. The frontal lobe is the area where long term memories are stored. The prefrontal lobe is involved with the intellect and largely determines our responses to circumstances in life. The parietal lobe deals with the somatic system which regulates our inner mental states as well as affecting temperature, pain, pressure, digestion and other automatic processes of the body. It is also the area of the brain governing what is referred to as "proprioception". This is our ability to perceive and visualize inner images and emotional states as well as orienting us spatially in our physical environment. The motor cortex regulates muscle control throughout our body. The sensory cortex controls physical awareness of bodily sensations. The temporal lobe regulates the spoken word. The temporal lobe also deals with auditory sensations and perceptions involved in comprehension and learning. The Broca's area is a part of the temporal lobe which organizes and modulates grammatically coherent speech patterns. The Wernicke's area is the area involved in speech comprehension. The occipital lobe deals with visual perception, recognition and memory.

The midbrain lies below the cerebral cortex. The midbrain controls and regulates states of conscious awareness and the unconscious state of sleep. It is the area of the brain that is also involved in speech, perception, memory and cognitive thought processes. It includes the pineal gland which affects sleep, and our interior sense of time. The hypothalamus regulates body temperature, controls our appetite, influences blood pressure, and other bodily functions and regulates the pituitary gland. The pituitary gland produces hormones that regulate many vital bodily activities such as our rate of growth, sexual development, digestion and other bodily functions. The thalamus is the major organ for the transmission of sensory stimuli coming into the brain. It is the initial control point for incoming stimuli and directs incoming stimuli to the various other parts of the cerebral cortex.

The hindbrain is made up of what is called the brainstem. It includes the reticular formation which is a group of nerve fibers deep within the brain that governs the activity of the entire nervous system.

The medulla oblongata directs the flow of information between the spinal column and the brain. It controls head movements, speech muscles, and regulates breathing and heartbeat. The pons regulates bodily spatial movements. It controls facial muscles, eye movement, and the muscles of the mouth, nose and facial expressions. The hippocampus involves the regulation of both short and long term memories and the control of our emotions. The basal ganglia regulates muscle control and is involved in physical movement. Just below the cortex of the brain lies the limbic system which functions to control and maintain emotional states. Charts 25-1 and 25-2, at page 173-174 identify the location of each of these major areas within our brain. This brief explanation of the structures and functions of the human brain is not an exhaustive or complete explanation. It is only to familiarize the reader with the most important functions of the human brain system. These major areas of the brain will be referred to in the following sections of this chapter in order to acquaint the reader with how the brain translates incoming stimuli into imagery which creates patterns of inner and outer realities making up the many aspects of our life.

The Structure of the Brain's Neuron Cells

The neuron cells of the brain are a specific type of brain and nervous system cell whose purpose is to receive and transmit various electrochemical stimuli from the five sensory receptors. These neuron cells act upon and conduct electrochemical stimuli to neuron cell bundles often referred to as neuro networks. The neuron cell consists of a main body called a "soma"; an elongated portion of the cell extending out from the body called an "axon"; and numerous tentacle type branches called "dendrites". The dendrites transmit electrical chemical energy between the various cells of the brain. The soma or cell body arranges the incoming energy in such a fashion that it can be interpreted by the human brain for the purpose of creating images out of the electrochemical bits of information being transmitted to the neuro networks. The information of the soma is transmitted to other cells via the axon which in turn transmits information to the surrounding cells by way of the dendrites. The neuron cell dendrites transmit information between cells by what is referred to as "neurotransmitters". A neurotransmitter is an electrochemical substance that acts to transmit data by electrochemical energy in the space between cells. The space between cells is called the cell "synapse". For a representational sketch of a typical neuron cell refer to Chart 26 at page 175.

There are many types of neurotransmitters operating within the brain system, and probably there are many more which have not

CHART 25 - 1

Anatomy of the Forebrain area of the Human Brain

THE CEREBRAL CORTEX AREA

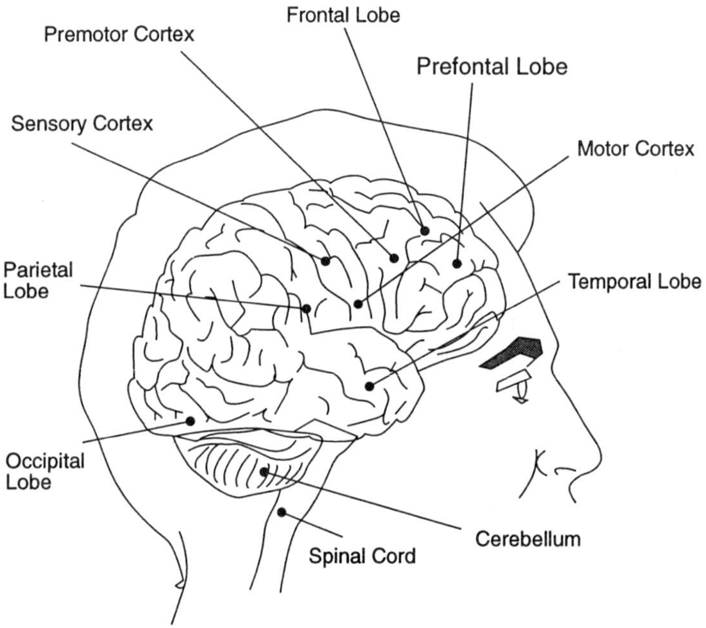

CHART 25 - 2

Anatomy of the Midbrain and Hindbrain Area of the Human Brain

THE MIDBRAIN AND HINDBRAIN AREA

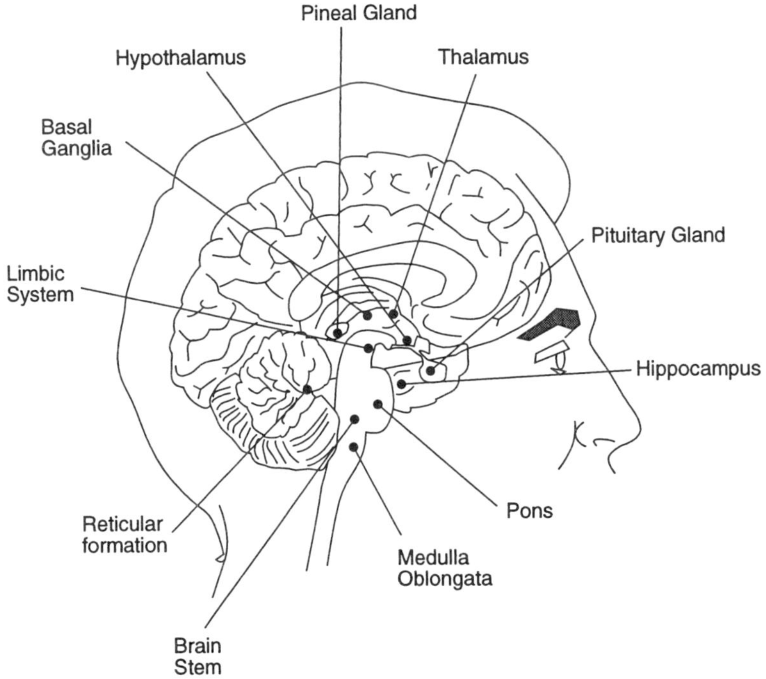

CHART 26

The Anatomy of a Neuron Cell

The Neuron Cells of the Brain are involved in converting sensory stimuli into Basins of Attraction which is then interpreted by the mind as analogs in creating impression of inner and outer realities

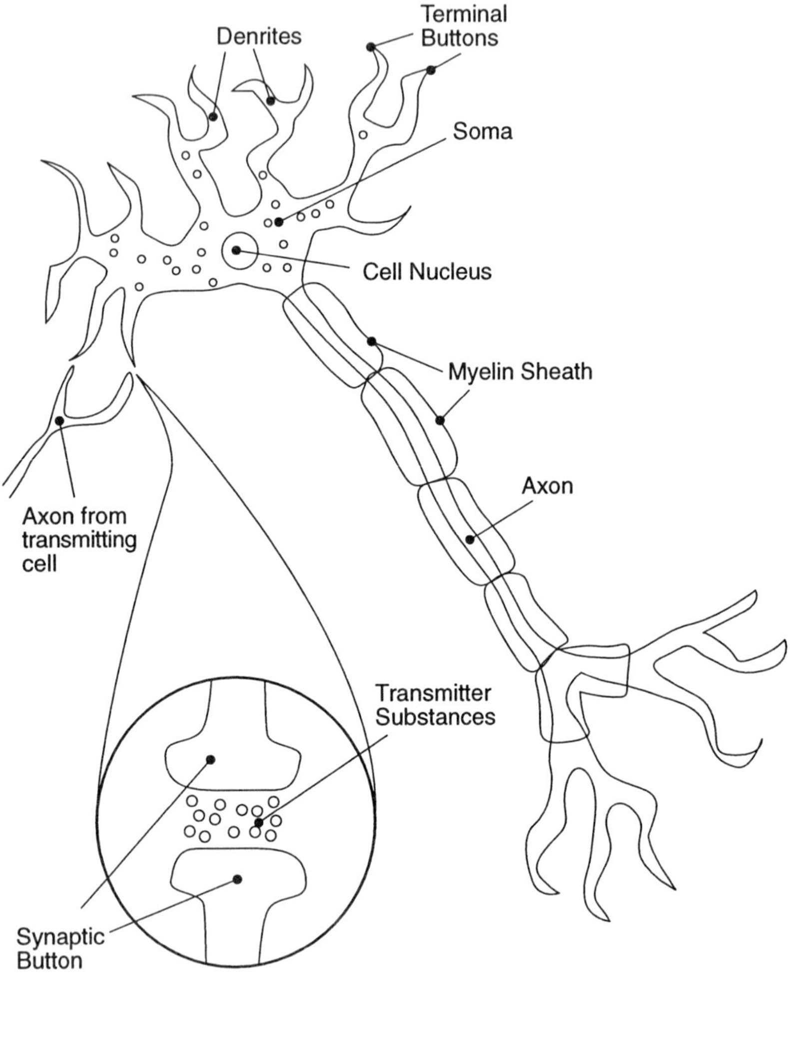

The electrochemical processes of the neuron cells take place at the synaptic cleft between cells. Sensory data is transmitted between cells via electrochemical substances called neurotransmitters.

yet been discovered. The reader has probably heard of some of the most generally known neurotransmitters. One such neurotransmitter is called acetylcholine which acts as an inhibitory or retarding agent which essentially slows down the activity between neuron cells. Another neurotransmitter, norepinephrine, acts to increase neuron cell activity of the brain. Adrenaline is a type of neurotransmitter which has the effect of increasing cellular activity and is known to enhance the " fight or flight" defense mode. Dopamine has to do with the regulation of body movement and personality and mood states. Serotonin affects mood states of the mind and is often used as a sleep aid. Endorphins act as pain killers and are activated by acupuncture treatment. Intensive exercise produces increased levels of endorphins which cause what is know as "runners high". Gamma Aminobuytyric, or GABA for short, affects bodily movement. Substance P regulates levels of pain in the body. Glycine is an inhibitory transmitter and slows down activity within the brain. There are many other types of neurotransmitters, but this brief description covers the major ones.

The use of psychotropic drugs has the effect of altering the electrochemical processes of the brain which, in turn, alters the energy and images of the secondary belief systems of the mind to a degree that life is experienced as more tolerable. The person experiences much less mental and emotional pain. However, the underlying negative energy forces forming the primary belief system structures are not extinguished or removed by drug therapy. Consequently, when drug therapy is terminated the underlying negative energy and images will often reappear in the personality and operational systems of the mind. When they do they will again disrupt the secondary belief systems which caused the pain and suffering in the first place. The reason for this is that drug therapy only rearranges the electrochemical impulses regulating sensations of pain because the drug is formulated to only deal with the pain and not to rearrange the underlying structure of the neuro networks giving raise to the pain and suffering. Once the prescription drugs are removed the source of the pain is often reexperienced.[3]

What is needed is a drug which not only acts on the stimulus of the pain itself at the level of the neurotransmitters but also removes the underlying source of the pain which is represented by the actual arrangement of the underlying basins of attractions and emotional analogs of the subconscious mind. The problem with such

3 Often drug therapy in combination with cognitive therapy brings about permanent relief from chronic depression and anxiety for the reason that when the drugs alleviates the symptoms of the depression and anxiety the person is able to rearrange their neurological responses in life giving rise to their symptoms through effective cognitive therapy.

a drug would be that in removing the underlying source of the pain it would also remove the related aspects of the primary personality and operational programs of the mind and the attendant short and long term memories associated with the source of the pain. Such is the case with respect to the use of electric shock treatment in connection with severe depression and anxiety. The result, while beneficial, often leaves the patient with a rather flat affect, bland personality and loss of memory of significant events in his life. The reason for this is that the shock treatment, while effectively deals with source of the pain of depression, at the same time results in disrupting the subconscious analog patterns forming the related personality and operational systems of the brain and mind.

How the Brain Organizes Incoming Stimuli

The process of creating our outer physical realities, whether it is how we look, how we act on the job, or how we relate to others, all have to do with how we organize incoming stimuli from our outside world within the existing neurological landscape of our brain. Incoming stimuli is never recorded in the neocortex of our brain like an exact copy of a photograph of the external world. Rather it is simultaneously received and interpreted by the existing landscape of the brain. If the existing landscape is negative this will have a negative impact on incoming stimuli representing the current events in life. These events will be interpreted as being negative in nature. All stimuli coming into the brain are really neither positive nor negative, they are simply neutral. It is our mental interpretation of such events which makes them either positive or negative. This is not to say that bad things don't happen to good people. However, we always have the capacity to determine how we will respond to all events encountered in life, even so called bad ones. Its like the story of two young boys, one positive and one negative. The boys were placed in separate rooms. There were literally hundreds of toys in the room with the negative boy and horse manure in the room with the positive boy. When the negative boy was asked why he wasn't playing with all the beautiful toys he responded that he felt lonely and was just too confused to make up his mind which toy to play with first. When they looked in on the positive boy he was laughing and throwing horse manure all over the place and having a really good time. When asked why he was so happy he replied that with all of this horse manure there had to be a pony in there somewhere.

In order to fully understand exactly what was going on in the life of these two boys we need to take a closer look at how the neurological processes of the brain create realities in life. As I said, incoming events are never recorded exactly in the manner they take place.The brain simultaneously identifies and directs the incoming stimuli of exterior events to their corresponding internal basins of attraction subserving the primary personality and operational programs of the mind. The incoming event is recorded in a manner which is consistent with the landscape of the brain's existing basins of attraction. Again, if these networks are negative the incoming event will be interpreted as being negative. What has happened is that the incoming sound, sight or kinesthetic stimuli from the external world is converted into the electrochemical impulses of the neurotransmitters. It is these neurotransmitters that activate bundles of neuron cells making up the basins of attraction. These newly organized basins are then matched up through their energy and matrix configurations with the appropriate existing basins and together they form a composite pattern of the external world. It is this composite pattern which is then interpreted by the subconscious mind as being either a positive or negative event in the life of the individual. This inner neurological response to the outer event is often referred to as the brain's somatosensory system response. This somatosensory system response, or simply "somatic system response," is the brain's way of providing an internal emotional response to the external events in our life.

How the Human Brain Organizes Outer Physical Realities in Life

It is this internal somatic system rather than the external sensory receptor system or the conscious cognitive mind that primarily determines how you process, interpret and experience your outer physical world. If your internal somatic system is dysfunctional your interpretation of your outer world will be largely influenced by this negative internal system which will give you a negative interpretation of the incoming signals from your outer world. What is actually taking place is that if the underlying somatic system is negatively programmed it will not only interpret the incoming signals as being negative, but it will also create negative neurological basins of attraction. These newly created basins will be negatively arranged based on the composite mental picture composed of the negative internal programs and the incoming signals which are negativity interpreted. It is these negative basins of attraction that, in turn, govern the operation of the sensory receptors that constantly scan the exterior landscape of your physical world. This scanning process of our outer world organizes for us a sense of a meaningful reality and our relationship to it. Refer to Charts 27-1 through 27-4 at pages 180-183 for illustrations of how these basins of attraction and their attractor pathways operate in arranging stimuli into either positive or negative mental imagery.[4]

The scanning process of our sensory receptors seeks out relevant elements in our exterior environment in order that we may make sense out of them. If our operational basins are dysfunctional they will cause the sensory receptors to be biased in favor of picking up on negative events in life instead of being calibrated to read the exterior landscape as being essentially positive. What is happening is that the entire system is geared to picking up and interpreting negativity in the exterior landscape in order to reinforce this inner negative system rather than searching out any positive elements in the exterior landscape. Conversely, if the interior basins of the neurological processes of the brain are essentially positive they will cause the inner somatic system to be essentially positive. This, in turn, causes the sensory receptors to automatically scan for positive elements in the exterior landscape in order to reinforce a positive internal system.

This last point is probably one of the more important points in this book. *The polarity of the energy of the brain, whether positive or negative, will subconsciously drive the entire system to interpret its physical environment as being essentially a positive or a negative one.* Few people understand this or know how to take effective action to change a negative neurological brain to a positive neurological brain in order to create positive realities in life. Once you fully understand exactly how this process works you will start to implement it in your life in order to create positive realities in all areas of your life. You will never intentionally hold a negative feeling or thought toward yourself or others under any circumstances ever again! For if you do you will be hurting only yourself and the long term damage you do to yourself can sometimes be devastating!

On the subconscious level of the analog system, like analogs attract like analogs. On the neurological level it is often stated that neuron cells that fire together wire together. Remember, on the neurological level and on the subconscious level negative patterns attract negative patterns. Unfortunately the brain and subconscious mind do not judge whether a thing or event in life is either good or bad. They simply operate on the similarity of their patterns in the neurological and analog systems of the brain and mind. They do not exercise conscious rational thought in arranging life realities. They simply arrange realities based on the neurological bias of the basins of attraction and the values of the subconscious analog system. Accordingly, if the neurological processes of the brain are dysfunctional the subconscious mind will have a built-in bias for continuously

4 For a discussion on basins and their neuro pathways refer to Mark E. Furman's book entitled, *Minds in Motion, The Human Performance Technology for the next Millennium.*

CHART 27-1

How the K/V/A and other stimuli from the sensory receptors feed data into the neuro networks of the brain for processing into mental imagery

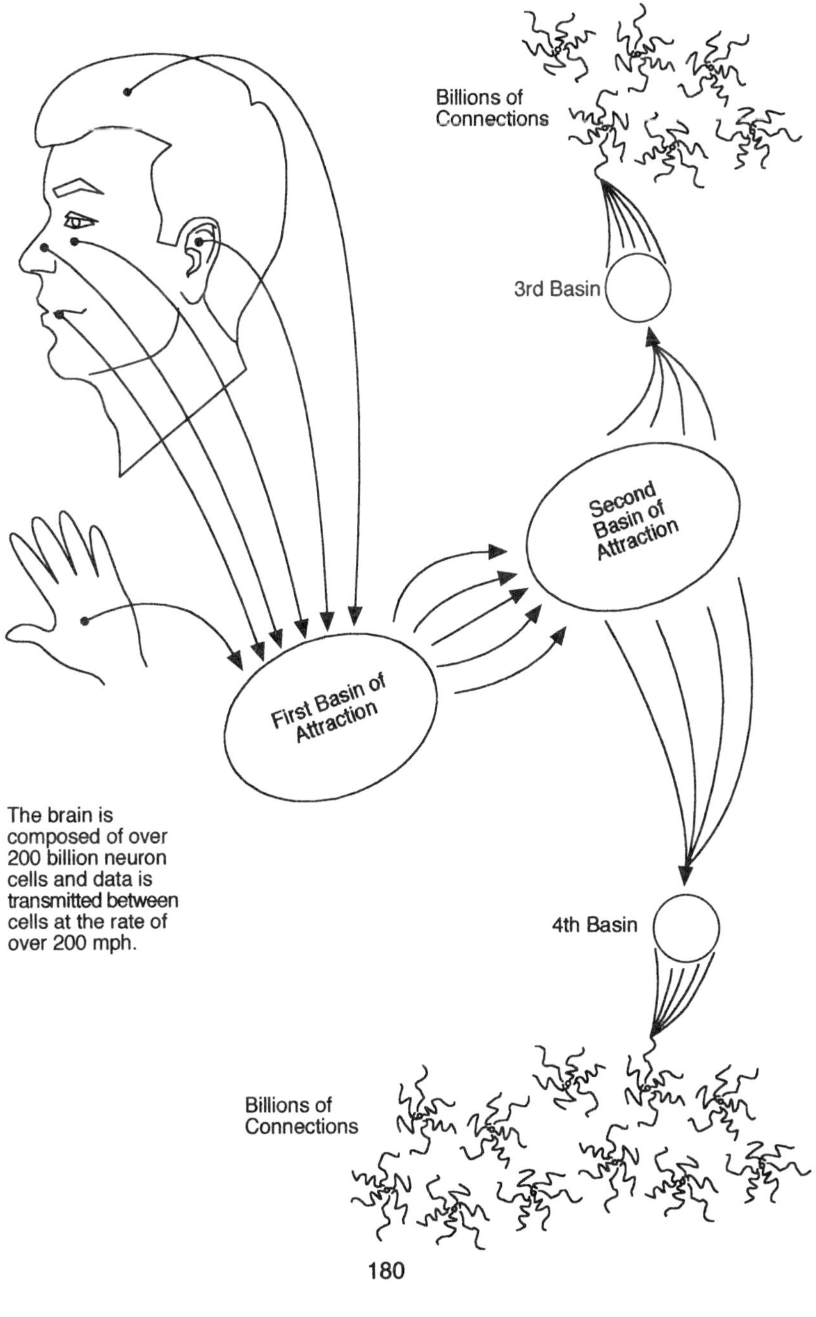

CHART 27-2

How a Positive Basin of Attraction and Its Neuro Pathways Collapse a Negative Basin

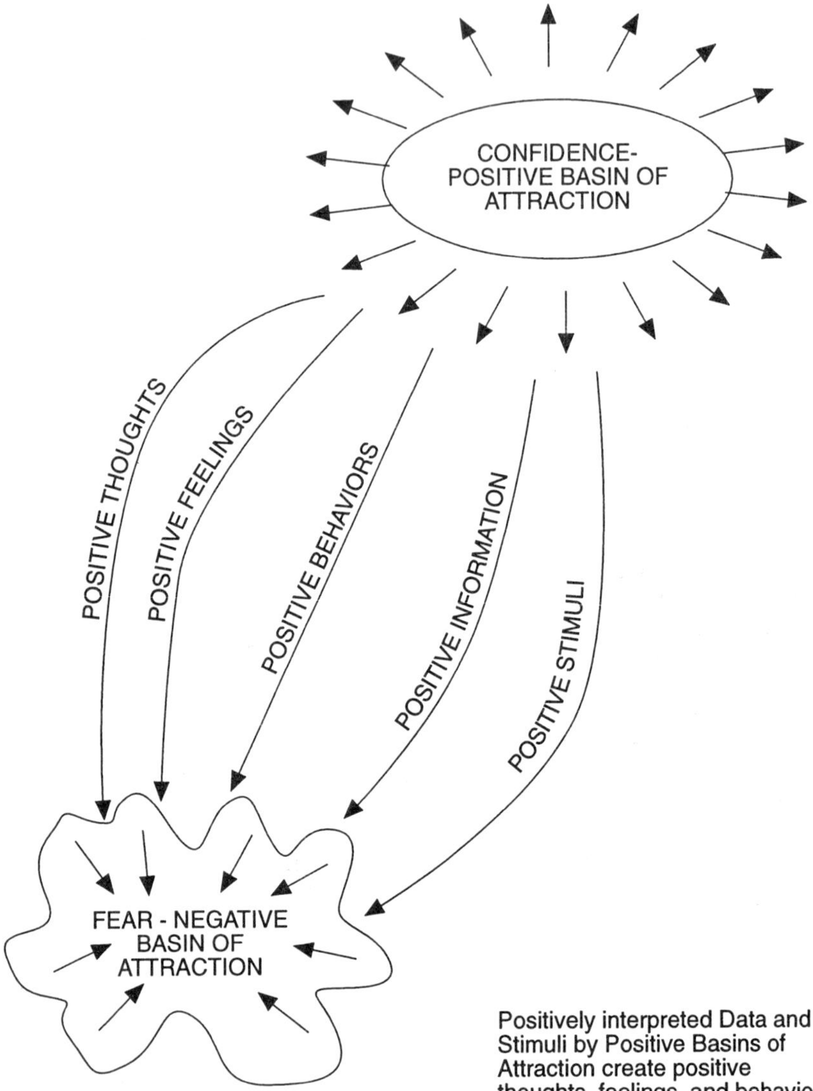

Positively interpreted Data and Stimuli by Positive Basins of Attraction create positive thoughts, feelings, and behavior that destabilize Negative Basins of Attraction and collapse the Basins of Attraction and collapse the Basin's ability to create negative realities.

CHART 27-3

Negative Neuro Network Pathways Connecting Basins of Attraction

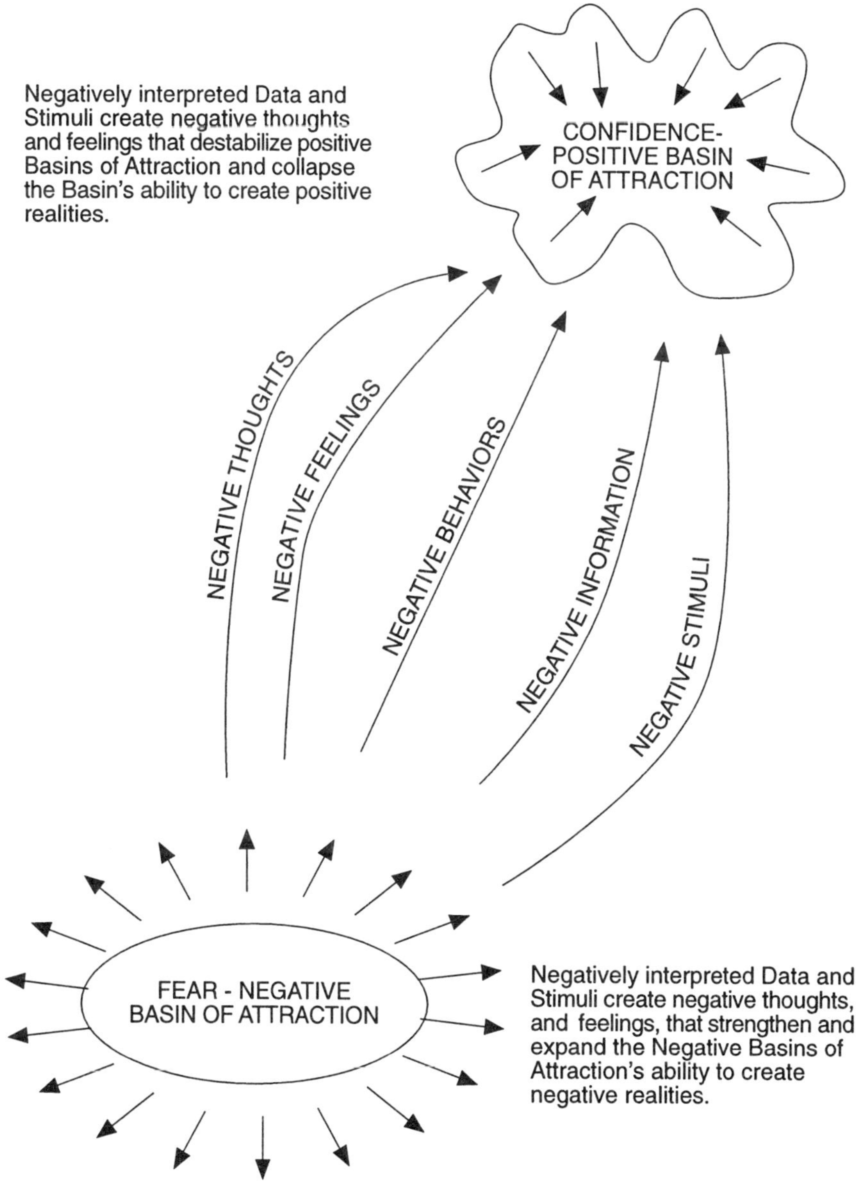

CHART 27-4

The Effects Disrupting Negative Neuro pathways Have on Destabilizing Negative Basins of Attraction

Disrupting negative pathways leading to or from a negative Basin of Attraction disrupts, destabilizes, and collapses the Negative Basin.

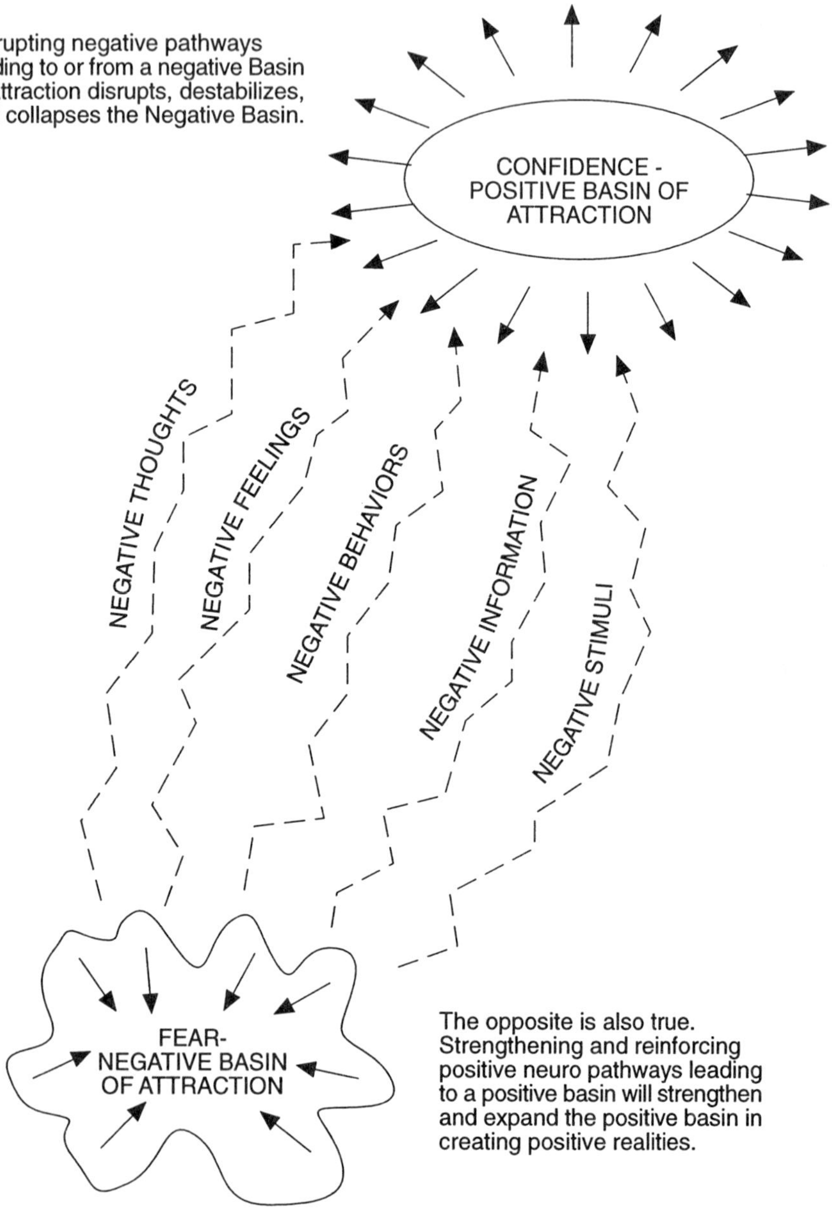

The opposite is also true. Strengthening and reinforcing positive neuro pathways leading to a positive basin will strengthen and expand the positive basin in creating positive realities.

scanning for negatively charged elements in the exterior physical landscape of life. Consequently, the inner somatic system not only interprets events in life as being negative but it also programs the sensory receptors to pick up any existing negativity in the environment in order to reinforce a negative spin you have subconsciously placed on life in general. What you have is a series of closed sensory cybernetic loops that operate on the principle of feed backward and feed forward propagation of stimuli throughout the entire system which reinforces a person's negative or positive outlook on life. *The human mind and brain not only interpret our experiences of our outer world, they actually create that outer physical world!* It is this personally created outer physical world which we refer to as your personal reality in life. Refer back to Charts 8-1 and 8-2 at pages 92-93 which show how these continuous cybernetic loops operate in creating realities in life.

Far too often when we fail to understand this organizational structure of the brain we endeavor to change a negative life into a more positive life by reorganizing the outer physical elements making up our outer physical world. Alternately we will attempt to change things by endeavoring to reinterpret things through the mental process of our conscious cognitive reasoning. The first process is seldom ever effective. We are all familiar with the old adage that when you move from one place to another you simply take all of your old baggage with you. The second process holds limited success. The reason for this is that issues in life creating your problems often have their origin in the neurological structures of the brain and in the analogs of the subconscious mind. It is here that our problems need to be addressed in order to obtain permanent effective relief.

The MetaCreation processes are geared to effectively remove negativity at the neurological level. We use the Metacreation processes to create alternative positive realities in life by repatterning the brain and mind. Accordingly, the processes are effective because they are based on the way the neurological functions of the brain program the analog system of the subconscious mind. They take advantage of the fact that the subconscious mind searches the exterior landscape for characteristics which reinforce the positive patterns of the internal structure of the brain and mind. The subconscious mind accomplishes its task by scanning the exterior landscape in search of complimentary patterns to match the internal landscape. These complimentary patterns in the external landscape are made up of their representational submodalities. Remember, it is submodalities which make up the features of the analog patterns of the subconscious mind. It is these complimentary submodalities which the subconscious mind searches for as it scans the landscape of our exterior world. It is the analog patterns of these submodalities which we use in order to turn a negative system into a positive system in order to create more positive realities in life.

We talk in terms of the subconscious mind sorting through unlimited possibilities in order to bring to conscious awareness the most appropriate people, places and things needed to manifest your desired realities. We are, in fact, referring to this process of bringing to conscious awareness those analog patterns of the submodalities representing the desired realities. The result of all of this is that the inner somatic system, the neurological basins of attraction, the subconscious analog system, the sensory receptors picking up the exterior signals, your conscious cognitive thought processes, your actions and behavior, and your actual physical reality itself, are all simply different aspects on the one and same mind/brain system organizing the many and different realities in life! For a description of how all of these systems work together synergistically to create inner and outer realities in life refer to Chart 28 at page187. This chart brings home the fact that by changing any one of over thousands of variables in life you affect changes in all of the other variables resulting in changes in your realities. This phenomena is characteristic of the morphological self organizing system of the human brain.

Therefore, your personal outer reality consists of your response to the many events taking place in your outer physical world. However, your personal reality does not consist of the entire outer physical world. It only consists of that small portion that you become personally aware of because its incoming stimuli cause a response in your brain in some fashion or another. The rest of the world is just that, a physical world, but not necessarily part of your own personal reality. For example, if you just bought a new Ford Taurus automobile your personal reality would now include your new car and not all of the other thousands upon thousands of Ford Tauruses on the road. However, if you happened to hit another Ford Taurus on the way home from work at that very instant it just became part of your own personal reality. Whether it is a totally negative event in your life depends upon your personal reaction to the accident. Accordingly, your somatic system responds to and interprets each and every event that takes place in your world of which you become personally aware.

How you can effectively reinterpret your outside world reality from one of a self limiting, painful or negative world to a positive, prosperous and enjoyable world has to do with two major organizational systems of your brain and subconscious mind. The first major organizational system has to do with how the brain itself organizes and processes incoming sensory signals into basins of attraction making up the existing landscape of the brain. This system is referred to as the neurological processes of the brain. The second major organizational system has to do with the manner in which the subconscious mind interprets the many patterns created by the neurological basins of attraction making up the physical structure of the brain. This system is referred to as the analog system of the subconscious mind. In this Chapter 5 we explore how the neurological processes and structures of the brain contribute to creating realities in life. In Chapter 6 we deal with how the analog system of the subconscious mind operates in creating those realities. For it is through the process of rearranging analog patterns and pathways by strengthening, changing and weakening their electrochemical energy that we create, alter and dissolve realities. Once you understand the processes of the brain and mind and how they work together you can start to effectively take control over these processes. By taking control you will be able to not only reorganize the neurological structures and functions of your brain but you will also be able to reorganize the analog structures and functions of the patterns of your subconscious mind in order to effectively create for yourself a more prosperous and meaningful life.

The balance of this chapter deals with the neurological processes of the brain in creating reality. We first explore how the brain translates visual stimuli into neurological patterns that can be interpreted as analog images by the subconscious mind. We will then look at how sound waves are converted into auditory patterns that are interpreted by the subconscious mind as linguistic and auditory expressions, such as musical tones which are meaningfully interpreted. We will then look at how we create long and short term memories. Next, we will explore how the brain creates a sense of inner time as well as how the brain and mind create and keep track of outer time and space in which events take place in our physical three dimensional world. We will look at how the brain creates and interprets speech patterns. Finally, we will explore ways in which incoming stimuli can be reorganized by our brain and reinterpreted by our subconscious mind in order to create for us a more prosperous and meaningful life.

How Our Brain Organizes Visual Images

Incoming visual stimuli take approx. 65 to 100 milliseconds to organize a visual image in the visual cortex area of our brain. Incoming visual images immediately start to fade and need to be reinforced every 250 milliseconds. This rapid organization of visual images allows a person to continue to closely inspect an object without

CHART 28

By changing any aspect of your life changes the imagery of your inner and outer realities which in turn creates a better life for yourself

Neurological Stimulus Changes in the sensory stimulus consisting of (1) data, (2) stimuli, (3) thoughts, (4) feelings, (5) behavior and (6) past experiences, changes the directional flow, intensity and output of the energy creating your mental images which changes your inner and outer realities.

Sensory Submodelities Changes in the (1) intensity, (2) pressure, (3) tone, (5) visual effects, (6) sound, and (7) frequency, of the sensory submodelities causes changes in the sensory modalities of the mind/brain resulting in changes in the images creating your inner and outer realities.

Sensory Components Changes in the (1) auditory, (2) visual, (3) taste, (4) smell, (5) tactile, and (6) kinesthetic sensory modalities of the mind/brain cause changes in the mental patterns of the mind resulting in changes in your inner and outer realities.

Time and Space Changes in the time sequencing of your realities within your mental and emotional states of a past, present and future, and changes in the placement of realities in space causes change in the timing and spatial relationships of your inner and outer realities.

Mental and Emotional Patterns Changes in intensity, frequency, or positive or negative modes, of your mental and emotional patterns of your mental imagery create changes in your inner and outer realities

Thoughts and Beliefs Changes in your beliefs about yourself, about your career choices, spiritual beliefs, relationship beliefs, beliefs about sexual attitudes, financial beliefs, community and social beliefs, family beliefs, and other beliefs concerning self and others, create changes in your inner and outer realities in life.

Physical Wellbeing The type of food you eat and life style you live affects, either positively or negatively, your mental and physical energy which in turn affects your thoughts and emotions and your physical health and wellbeing.

Ego Personality Overcoming ego personality expands your cognitive structure thereby expanding consciousness and self awareness.

Objective Reality By changing your knowledge and understanding concerning objective reality changes your awareness, perceptions and responses to life which allows you to create more expanded life enhancing realities.

Actions and Behavior Changes in your behavior, actions, life style, vocabulary, acquired acknowledge, appearance, social mannerisms, conversation skills, business aptitudes, education and leisure skills create changes in your inner and outer realities.

The purpose of this chart is to illustrate that changing any one of the possible thousands of elements influencing the creation of the mental and emotional imagery making up your inner and outer realities in life, can potentially change your mental and emotional imagery, which in turn changes your inner or outer reality in life. This chart dramatically illustrates how important it is to take control over all aspects of your life so that you will be able to take control over your life's realities and create a better life for yourself.

creating an after image or overlapping double images of the object. The brain is able to hold the image in a relatively stable position as we physically move our head and body around in order to inspect an object from different angles.[5] Our two eyes being separated by a distance of approximately one inch create the effect of stereoscopic vision. It is this stereoscopic vision that creates a 3-D image which gives us the capability for depth perception. Depth perception allows us to judge the distances between us and physical objects located in three dimensional space. Visual images are not exact copies of external objects but rather are formed out of the combination of both the visual memories stored in the vestibular system and the new incoming images which together form a composite image organized by the somatic system and projected onto the visual cortex area of the brain.

Once a visual image is created we can then close our eyes and mentally move the image around in our inner visual field in the visual cortex area of our brain. Through this process of mental visualization an image can be enlarged or made smaller, projected deeper in space or brought in closer in our inner visual field or viewed from different angles. This process is the result of what is referred to as the association effect of the bimodal and trimodal neuron cells of the prefrontal cortex. These specialized cells can organize information coming from two or three sensory systems, such as information coming into the system via the auditory, visual and kinesthetic sensory receptors. Shifting inner metal images in this fashion can be used to dramatically alter the way we visualize a mental image which, in turn, affects our emotional response concerning the image.[6]

It is the brain stem which creates a type of inertial guidance system that coordinates motor, kinesthetic, visual, and auditory movements in physical space and in linear time. In this manner we can easily achieve a sensation of physical bodily presence and movement in our three dimensional world. It is this brain stem type of inertial guidance system that allows us to create internal representations of our external world. Again, all of these internal representations consist of the combinations of the existing neurological memory patterns stored in the neocortex area of our brain and the many incoming visual, auditory and kinesthetic images coming into the system via our sensory receptors. Consequently it is the combination of our auditory, visual, motor, tactile, and kenesthetics senses which, operating in combination, create for us our outer physical placement and movement in time and space.

In effect we mentally create our own personal sense of our

---

5 Mark Furman, op. cit. p. 26.

6 Ibid p. 31.

physical presence in the time and space of our world. We do not live separate and apart from that world. Nor does that world exist separate and apart from our sense of a personal self existing in that world. In effect our personal physical world and our sense of self living in that personal world are simply two aspects of the one and the same mind/brain self organizing system creating realities in that world. Objectifying our personal world and viewing it as existing separate and independent from us causes us to lose a significant degree of control over that world thereby restricting our ability to mold and form it in a manner more to our liking.

How Our Brain Organizes Sounds

The left auditory field processes incoming sound waves into basins of attractions located in the neocortex area of the brain. The brain processes sound wave stimuli into sequential tones and discerns variations in the rhythms of the various incoming tones and sounds. The right auditory field processes incoming sound waves into analog components that can be easily recognized as the symbols of language by the brain and mind. These two aspects of sound are then combined into a composite pattern which not only creates a meaningful linguistic expression but also provides the tone or inflection of the spoken words. This composite pattern allows us to interpret the manner in which language is delivered as well as understand the spoken words themselves.[7] In this way you are able to determine whether something is spoken in anger or in love as well as all of the other possible variations in between. Sounds when combined with associated visual images, and tactile or kinesthetic feelings, give us a sense of reality of the many objects and activities making up our life. That is why when you change the submodalities making up the auditory, visual and kinesthetic images formed by your brain you change your experience of these images, and thus you change the realities you experience in life.

How Our Brain Creates Memories

The frontal cortex area of the brain acts as a working memory field which is used when processing incoming stimuli and correlates it with images in the existing long term memories that reside in the neocortex area of the brain. The brain accomplishes this task by indexing past memories by their location within the frontal cortex. This is the reason when you tilt your head or eyes slightly upward and to the left you are able to access a memory that doesn't readily come to mind when looking straight ahead. Whereas you tilt your head and eyes slightly up and to the right to mentally construct a new image of something you just heard or read about but are not sure what it might

7 Furman, op.cit., p.35.

look, feel or sound like.[8]

An interesting exercise is to think of something enjoyable and with your eyes closed tilt your head in the direction of the image and increase all of its submodalities. That is increase its color, location, clarity, size and make it three dimensional. With your eyes still closed tilt your head in the direction of something which is not very pleasant and repeat the process by adding submodalities to make the image more real. Now look back and forth between the two images and feel the emotional differences you experience when mentally viewing each internal image separately. This difference in emotional states is the result of the pyramidal cells of the frontal cortex stabilizing each image in your inner visual field. When each image is stabilized in this way you automatically experience the submodalities associated with the image including emotional states. This is the result of the inner somatic system acting upon the inner imagery you have just constructed from your short or long term memory fields. In order to change a negative emotional state, if it is a minor one, simply change the location of the mental image in the inner visual field and remove its submodalities, that is, reduce its size and reduce its clarity. What you are doing in effect is destabilizing and dissolving the neuro networks and neuro pathways making up the composite image you are experiencing. As you destabilize and dissolve the submodalities making up the image you also remove any negative emotions associated with the image. Removing major negative emotions associated with major trauma in life takes something more than just destabilizing negative basins by simply reducing its submodalities in this fashion. How to effectively deal with major trauma is explained in Chapter 8.

How Our Brain Creates a Sense of Time and Space

The basal ganglia area of the brain has the ability to create inner sequencing of periods of time to match our conscious awareness of time interval which we measure in seconds, minutes and hours. It is able to do so by using the neurotransmitter dopamine as the modulating substance similar to the manner in which an hourglass slowly measures out the stream of sand in order to measure the span of one hour. For instance, you have experienced the effect of just prior to going to sleep at night you thought to yourself that you wanted to wake up at a set time in the morning, such as 7:00 a.m. You found that you were able to wake up at exactly that time or just shortly before or shortly afterward. The reason for this is that the measured dosage of dopamine kept the level of acetylcholine high enough to allow you to sleep to the appointed hour. Through its hourglass effect the brain starts to reverse this process and as you approach

8 Ed & Marianne Reese's 1996 NLP Practitioners Workbook.

the time to get up the level of norepinephrine increases. This, in turn, increases the activity of the brain causing you to wake up at the appointed time.[9]

Our awareness of outer sequencing of time operates in a little different fashion. At about the age of four to six years old we become aware of the sequence of 24 hours in a day and that there are seven days in a week and that a year consists of 365 days. Historically we could have sequenced our daily lives in different time intervals. For instance, the early American Indians used the seasons of the year rather than calender days as the most important time intervals in their lives. The season for sowing and harvesting crops was more important than keeping track of individual hourly time periods and days. Since the early American Indians didn't have wristwatches keeping track of the seasons was easily accomplished by observing the phases of the moon. Once we learned how to tell time by keeping track of the passage of 24 hours in a day our conscious mind was able to provide the brain with the necessary stimuli needed for the brain to arrange the patterns of its neuro networks to recognize internals of time based on our clock and calender measurements. Similarly, the neocortex also arranges measurements in space by forming neurological basins of attractions and attractor pathways corresponding to measurements based on the meter stick and yardstick in order to allow us to measure distances between us and objects in three dimensional space.

Consequently, the creation of time, both inner time and outer physical time, are subjective processes of our brain and mind. It is true that the time of one year, consisting of 365 days, is measured by the period in which the earth makes one complete revolution around the sun. However, the inner experience of time for most people is considerably different than outer physical time as measured by our clocks and calenders. Depending on your age you internally experience time intervals differently. Young children often find time to drag by slowly, particularly when they are traveling long distances. How often do parents hear the cry from the back seat of the car, "Are we there yet?" For an 8 year old one hour often seems like an eternity. On the other hand, older people find that time seems to be slipping away much faster than when they were young. Doctors have found that mental disorders distort our perceptions of time. Schizophrenics often tend to greatly overestimate one minute time intervals by several minutes or more. Clinical depression tends to cause people to underestimate time and one minute may be estimated to be only 40 to 50 seconds.

9 Furman, op. cit., p. 81.

Time distortion by the human brain and mind is due to several causes. First, it can be related to distortions in the neurotransmitters themselves which govern the regulation of our sense of internal time. As you distort the brain's levels of norepinephrine and acetylcholine you distort the brain's operating system for tracking internal time. Also time distortion can be the result of the operation of our own inner somatic system. When we age our somatic system creates a sense of finality of life itself which in turn creates a sense of urgency. Part of the problem is that when we age and retire we often do not replace the important activities of a career and family with other meaningful pursuits in life. Consequently, we are left to contemplate a future where the only next major event to look forward to is death itself. But if a person has lived a full and complete life and is still involved in meaningful activities this sense of urgency we all face the closer we approach the end of our life is not as dramatic, and as a result inner time acceleration is not as pronounced.

The major problem, however, is that once the brain organizes time and space through the regulation of our neurotransmitters to form an impression of our three dimensional world we then came to believe that such measurements in time and space were somehow permanently fixed as the measurement of our exterior world. We consequently objectified time as well as space and came to believe that time and space existed outside and independently of us and largely outside of our direct control, when in fact time and space is simply a mental construct of the brain and mind! This is not to say that these neurological coordinates for measuring time and space do not serve a very useful purpose in ordering our daily life, for they certainly do. But as we objectify time and space we lose effective control over the events in our life taking place in time and space. Since time and space are organized by our brain and not organized by anything outside of us we have far greater potential to control both past time and future time than we ever thought possible. When we come to recognize this fact by understanding how time and space are organized by the brain we can then destabilize a negative past which does not serve us very well and organize for ourselves more enjoyable positive future realities.

How Our Brain Forms and Comprehends Speech Patterns

The Broca area of the brain initially organizes grammatical coherent speech patterns. The Wernicke area is involved in speech comprehension. If a person's speech muscles and auditory cortex are fully functional but the Wernicke's area, or the temporal lobe of which it is a part, is severely damaged the person will not be able to effectively understand what is being said although he is able to clearly hear what is being said. The motor cortex area of the brain controls the vocal cords' muscles for the purpose of forming sounds into understandable speech patterns. If the Broca area is significantly damaged, even if the motor cortex is fully functional and is capable of forming speech patterns, a person will not be able to do so for the Broca area is no longer able to organize grammatical speech patterns for the motor cortex to use in forming meaningful dialogue.[10]

You may be wondering at this point how it is that the brain is able to form and understand literally tens of thousands of word patterns which form an almost endless variety of phrases and sentences. How are we able to organize what is said, what we see, or what we read, much less understand it all in any meaningful way. Mankind's ability to formulate a language and understand its meaning has been the most important single factor in the development of human intelligence. It is what allowed mankind to evolve far beyond any other living primate. The development of a spoken language, probably more than any other single factor in human history, forced, at an unprecedented level of evolution, the rapid development in human consciousness, the associated neuro networks of the brain and the analog systems of the subconscious mind.

The complexity of the neuro cortex of the brain in organizing speech patterns expressing as human thoughts articulated in a spoken language understood by others almost defies comprehension. In fact, it wasn't until the last several decades that the neuroscientists fully understood the internal brain and mind processes in organizing intelligible language patterns which are understood by others. The complexity of this process is readily apparent when you consider that the human brain has to recognize and organize 79 different phonetic speech patterns into literally tens of thousands of words by using only 26 letters of the English alphabet. It would be like switching all the trains in the country from one central control panel and then doubling up all of the various switching possibilities to be handled by each switch and not crash any of the trains in the process. It could be done, but you better have a very sophisticated control system. The brain evolved into such a sophisticated system. It has almost unlimited capacity in what is referred to as its recurrent neuro networks. These networks are arranged in similar fashion to an immense parallel processing system. This system is described in great detail in the following chapter. For our purposes here it can be understood in the way it encodes the incoming phonetic sounds to form words and sentences. It is able to not only recognize words but it is able to recognize specific words in the context of the phrase or sentence in

---

10 Paul Churchland, *Engine of Reason, Seat of the Soul,* pp.133 and 134.

which they appear.

The brain forms a neuro network to represent a newly formed word. It also recognizes the sequence of the word patterns in a sentence and organizes a neuro network pattern to represent the word in the context of the sentence.The brain maintains in memory numerous closely associated neuro networks representing slight variations in the sequence of word patterns in a sentence. In this way the brain provides us with a specific basin of attraction to represent a specific incoming word pattern which is closely associated to the existing neuro networks representing previous variations of the incoming word patterns.[11] An example of the slight variations in similar sentences would be "Dog who chases cat who chases mouse" versus "Dogs who chase cats who chase mice" versus, "Dogs who chase cat who chase mice". I could go on with more variations of this simple sentence structure but you get the point. The brain's neuro networks, in effect, double up in the usage of its basins of attraction by using almost the identical neuro networks to form and track these slightly different sentences. The technical terms for this process is called vector trajectories forming word patterns in the vector space of the recurrent neuro networks forming the word basins of attraction.[12]

The differences in sentences are noted by the brain by the creation of sightly different neuro trajectories and slightly different variations in the neuro networks of the overlapping basins of attraction forming the original sentence structure.This system of neuro networks representing slightly different variations in word patterns through the creation of slightly different trajectories and slightly different neuro networks to represent new incoming variations allows the brain to both quickly and efficiently organize new word patterns into meaningful language. By retaining these earlier patterns in memory it is able to assign incoming speech patterns directly to similar basins of attraction representing existing speech patterns. By using the existing speech patterns in this way it can easily build upon what it already knows and needs only to account for any new variations in the incoming stimuli by creating new trajectories and new vector spaces within the existing neuro network patterns.

In simple layman's terms what I am saying here is that the brain provides for differences in word combinations by not only arranging the organization of the pattern of the basins of attraction themselves but by also arranging the various paths used to connect basins representing word patterns. The motor cortex controlling the vocal cords can then express new patterns of speech by interpreting

11 Churchland, op. cit., p.142.

12 Ibid p. 142.

the word patterns represented by the newly organized neuro network patterns. In this way the brain is able to account for numerous slight differences in similar sentence structures. You can see that this system of organizing neuro networks to represent language patterns does not use fixed rules to keep track of almost unlimited variations in common speech. If it did the searching and matching process through the use of complex rules would be considerably longer and more complex than the method used by the brain's existing parallel processing system. The system transfers variations in speech between neuron cells at the rate of over two hundred miles per hour. However, neuroscientists track this process in terms of milliseconds. You can see that this is a very efficient way to keep track of the thousands of word and sentence variations we hear each day.

A child doesn't just memorize vocabulary in order to understand the spoken word but learns his native language through the repetition of the language patterns spoken first by his parents and then by others during his school years. The child's brain is able to rapidly build upon its existing speech patterns new networks to represent the new increasingly more complex speech patterns it hears each day. Recent studies have indicated that in fact you can't stop a child from learning his native language, short of locking him in a closet all day. Children are not just copy cats. They don't just memorize word patterns, but their brains quickly learn the underlying organizational structure of the English language itself consisting of the 26 letters organized into the 79 phonetic sounds. The brain then organizes all of the word patterns it receives into neuro networks representing these phonetic sound patterns. In this way a child is able to quickly recognize and then organize these sounds patterns into words and sentence. Ultimately the child learns to recognize approx.10 new words a day until they are about 30 years old. At this rate the adult brain can recognize and process a vocabulary of from 80,000 words to over 100,000 words. Exactly how we come to understand the meaning of all of these words, and then by using the intention of our thoughts based on these linguistic patterns, organize and express realities for the purpose of creating a meaningful life for ourself is discussed in Chapter 6.

How the Brain Discreates Negative Outer Physical Realities

By now you are probably wondering if we create our outer realities by using the power of our brain and mind to create mental images of the thing desired, how do we reverse the process? How do we discreate physical realities already in existence which we would rather not continue experiencing in our life? If you think about it for awhile the answer becomes rather obvious. You simply reverse the creation process, by bringing the conscious mind, through the intentionality of thought, to the basin of attraction in which the creation was initially formed through the neurological connections of the brain and form a new thought which would eliminate the creation from your life. The energy of the conscious mind and the energy of the basin's field align with the image of discreation. The analog system of the subconscious mind then acts to bring to conscious awareness the most appropriate way to eliminate the creation from your physical environment. This, of course, doesn't mean that the physical object simply disappears into thin air any more than did the object appear out of thin air. Infinite possibilities within the the analog system of the subconscious mind bring to conscious awareness the simplest most appropriate means possible to physically eliminate the object from your life in accordance with physical law.

We all experience this process of discreation everyday of our lives. We are just not consciously aware of how the process works, or that it even initially takes place in our life at the neurological level of our brain. For example, you may want to get rid of an old car which is no longer of much value or convenience to you. You might think that you can trade it in on a new car, run an ad and sell it through the newspaper, give it away, donate it to charity, sell it for parts, abandon it, or dispose of it in many other ways. You might not be aware that the process of discreating the old car initially takes place at the neurological level of the brain at the level of the basin of attraction giving rise to the thought that you decide to dispose of the old car. At this level of activity of the brain the neurological synapses of the neuron cells send out electrical chemical stimuli to the appropriate basin of attraction which the subconscious mind then interprets as thought, emotion, or behavior which in turn the conscious mind acts on in the most appropriate manner to carry out your wishes in disposing of the old car.

What has happened is that you are only consciously aware of the fact that you decided to get rid of the old car and decided that the best way to accomplish this, for purposes of this illustration, is to either trade it in on another car or sell it outright. You are not aware of the fact that the decision to dispose of the old car first started at the neurological level of the brain at a specific basin of attraction. You might say to yourself "What difference does it make as to the internal processes of the brain and mind so long as I got rid of the old clunker?" The difference is that the more aware you are of what is going on at the primary level of your brain and mind of how you make decisions in life the more you are able to make wise or effective decisions in life. Far too often people remain unaware of how and why they make the decision they make in life. This is particularly important if the decisions you have been making bring about a lot of lack, limitation and suffering in your life.

How Conscious Thought Affects Neurological Basins of Attraction

All thought and its corresponding feelings regardless of the content of the thought, whether positive or negative, create one of three different states at the level of the neurological activity of your brain. A thought, (1) creates or dissolves basins of attraction, (2) strengthens or weakens basins of attraction, or (3) creates cognitive dissonance or mental confusion between two or more basins of attraction. The reason for this is that thought will alter basins and their connecting pathways by changing, strengthening, or weakening the electrochemical energy of their neurotransmitters thereby creating, strengthening, altering or dissolving realities in life. Therefore, if you wish to reinforce a positive habit or behavior, or reinforce a mental or physical creation you might want to think of all the possible reasons why it serves some useful purpose in your life. On the other hand, if you want to modify or change behavior, or modify or change a mental or physical creation, you would want to think of all of the reasons why such a change would be in your best interest. You may want to make a written list of all of the reasons to support your mental creations and periodically review it. If you can read the list from time to time and experience strong submodalities of feelings and emotions when doing so all the better since thoughts, emotion, behavior, information and stimuli all reinforce the creation desired. For this reason goal setting and an action plan to achieve specific goals are highly recommended by motivational and self help teachers. All thought at the neurological level of the brain either creates or dissolves basins of attractions or creates confusion. However, all thought at the level of the mental activity of the subconscious mind through its energy and images simply interprets the brain's basins in order to create patterns of images. Consequently, the brain organizes neurological stimuli, the subconscious mind creates analog patterns, and the conscious mind interprets the many analog patterns in order to create a meaningful life for you.

Cognitive dissonance and mental confusion are not always as bad as they sound for the reason that the mind will start to scan the horizon of the mind/brain landscape in an effort to resolve the confusion and make sense out of the existing confused patterns of thought it is experiencing. Some of your most creative ideas can arise out of cognitive dissonance. The reason for this is that confused states of mind constitute what is called phase transitions in the basins of attraction. That is, the neurological synaptic pathways are in phase transition. In phase transition they start to open, loosen their electro-chemical connections and tend to become fluid. Their energy sets up dynamical tensions within the system in searching for a solution to the confused state and a restoration of phase stability within the system. In phase transition the brain is more amenable to the creation of new alternative attractor pathways leading to new creative basins of attraction giving rise to original and imaginative solutions or resolution to the initial mental confusion. For the most part, however, both the basins and attractor pathways remain, more or less, stable in order to continue to give you a stable experience of a solid outer physical reality in life.

One downside to confused states of mind is that instead of being open to new and innovative solutions you become anxious and fearful for not having an immediate resolution to the confusion you are experiencing. In this case your brain starts to produce more norepinephrine and less acetylcholine. The entire neurological structure tends to "lock up" and instead of creating new attractor basins moving toward solution the entire neurological system become firm, or structurally rigid and not very amenable to changes in this anxious state. What you need to do if you are anxious or fearful and can't seem to make any reasonable decisions or find resolutions to your confusion is to divert your attention from the problem. You do so by occupying your mind with relaxing thoughts and activities such as listening to relaxing music or distract the mind/body by working out or by pursuing other relaxing activities in order to divert cognitive mental activity from the problem. This will eventually unlock the rigid state of the mind/brain structure by reversing the levels of norepinephrine and increasing acetylcholine. This will allow you to find the solutions you have been seeking. Episodes of periodic confusion leading to phase transitions are a classical characteristic of any complex self organizing system such as the human brain. By relaxing you are able to find your way out of the morphological maze of this dynamical self organizing system.

Some Differences Between States of Inner and Outer Reality

Our inner realities are made up of our interpretation of our inner mental states whereas our outer realities are made up of our mental interpretation of our outer physical environment. Both, however, consist of our conscious interpretation of our brain's neurological basins and of our subconscious mind's analog patterns. It is these neurological basins and the analog patterns which reflect the inner and outer imagery making up the composition of our lives. If a person were to say that he is happy, peaceful and calm when in fact his internal operating systems are composed of significant negative energy such a person would be said to be in a state of denial. Similarly, if a person were to say that he is a millionaire and at the same time he hardly has enough money to pay his bills he would be characterized as being irrational. Simply affirming a positive state of mind or a positive state of external reality doesn't necessarily make it so. Nothing I said about creating states of inner or outer realities should be interpreted as meaning that by a sheer act of will power alone you can create internal or external realities in life.

In the case of the person who seeks to be happy in life he needs to first clear significant negativity from the prime directives operating at the subconscious level before he will be able to create states of internal happiness. Similarly, with respect to the person who desires to manifest a million dollars in his life he also needs to remove any negative impediments which might be blocking his prosperity on the level of the prime directives. Then he needs to program the subconscious mind to sort for those people, places or things which will bring to conscious awareness, in one form or another, his prosperity. The first person might affirm that his happiness is a state of mind which is presently unfolding in his life as he consciously removes any negativity which is blocking his happiness. Such a statement, if acted upon, would be rational. Similarly, if the second person were to state that prosperity is presently manifesting in his life as he removes any mental blocks to his prosperity and as he programs his subconscious mind to scan the landscape of his life in order to bring to conscious awareness those things needed to make his life prosperous. This also, if acted upon, would be rational.

How Physical Laws Affect the Creation of Outer Physical Realities

Your physical manifestations only come about in a natural way consistent with the way physical laws govern the way material things operate in this world. Quite often in applied mind science literature you will come across the statement "if you think that you manifest your physical realities in life let's see you walk through a brick wall or materialize a Mercedes in your driveway." Such statements clearly miss the point. I never maintain that you walk through brick walls or materialize Mercedes contrary to physical law. I simply say that the brain and mind have the capacity to bring to conscious awareness those things in life which allow you to create the type of life you desire for yourself. In the case of the wall, the subconscious mind will bring to conscious awareness the most appropriate means to remove the wall if it is a block to progress in life. Similarly, the subconscious mind will search for the most appropriate way to manifest the Mercedes consistent with physical laws. However, if you do not believe that this is possible, that belief also becomes a program directive of the mind and will be acted upon in order to prove you right in not being able to create those things you seek in life. This has been a major limitation of the historical Cartesian way of thinking. That is, when we view our world as existing outside of us and outside our control in fixed time and space we lose a considerable degree of control over that world.

The old Cartesian way of thinking fosters the idea that if we are to dominate the environment we must take physical control over every aspect of the environment and force it to respond in the fashion we desire. This way of thinking has resulted in the exploitation of the world's natural resources, global pollution, and a whole host of other social and environmental problems. MetaCreation postulates that you need only to take control over the mental processes which create your environment. You do not need to take conscious control over every aspect of the exterior environment. Rather you need to cooperate with your environment and live in harmony with it. *What I am simply saying is that in order to manifest your good you need only know what it is that you want out of life. You do not need to know all of the details of how it comes about beyond what you might naturally be expected to do as circumstances reveal what needs to be done.* You need only to stay positive and open to your good. The subconscious mind can sort endless possibilities to manifest your desires in the most natural way possible. If you endeavor to take control over every last detail of how your realities are to manifest in your life you are no longer relying on the awesome power of the subconscious mind to sort endless possibilities for you. You have consciously chosen to make such selections yourself. This is not to say your selections are right or wrong. It is only to say that your conscious selections are limited as compared to the unlimited power of the subconscious mind. By trusting the MetaCreation process you are able to let go and allow the greater power of the subconscious mind to sort it all out for you.

Finally, a word on physical law governing the natural order of things. Remember, quantum physics expresses the nature of how things are created and come into physical manifestation, whereas physical law governs how things operate in our physical world once created. While often there seems to be a conflict between quantum reality and physical reality, in fact there is none. The one expresses the creation of all matter, the other expresses the operation of all matter. They both express different aspects of the same thing, that one thing being matter arranging and rearranging itself into thousands upon thousands of patterns of reality. Similarly, the brain and mind when creating mental states organize the states on the atomic and molecular level of the brain and mind in accordance with quantum reality. Once organized on the neurological level its manifestation in outer reality is governed by the operation of physical law.

Remember, at the quantum level of the origination of physical matter, whether subatomic particles or massive celestial bodies, time and space are aspects of the creation process itself. Time and space do not exist as separate, independent and fixed objective states in which matter physically resides. The very act of creation itself creates the time and space of the creation of all matter. Similarly with respect to the creation of any mental state or an awareness of any outer state, as to the act of creating such mental states, time and space are not independent coordinates in which mental states arise. All mental states whether inner states or outer states are created independently of the fixed time and space of our three dimensional world. *While quantum physics has proven the old Cartesian way of thinking wrong in how its applies to the physical world it has not yet done so with respect to our mental world where we create all realities in life! It hasn't done so for quantum law as presently set forth by quantum physicists does not address the origination of mental states.* The result of all this is that we do not create a life for ourselves in a fixed three dimensional world in fixed time and space which is objectified and over which we have very little direct mental control. Unless, of course, you want to continue to believe, see and experience it that way!

Summary on the Reality Creation Processes by the Brain

The major impediments to the creation of outer realities consist of either (1) doubt as to your own ability to create in this fashion, or (2) the belief that the creation process itself as spelled out here doesn't really work in this fashion. There are literally thousands of reasons which fall into either of these two broad categories of why the creation process doesn't really work. Like the creation of inner realities, you need to train yourself to catch these negative neurological thought stimuli at their source and either eliminate them or redirect them to positive basins of attraction in order to effectively create positive outer realities. Once you know how to work the process and have experienced the process of creation in your life your belief in the process strengthens which, in turn, strengthens the creation of basins of attraction. You start to create outer realities in the conscious awakened state as effectively as you are able to create inner realities in the relaxed state.

You need to understand that all personal realities, both inner and outer, are the creations of the neurological processes of the brain and are organized and projected into the visual neuro cortex field of the brain. We have seen that in order to make this critical element of the creation process an operational part of your conscious mind you need to expand your cognitive structure by accepting it as the operative process by which we create all personal realities. This is probably the hardest step to take in the creation process because there is nothing in your cognitive structure up to this time which would indicate that this is in fact the way the brain and mind create realities in life. You need to believe in it in order to make it an operational part of your mental cognitive structure before your brain will act on any new programming instructions based on these new beliefs. Simply stated you need to believe it in order to see it and not the other way around. The nature of our inner and outer realities in life and how we create them might be best described in the context in which they were first revealed to Tommy Mindful during his trip to Superstition Mountain where he encountered the Mystic Shaman.

Tommy's Odyssey at Superstition Mountain - How the Shaman Explains the Nature of Reality

It was still very hot at 9:00 o'clock at night when Tommy's plane touched down at the Phoenix International Airport on Wednesday, August 18, 1997. Tommy quickly retrieved his 50 pound backpack at the baggage claim area, rented a car, and arrived at Apache Junction just before midnight. He arose by 4:30 a.m. the next morning and headed east on U.S. Hwy 60. About 45 miles out of town he turned left onto Scorpion Gulch Road, a twisting and turning dirt road which took him toward the Superstition Mountains. It was still dark and he was barely able to see the arrow pointing to the Peralta Trailhead. He drove on for about another mile to the trailhead parking lot. He locked his car, hoisted his backpack to his shoulders, secured his waist band, pulled down on the brim of his safari hat, and headed north along the base of the Peralta Mesa.

As the sun came up he could see off to his left running parallel to the canyon the Dacite Mesa which ran north as far as the eye could see. Throughout the canyon grew magnificent saguaro cactuses which looked like giant organ pipes strewn about at random on the canyon floor. By midmorning he stopped to rest on the shady side of an outcropping of rocks with some pinon pines growing in the shade. It was just too hot and there wasn't enough wood around in such a desolate place with which to build a campfire. He made do by just having a couple of long swigs of cold water from his canteen, a power bar, a couple of dried biscuits and some trail mix. After only a 15 minute rest stop he started out again for he realized that by early afternoon it would be much too hot to hike in the blazing sun. Temperatures at this time of year can easily rise well above 110 degrees in the desert.

After hiking roughly parallel to the Mesa on his right for about 20 miles he took another siesta on the south side of what looked to be an old burned out pueblo with only two walls still standing. At was now 2:30 p.m. and it was getting just too hot to hike in the blazing sun. He dropped his backpack to the ground and used it as a back rest and after a lunch of water, nuts and precooked rice he took a nap for he was getting pretty exhausted from the long day's hike. He awoke at about 4:00 p.m. stretched, took a couple of swigs of water hoisted his backpack to his shoulders and again headed north in the general direction he had been going all day. Although all day he had been hiking at a rather vigorous pace he did not slow down although his steps became more deliberate. He seemed like a man driven. Was it the desire to find the treasure of the Lost Dutchman Mine or was it something deeper and more meaningful that was driving him?

He heard the cry of a crow circling far overhead. A rather haunting "caw, caw, caw", like the cry of the very lonely, the very desperate. For the first time he became consciously aware of seeing that crow all day long but hadn't paid any attention to it. Tommy became aware that the crow was in some way leading him toward his destiny. The crow continued his lonely call in the distance "caw, caw, caw" as though the crow was encouraging him to follow, and at the same time warning him of some unknown premonition of impending doom that lay ahead. He felt that he was somehow a part of a greater drama being played out in these desolate foot hills of the Superstition Mountains. Was it to be a test of his courage, was he to finally face his deepest fears, or was it just his imagination brought on by the heat of the day and fatigue from the long hike?

By 9:00 p.m. he was still hiking. But now it was cooling off just enough to make the hike barely tolerable. He was not sweating as much, but in the dry Arizona desert sweat never accumulated on your shirt or skin but immediately evaporated. The sun was beginning to set but there was just enough sunlight left for him to see in the distance about a half mile up ahead what looked to be a small adobe hut built low to the ground with a few outbuildings and a small corral attached to the dwelling. There was a light in the small window so the place was inhabited. As he approached the hut about 10 minutes later a small mongrel dog ran out to meet him and was yapping up a storm. This alerted the occupants who probably had not had any visitors in months and out of curiosity they came outside to see who was coming down the trail. Tommy was met by a native American Indian who he later learned was of Apache descent and a woman who was his wife, and an old man.

They invited him to stay the night in the small corral next to their hut. They shared their supper of beans and rice with Tommy for they wanted to hear what was going on in the outside world. They had a radio with very bad reception, and no T.V. Tommy thought to himself that this might be more of a blessing than a curse. The next morning they filled Tommy's canteens with fresh water from their deep well and warned him that the terrain he was heading into became increasingly hostile and even dangerous. When asked they wouldn't explain what they meant by dangerous. There were many stories of prospectors and other treasure hunters just like himself looking for the treasure of the Lost Dutchman Mine who went up into those mountains and never returned. They called out to him to be extra careful as he headed off again in a northernly direction. No sooner than he left the hut of the Apache Indian did the crow appear again leading him in a northernly direction.

This day the going was a little slower since the terrain became increasingly more rugged with a slightly uphill climb whereas yesterday was mostly flat with less gullies and rocky knolls. Because he wasn't making as good a time this day, he ate his lunch as he hiked and by 2:00 p.m. it was just too hot to hike on. However, the crow was just ahead and circling to his right over what appeared to be the entrance to a canyon cut deep into the Mesa. He pushed on in spite of being very tired. Within a half hour he was at the entrance to the canyon. At the base of the canyon there appeared to be the ruins of an old prospector's hut and old rusted mining machinery. The hut would be a good place to rest until the heat of the day started to subside. As he sat munching some crackers and cheese the crow alighted within 10 feet of him and started its "caw, caw, caw" cry again. Tommy tossed a cracker to the crow. The crow quickly moved in to retrieve the cracker and flapped back several feet before eating it.

By 4:30 Tommy started hiking north away from the canyon when the crow buzzed him several times and then started up the canyon. Tommy tried to ignore the crow for his maps all indicated that he should continue north towards the Peralta Peaks which he could see in the far distance. There he hoped to find the Lost Dutchman's treasure. Every time he started north the crow would circle back, buzz him several times and then head up the canyon. Tommy finally thought to himself "O.K., you old crow lets see what's up that canyon." He followed what looked to be an ancient dried up riverbed for about 10 miles winding up the canyon floor. The crow was now always about 50 feet ahead of him.The hike up the canyon became increasingly steep and rugged. Since he was hiking in the shadows of the steep canyon walls, this helped cut off the direct glare of the afternoon sun beating down on him. Also there was a slight breeze coming down the canyon off the Mesa above which made the hike almost pleasant, although still very hot.

It was almost 9:00 p.m. and he must have been about two thirds up the canyon when the canyon floor seemed to level off and widen out. The canyon opened out into what might have been a very large meadow at a much earlier time when fresh water flowed in the dry riverbed. Maybe it was a large gathering place for native Indian ceremonies in the dark distant past, Tommy thought to himself. The crow was now flying low on the horizon and seemed to drop down at a location just up ahead at what appeared to be a well maintained native Indian adobe lodge. Against the north canyon wall behind the adobe lodge was a well preserved three story pueblo community. There appeared to be lights flickering in some of the windows and in the shadows he could make out the shadowy figures of men, women and children working, talking and playing. Opposite the pueblo about 50 yards out in the middle of the canyon stood the well maintained mud and clay adobe lodge. As Tommy drew closer to the adobe lodge there was a campfire burning brightly and on the other side of the campfire stood a figure dressed in old ceremonial clothing and what looked to be an old coyote skin shawl or hood over his head that blocked Tommy's view of his face.

As Tommy drew up to the burning fire the figure gestured for him to sit down. Tommy dropped his heavy knapsack to the ground and sat on the log nearest him. He was so tired from the day's hike up the canyon that he slumped to the ground and used the log as a back rest and stretched out his legs alongside the fire pit. The hooded figure said nothing but was mixing a brew in an old miner's coffee pot on the fire. When the brew was done the hooded figure poured two tin cups full of the hot brew and handed one to Tommy. While it tasted rather bitter it had a soothing effect upon him. He started to feel very relaxed, better then he had felt in months. It seemed to be a combination of the warmth of the campfire, the brew like tea, the presence of the hooded figure and the energy which seemed to permeate the entire area. Whatever it was he could care less for a strange peace and quiet came over him which he had never experienced before in his life.

His thoughts drifted back to those younger years growing up with Ricky, how Kenny and Stella, their parents, were always fighting and Kenny would always walk out and not show up again for a couple of days. The last time, of course, Kenny walked out after a particularly bad fight with Stella and never came back. Tommy was only 6 years old at the time and he really missed his father. Although Ricky was only four years older it was Ricky who took care of Tommy, and in a way Ricky became Tommy's substitute father, for Stella was rarely around much after Kenny left. When she was there she was often so drunk, or hung over, that her presence didn't make much difference in their lives one way or another. As he was relaxing lost in his own thoughts he looked over in the direction of the hooded figure who stood opposite him on the other side of the fire sipping his brew. Tommy sat up with a jerk for he couldn't believe his eyes. Around the neck of the hooded figure on a leather strap hung the symbol of the triple cross!

Off and on all his life he had heard about the triple cross but never knew what it meant nor exactly what it looked like! He first remembered hearing about it when he was a young man in college. He recalled he went to have a Tarot Card reading by a local fortune teller. She was the first one to reveal to him that he was under the spell of the triple cross. He asked her to interpret what the spell meant and she replied that while she understood the meaning of the triple cross in Tommy's life she was forbidden to reveal its mysteries to him for he must personally endure the lessons of the triple cross for himself. She did state that she could assist him in his quest by introducing him to eastern style meditation. She advised him that in addition he should follow a pure vegetarian diet and practice Hatha yoga exercises daily in order to ward off the worst of the evil effects of the spell of the triple cross. These practices and diet would put him in good stead in facing the future trials in his life which he was sure to face as a result of the spell of the triple cross. Some 10 years later a vedic astrologer read his astrology chart and stated that it revealed that Tommy was under the influence of the triple cross. He also stated that he could not reveal its significance to him for he must face his own trials in life in order to fully understand the deeper significance of the sign of the triple cross. Others in his life had mentioned the triple cross, but no one ever explained its significance to him.

Since Tommy's early years were ones of hardship he always assumed that the triple cross stood for his poverty, lack of love, and the loneliness he experienced most of his life. He remembered it was the lack of love and loneliness which weighed most heavily upon him. His thoughts drifted back to when he was eight years old in the third grade. It was Valentines Day and all the kids in his class were to bring in their valentines and at recess they were allowed to walk around the room and place their valentines on their friends' desks. They would then return to their own desk and the teacher allowed them about 10 minutes to open their own valentines. Tommy was excited about the prospect of receiving his valentines and so hurriedly handed out all his valentines and returned to his desk to open the valentines he received.

When he got back to his desk there were no valentines on it. Not one single solitary valentine! He just sat there with tears welling up in his eyes. No one cared enough to give him even one valentine after he handed out valentines to all his classmates. Suzy Lee Thompson, the prettiest and most popular girl in the class, sat directly across from Tommy. Suzy Lee looked at her huge collection of valentines. There was so many they were literally falling off the edges of her desk. She then looked over at Tommy and saw that he didn't even get a single valentine. Suzy Lee took the top valentine off her pile, one with a big red heart on it and tore it in half and kept the bottom half which was signed by a classmate and took the top half and wrote on it "To Tommy Mindful, be my Valentine, Suzy Lee." She handed it to Tommy with a smile. He was so happy to receive just half a valentine. He went home that day and felt that this was the best Valentines Day he ever had for he received a half valentine from none other than Suzy Lee, the most popular girl in his class. He treasured that valentine, and later on in high school he added it to the collage he created in art class symbolizing the most memorial events of his grade school years. To this day that collage with Suzy Lee's half valentine still hangs on the rear wall of his office.

And now he had come face to face with the triple cross that had placed a curse on him all of these years. The mystic Shaman started to speak to Tommy. While Tommy could clearly hear his voice inside his head the Shaman's lips never seemed to move. The Shaman stated that the mystery of the triple cross was locked up in the meaning of the three horizontal bars. The triple cross was neither good nor evil but simply represented the energy and images of all of the realities that ever permeated the universe from the dawn of time. The lower bar represented the energy of fear and hatred and was formed out of the negative polarity of its energy. The top bar represented the positive energy of love and beauty. The larger middle bar represented the combination of both the positive energy and the negative energy which together arranged and rearranged themselves into the matrix configurations forming all matter found throughout the entire universe. The mystic Shaman went on to state that it isn't the energy nor matrix configurations that creates good and evil but mankind's own positive and negative thoughts, desires, behaviors, and actions in creating realities out of the polarity of the energy and matrix configurations of the atoms and molecules arranged at the quantum level of their brains and minds that creates all of the happiness and all of the misery in the world!

The Shaman went on to state that the vertical bar represented mankind's ability to move his awareness from one level to the next higher level up in consciousness in order to synthesize his energies from a higher perspective and in doing so be able to overcome obstacles in life and create expanded more meaningful realities.[13] Once a man comes to understand and know this and learns to take control over the processes of his brain and mind he will, at long last, be able to create a world of prosperity, love and happiness, and by doing so direct the course of his own destiny and the destiny of all mankind. The Shaman concluded by stating that when you transcend the duality of good and evil, right and wrong, and love and hate you will have truly become a master over your own destiny and can at will collapse the entire energy field represented by the duality of the energy of the triple cross. You will then be able to not only create your own realities in life but you will be able to experience the source of those realities, the very source of your own being!

Tommy sat there quietly for awhile contemplating the astounding revelations of the Shaman. He then watched out of the corner of his eye as the Shaman walked slowly into the adobe lodge and momentarily reappeared carrying a very old leather bound manuscript or book of some sort. Carved into the leather cover of the old manuscript were the words "Ancient Gnostic Gospels." Tommy stood up. Without saying anything the Shaman handed the old manuscript to Tommy and indicated that Tommy should open it. With a random gesture Tommy opened the book at about the middle and he started to read aloud from the ancient text which referred to Saul of Tarsus of ancient biblical times. It is Saul speaking as Tommy read the following:

> "To keep me from becoming conceited because of these surpassingly great revelations I have just received, there was given me a thorn in my flesh, a messenger of Satan, to torment me. Many times throughout my wretched life I pleaded with the Master to take it away from me. But he said to me, 'My grace is sufficient for you, for my power is made perfect in your weakness. In every minute of every hour I am there with you for in your darkest hour I abide within you and you in me for before you were in bondage but now you are set free'.Therefore I, your humble servant, will boast all the more gladly about my weaknesses, so that the Master's power may rest on me. That is why, for God's sake, I delight in my many weaknesses, in insults, in hardships, in pain, in ignorance, in difficulties and in persecutions. For when I am weak, then I am made

13. Tommy later called this process "cognitive restructuring of the mind" which is described in Chapter 9.

strong."[14]

As Tommy read these passages they had an extremely profound effect on him. He dropped to his knees and started to cry out loud in pain for the meaning of his entire life was being revealed to him at that very moment. Suddenly he was filled with a very strong feeling of a presence that was so overwhelming that he felt that he was literally being torn out of his body and lifted into an altered state. He was filled with immense feelings of remorse, forgiveness, compassion and illumination. At the deepest levels of his brain and mind, at the quantum level where his brain and mind organized every reality he had ever known in life, he was being transformed from suffering and pain to illumination and love. A Voice started to speak to him and this is what it said:

> "Yes, of course, I was there during each of those dark, lonely and neglectful nights. For I am both everywhere and all at once. Yes, in an instant I could have brought you the love and peace you so desperately sought all your life. Yet, you did not call out to me to reveal to you the nature of your neglect nor the true nature of love itself. For you will ultimately find that all love starts with self love. Had you just called out I would have responded in an instant. I have given each one of you free will to search out and find your own truth and to choose your own path, your own destiny in life. It is not in my nature to intercede in your daily affairs unless you sincerely desire my help and guidance. Had you done so earlier I could have given you the means to stop the neglect and loneliness in an instant. I could have surrounded you with the light of my illumination and loving presence so that the pain of neglect would not have had such a traumatic and devastating effect upon you, and you would not have had to suffer as much as you did all through these years."

> "You did not realize that I understood and felt your pain as deeply as you did, and I could have also hated your parents and pitied you to the same degree that you hated them for I love you so much. You do not realize it is not in my nature to do so for I love your parents, Kenny and Stella, as much as I love you. For I express all that

14. A rough paraphrase of this very ancient text can be found in the more modern version of the Holy Bible, New International Version, II Corinthians, Verses 7 through 10.

love is, for I am Love.You needed to only call out to me like you have been doing just now. But, you chose instead to curse your parents and pity yourself all these years. And the consequences of cursing your parents, regardless of what they might have done to you, brought down on your head hot coals of hatred, anger, and anguish, and you have needlessly suffered through all these years.

You suffered needlessly out of your ignorance, your lack of awareness and understanding of what life is truly all about, and of what your were doing to yourself. However, my grace is sufficient for you every minute of every day. It will sustain you in every situation you will ever be faced with in life. It will strengthen you to met any hardship and to endure any difficulty and it will ultimately lead you and guide you and give you all of the answers you will ever seek in life. For you will attain illumination and be released forever from the bondage of fear, hatred and self pity. For now you have asked and I have responded, for I always respond to any prayer, petition or supplication sincerely prayed by those who dearly seek to know the truth about their own reality, their own destiny, and ultimately the source of both."

The Voice went on to say:

"You now have been given the answers you were so desperately seeking all these years. Now I will also give you something that you did not ask for. Henceforth you will be able to unconditionally love your parents and yourself like you never were able to love before because you were blinded by your own hatred, anger and self pity."

With that statement Tommy was filled with a tremendous feeling of compassion and love for his parents and for himself. It extended outward to include an unconditional acceptance and deep love for his wife Claire, his two children, Anthony and Philip, his older brother Ricky, and for Betty and Louanne, and all the other people in his life.These feelings were literally indescribable. At long last Tommy had received the answers he had so desperately sought all of his life and now he was at peace! After a time, he really didn't know exactly how long, Tommy was able to compose himself. At last Tommy had received the answers he had so desperately sought all of his life, and now he was at peace! Tommy was not very religious, at least not in the traditional sense. He quit the Catholic Church many many years ago and since then seldom went to church and didn't read the Bible. He never gave much thought to God and what role He plays in life. However, out of his pain, Tommy was always searching for answers to life's mysteries and for meaning in the suffering and pain he had experienced all of his life.

Now he came to understand that if you connect yourself to the energy of your subconscious mind, the same energy that pervades the entire universe, its guidance and direction can be awesome. If you are religious and belong to a church you might characterize what Tommy was feeling as the grace of God working in his life. But in whatever manner you choose to characterize it, the point is that Tommy came to realize, in that moment in the canyon, that we all are a part of a greater whole, that this greater whole can be accessed in order to actively assist you in all areas of your life. In his own life Tommy referred to this process of accessing this energy through the power of the subconscious mind as the ability to draw upon those aspects of the field of consciousness of the greater whole, of which we are all a part, in order to create for yourself an entirely new life with new purpose and new meaning. Tommy characterized these aspects of life at the level of creation of the subconscious mind as infinite potential and infinite correlations with infinite possibilities. At the level of self identity without attachment or ego identification to the things of this life he characterized them as wholeness, perfection and completion. However you are free to characterize these aspects of life in any manner you wish. They always work regardless of what label you put on them, or what label they go by in your life.

It was after this episode that Tommy was able to travel back in time and space and take the negative energy and images of those early neglectful childhood days, which were lodged in the neuro cortex level of his brain, down to zero. He was able to then reframe them to more uplifting and supportive life energy and images. Today Tommy can truly say that he unconditionally loves his parents! He now sees Kenny, his father, as a loving and kind man who supports and encourages him to succeed in life and he also unconditionally loves Stella, his mother and sees her as also telling him that he can achieve whatever it is that he wants out of life. Whenever anyone says anything negative about them he now simply responds with one of four statements. First, they did the best that they knew how. Secondly, they didn't know any better. Third, you have to love them in spite of everything. And finally, God bless them wherever they may be. Then get on with your own life.

Tommy, finally, learned that we all have something important to say about our life. We all have a right to be here and to say it. We have an inherent right to stand up and to be heard. What each one of us has to say is very important. We have a right to the fullest expression of who and what we are. We all have a right to live life to the fullest and to enjoy life. We all have a right to seek our own answers and to find our own meaning in life.Tommy finally realized that his parents, wherever they are, encourage him in whatever he undertakes to do. In fact, they have become his greatest teachers in life. If it were not for them he would never have been able to learn the lessons that he had learned. Tommy came to realize that some of these lessons were hard and at times very painful. It is in this very suffering borne out of his weakness that lies his strength. Tommy learned that from the greater whole of which he is a part he is sustained each and every day of his life. Life has, at long last, truly become meaningful and beautiful. Tommy can truly say that he has found his way home at last. For his home is not the small village in northern Minnesota but in his heart. Tommy had coined a phrase that captures this beautifully "Undeserved pain holds within itself the highest redemptive values in life!"

Tommy realized that this process of creation involves not only changes in the belief patterns of your personality and performance programs of the mind in order to create changes in your inner and outer reality. He came to also realize it necessitates a change in the neurological structures of the brain and analog systems of the subconscious mind from which your very thoughts themselves arise. Unlike what others might say you can truly change your childhood past, not only reduce its negative hold on you, but actually reframe it and change it totally. In the process you can literally recreate an entirely new you, with the type of personality and operational models for living life successfully that you choose for yourself. Its all up to you. As you change your orientation in life you come to realize that the outer space of your reality and the time sequencing of your life all originate within the neurological basins of the brain and the analog processes of the mind. Such a change allows you to finally realize that in fact it is you who is creating your life realities. Experiencing the creations and discreations of life through the use of the MetaCreations processes reinforces the perceptions within your mind and brain that it is you who is creating realities. At long last you can finally come to understand the theory and apply the techniques in order to create the type of life you never before dreamed possible.

It was mid-morning before Tommy awoke. He sat up and only

saw a depression in the ground that might have been the remnants of an old fire pit where the blazing fire was the night before but the old fire pit didn't look like it has been used for many, many years. There was an old rock foundation where the adobe lodge might have stood and behind the house there was no evidence of the cliff dwelling, only some foundation stones scattered around the ground and the exposed caves in the canyon wall which might have been the interior rooms of an ancient cliff dwelling. There was no Shaman to be found, no crow, no sacred scriptures, no tin cups and coffee pot, or any other evidence that anybody had lived in this canyon for quite some time. Yet the energy of the place was as powerful as it had been the night before.

Tommy thought to himself that maybe he had just imagined the events that took place late last night. Maybe the night desert air was having an effect on him. Maybe he was hallucinating borne out of fatigue from the long hike he endured these last few days. While he was contemplating these things a glittering light kept flashing across his eyes. It appeared to be coming from a low hanging branch of the nearest pinon pine about 10 feet away. Tommy got up off his bed roll and walked over toward the pinon pine branch to see what was causing the glint in the sunlight. There hanging on the branch on a leather thong was the golden triple cross! Tommy took it off the branch and slowing, almost reverently, hung the triple cross around his neck. Maybe he wasn't dreaming after all. Maybe everything took place just as he had experienced it. He thought to himself, "What is reality anyway? Does reality consist of only what you see, hear or touch, or does reality not also include what you dream, feel, envision, hope for or sense might be happening in your life? Does it include your experiences of the past and your hopes for the future? Are these also subject to change as you yourself change? Is reality composed not only of your own energy but the energy of those other people and physical objects in your life as well as the energy of the entire universe?"

Tommy was beginning to realize that maybe he had been defining the reality of his life in much too narrow terms. Maybe the origin of his physical world was not really outside himself and outside his control. Maybe the Indian Shaman was right after all that there isn't really as clear a demarcation line between our inner reality and our outer reality as we think there is. Maybe, just maybe, we do in fact create our own realities in time and space. Maybe we can go back and change our past as it personally applies to our life. Maybe, just maybe, we can mold and form our future realities out of the very energy and images at the quantum level of our brain and mind. Maybe space and time themselves are simply an aspect of creation and that as we create new inner and outer realities in life we also create the time and space those realities reside in.

As Tommy was contemplating these thoughts he proceeded to walk over for a closer inspection of the cliff dwelling he saw the night before. On the ground he could see the remnants of a very old foundation of what might have been the foundation of some ancient cliff dwelling. He traced the foundations stones up the canyon for a short distance and came upon a very small cave opening hidden in a grove of sage brush and pinon pines. He went back and got his flashlight out of his pack back. He squeezed under the brush and pines and crawled through the small entrance into the cave. Once inside, the cave opened up to a rather large ceremonial room whose walls were covered with ancient Indian hieroglyphics. There were etchings of buffalo, deer and antelope with Indian hunters giving chase. Tommy continued to inspect the cave drawings and at the far end of the room the last series of drawings was of the earth, moon and sun, depicting the center of the universe in mythological terms. To Tommy's surprise superimposed on the universe symbols was the sign of the triple cross! The Shaman said that the cross symbolized the arrangement of the energy of the universe arranging itself into endless forms of matter we all experience each and every day of our lives. Tommy realized that the mystical Shaman must have been a type of spirit guide or teacher of some ancient Anasazi Indian civilization which occupied the Superstition Mountains thousands of years ago.

When Tommy crawled out of the hidden cave he looked at his watch and it was already 2:o'clock in the afternoon. He felt that if he left immediately and walked at a steady pace he could make it back to the old miner's hut at the entrance of the canyon before sundown. With a gentle breeze coming off the Mesa down into the canyon and by going downhill all the way in the shade of the canyon walls he made it back in record time almost a full hour before sunset. He had plenty of time to fix a fully cooked supper using a trail mix of macaroni and cheese, hot biscuits with jam and herbal tea. He sat until the fire burned down contemplating the many events of the past few days. He was already thinking about the many new ways he could help his clients in not only overcoming mental dysfunction but also how he could facilitate them in empowering themselves in all areas of their lives so that their lives would become much more meaningful and enjoyable. He thought to himself, "isn't this what a really good clinical practice should be about, not just overcoming limitations of the mind, but rather empowering people to find their own true meaning and purpose in life so that they may enjoy life fully and prosper in all things". Tommy realized at that moment that a good clinical practice was not just about dysfunction, but really was about empowerment!

Tommy slept well that night and got an early start the next morning. With the sun mostly at his back on the return leg of his journey and with the gentle downhill slope he arrived back at the Indian adobe hut just before sunset. As Tommy approached the hut and corral he was met by the younger man and his wife. The old man was several feet behind them. The old man looked closely at Tommy and started talking rapidly in his native Apache tongue. He kept up this rapid dialogue for what was about 5 minutes but it seemed like hours to Tommy. Tommy could no longer contain himself and asked the younger man what the old man was talking about.

The younger man said that Tommy was wearing the symbol of the sacred talisman of the triple cross, that the talisman of the triple cross was the mystic Shaman's symbol of his magical powers. Indian legend passed down from generation to generation told of the Great One known in Anasazi times as "Great Eagle". Great Eagle prophesied that before the end of this millennia someone would come out of the east who would reveal the secrets of the ancient Anasazi and fulfill their destiny. He would usher in an era of a thousand years of enlightenment and prosperity. Legend has it that the timing of this man's appearance is referred to by the Anasazi as the journey of a thousand stars. This someone would ultimately unlock the secrets of the universe and unite not only the Seven Great Tribes of the West but also unite all the tribes and all nations inhabiting mother earth. Finally, he said that grandfather says that we will recognize that person when he returns from the sacred burial grounds in Diablo Canyon. He will be wearing Great Eagle's talisman, the sign of the triple cross. Grandfather says that you are that someone! Your real name is "Summa Eagle" the person the ancient Anasazi foretold thousands of years ago. "Summa Eagle" means "Greater Eagle" for you will embody even greater powers than the mystic Shaman himself, Great Eagle.

The young man explained that grandfather said that Great Eagle understood how it is that all matter is formed out of energy and he was able to change shapes and become the wolf, eagle, bear or crow and . . . . Tommy interrupted the old man at that point and asked him to repeat the part where he said that Great Eagle was able to change his shape and become a crow. The old man nodded his head yes. Tommy then explained that a crow led him into Diablo Canyon to the spot of the ancient burial grounds. The old man said it wasn't a crow. It was the spirit of Great Eagle. The younger man then stated that grandfather says that you have been sent by the ancient ones to fulfill the destiny of the ancient Anasazi and the destiny of the entire world. You alone can interpret for all of us the hidden meanings of the triple cross and bring illumination to the unaware peoples on this planet earth.

Tommy didn't know what to say or do for he was tired, thirsty and hungry from his long journey. The younger man sensed this and invited Tommy inside to rest awhile before supper. The woman killed the fattest chicken and roasted it over an open fire. She prepared a supper of roasted corn, snap beans, rice, corn bread and chicken. While Tommy was waiting for supper to cook the younger man pulled up some old floor boards and pulled out of the root cellar a jug of what could best be described as an aged mixture of barley wine. It was cool, very smooth, and very very potent. During supper Tommy tried to answer all of the old man's questions and tried to get in a few questions of his own. The conversation was very interesting but sometimes difficult to follow since the younger man had to serve as an interpreter, for the old man only spoke in his native Apache tongue.

Tommy did learn however that Diablo Canyon means the devil's canyon in Spanish. The ancient Anasazi didn't call it Diablo Canyon. The white settlers named it that because many prospectors looking for the treasure of the Lost Dutchman's mine went up that canyon never to return. The white man believed that the canyon was inhabited by evil spirits, whereas the Apache ancestors always knew that the canyon was the sacred burial grounds of an ancient Anasazi civilization. It was the spirit guides who protected the canyon, not the white man's devil. Since the Lost Dutchman's mine was within close proximity to these ancient burial grounds within the canyon the spirit guides made sure that no one ever found the Lost Dutchman mine, and the few that did never returned to tell about it. If they did they would have attracted hundreds, if not thousands, of adventurers into the area, and they would have surely desecrated the ancient burial grounds.

Their talks went on late into the night and after his third cup of barley wine Tommy was ready for bed. He picked up his backpack and headed for the door. The young man stopped him and offered Tommy the only bed in their small house. Tommy said, "No. He didn't mind sleeping out in the corral." However the young man pleaded and begged him to take the bed for if he did not do so that would displease the spirit guides since it would be interpreted as a sign of disrespect. Tommy thought that since he put it that way and since he was very tired, he would occupy the bed while the others slept on the floor. Tommy didn't know whether it was the comfort of a bed which he hadn't had in a couple of days or if it was the effects of the barley wine, but he slept like a rock and didn't wake up until he was gently shaken awake the next morning by the younger man. The young man said "Summa Eagle, wake up for the morning is late, breakfast is waiting and you have a long journey ahead of you, a journey of a thousand stars". Tommy didn't know much about the thousand stars bit, but the breakfast part sounded pretty good. Breakfast consisted of sweet tea, chapitas with ghee, and corn porridge with honey and goat's milk. Tommy was starved and had second helpings of chapitas and porridge. When it was time to go he showed his appreciation for their hospitality by giving the old man his stainless steel camping mirror. The old man would laugh when he saw his own toothless grin reflected back at him. He gave his Swiss army pocket knife to the younger man and he gave his nesting cooking pots to the woman since he wouldn't be needing them any longer. He also left the remains of his camping food except for 6 power bars which should hold him nicely on the last leg of his journey back to the Peralta trail-head.

Before he left the young man said that grandfather wanted to chant an old Apache ritual chant for good luck. Tommy said O. K. and the old man took the sacred talisman in his hands and chanted the ritual chant and then embraced Tommy. Tommy asked the younger man what the chant meant. The younger man replied that the old man was wishing Summa Eagle God's speed on his journey, a journey of a thousand stars. Tommy replied that he was only a good day's hike back to the trail head but that he appreciated the thought anyway. But the younger man said that Summa Eagle really didn't fully understand his destiny. God speed on the journey of the rest of his life, the journey he would lead the honorable ancestors and all the peoples of mother earth to their destiny, the journey that would ultimately fulfill the ancient prophesies. The journey in life that each one of us must take, the journey into the valley of death and to the mountain top of illumination and happiness, the journey of a thousand stars! With that Tommy bid them farewell and waved one last time as he disappeared over the knoll down the trail from the adobe hut of his Indian friends.

Tommy made good time and arrived back at the trailhead well before sunset. He placed his backpack in the back seat of his car and drove back down the dirt road leading to Hwy. 60 back toward Apache Junction. Within 30 minutes he was at the Last Chance Gas Station and Saloon. He called Lonni, the office manager, from the pay phone in the parking lot and said that she should call everyone for a 9:00 o'clock emergency staff meeting on Friday morning. He was going to change the office protocol and all of their therapy regimes. Lonni could tell from the excitement in his voice that something big was up. She too caught the excitement and started to ask lots of questions. Tommy would only say that he would outline it all on the plane home tomorrow and share it with her and the staff at the meeting on Friday morning and then hung up.

For the first time since his odyssey Tommy finally started to relax. He bought a cold soda pop, sat down on the porch under the awning outside and contemplated the week's events as he watched the sunset. He absentmindedly fingered the triple cross hung around his neck. He smiled to himself when he remembered that he always thought that the triple cross was a curse on his life and now it was empowering him. Then all of a sudden it hit him like a bolt of lightening! His destiny was now at hand; his life would never be the same again!

CHAPTER 6

The Structure, Function and Processes of the Subconscious Mind
The Analog Patterns of a Parallel Processing System

Introduction to understanding artificial intelligence

Some of the most far-reaching research in the area of applied mind science is currently being conducted by research scientists in the newly emerging field referred to as connectionism of the mind. Very little is known about this research outside the small but rapidly growing field of connectionist scientists using what is referred to as a parallel distributed processor. Remember, from Chapter 2 that a parallel processor is a very unique and sophisticated type of computer processing system. Earlier those working in the field of classical cognitive science primarily used analytical type processing of a binary serial computer which uses detailed programmable rules to simulate conscious thought. This earlier work encountered considerable difficulty in simulating human thought basically because human cognition did not closely track with the computer's programming capabilities using logical rule based processes.

In Chapter 7 on the origin of thought, you will see that in creating belief systems thought can be classified into four general categories. Only the second category classified as "logical thought" is anywhere similar to the results obtained from these earlier binary serial computers. These serial computers could not simulate human thought represented by the other three categories, namely, emotional thought, utilitarian thought and ritual thought, which includes thought patterns based on tradition and habitual thinking. The reason for this is that these other three categories of thought while not necessarily logical or rule based would be characterized as being rational. That is, human beings are fully capable of thinking and doing some really crazy or dumb things which you would not characterize as logical or even reasonable but are, nevertheless, very rational, at least to the person doing or thinking them. Even crazy people think that they are rational. These other thought patterns as well as the origins of thought are explored in detail in Chapter 7.

Primarily since 1989 some of these scientists have turned their attention to the parallel processing type computer, which processes data on an analog basis, and came up with some rather startling results. They found that analog processing more closely resembles the organization of human thought on the subconscious level of the mind as compared to these earlier classical cognitive models. In addition these parallel processors not only processed data similar to analog processing of the human mind, these computer models also embody characteristics of the human mind not originally envisioned by these computer scientists. Many of these other characteristics will also be addressed in this chapter.

This research holds profound implications for mankind for the reason that for the first time in history we have been able to find an operating model with significant possibilities for understanding the internal dynamics of the human mind. Up until this time science had to rely on the empirical evidence resulting from research using diagnostic tests and procedures involving actual human beings. This approach never allowed science to actually open up a person's mind and look inside, so to speak, in order to try to figure out exactly what was going on within the internal processes of the subconscious mind. Even if we were able to do so we wouldn't see anything since the organization of mental patterns and their interpretation into forms of human thought are dynamical energy processes not readily apparent from an investigation of the hardware. A distinct advantage of this parallel processor is that you can run thousands of hours of programs simulating human thought without the processor getting tired, or worse yet, bored or even resisting the process. While the data is not all in on the full potential of these parallel processors simulating human thought, the data that does exist is considerable and holds unprecedented possibilities for the future.

PDP - A system which models the subconscious mind

In Chart 6, at page 80 the first item "System Structure" shows that the conscious mind uses a system of cognitive structures to organize conscious thought whereas the subconscious mind uses analog processing to organize subconscious patterns representing thought. Cognitive structures are covered in considerable detail in Chapter 9. As indicated above the specific type of structural, that is, hardware system, used by the subconscious mind is similar to a "Parallel Distributed Processor".[1] We refer to a parallel distributed processor as a "PDP". What a parallel distributed processor is, how it functions, and what we can learn from it in connection with the organization of human thought is discussed in this chapter.

A PDP model is a type of parallel processing computer as compared to a binary serial computer. The PDP has no central processor. Our PC home computer is a type of binary serial computer which is controlled by a central processor. The PDP system is made up of a large number of neurolike processing units which operate on an analog basis. Each unit within the system has the capacity to com-

1 For an excellent discussion on PDPs see William Bechtel & Adele Abrahamsen' book, *Connectionism and the Mind.*

municate with its neighboring units. Each unit has a specific quality which is referred to as the unit's weight or value. The unit's value simply consists of variations in the polarity of the energy or matrix configuration of the unit. Depending on the positive or negative polarity of the unit's energy the units communicate with one another within the system by either attracting or repelling the values of other units. Units can communicate by locally affecting the value of one or two of its neighboring units or the system can activate numerous units in order to create complex patterns. The change in the units' value is often referred to as the units excitatory or inhibitory mode.

In this fashion the units organize themselves into patterns which are formed by the similarities in the relationships among units based on the similarities in their values. Units combine together to create patterns which themselves have their own unique energy and matrix configurations. With respect to the subconscious mind the characteristics of the patterns can be described in terms of modalities and submodalities. Remember, the modalities consist of the five sensory sensations of, taste, sight, smell, touch, and hearing. The submodalities consist of such things as sound, color, hard, soft, sour, sweet, and so forth. It is these submodality and modality patterns which the conscious mind interprets as representing the things making up our environment. Things in the environment were then labeled, that is, they were assigned a name, and in the process acquire a value judgment of whether or not they are beneficial or detrimental in life. Chart 3-A at page 63 sets forth descriptions of submodalities.

During the evolution of mankind when we started to think abstractly, we were able to move the process of thinking to higher order cognition. We not only were able to label our physical environment but we developed the ability to form and articulate abstract thoughts and concepts about that environment. This led to the process of reflecting upon why we think and perceive things the way we do. This ability is referred to as "second order thinking" where we not only know that we know something, but also we started to contemplate why we know that we know it. This field of inquiry is called "epistemology" the study of thought itself. Epistemology is the study of the processes and dynamics by which we formulate thought and not just the investigation or study of the content or arrangement of our language.

The process by which we started to label everything in our environment is referred to as pattern interpretations. In Chart 3-3, page 61, pattern interpretations is set forth as level 8. Chart 3-3 describes the functional levels of the brain and mind involved in creating inner and outer realities in life. Of all the functional levels of the mind level 8 "pattern interpretations" is probably the most critical one in the process of creating realities. It sets forth the basic format upon which the conscious mind ultimately determines the quality of life. How we interpret these subconscious patterns determines whether we will perceive ourself and our environment as being positive or negative. It is not the things in our life that make it positive or negative. It is the configuration of these internal subconscious patterns in life upon which we make conscious judgments which determines whether we experience situations in life as being either positive or negative. Consequently, it is these patterns at level 8 and the modalities and submodalities forming these patterns described at levels 6 and 7, to which we direct our attention in creating realities. When you consciously access these subconscious patterns and change their modalities and submodalities from negative to positive you start to change your inner and outer realities from negative to positive. In this way you can start to create a better life for yourself.

The units of a PDP are not preprogrammed in advance like a binary serial processor but rather the PDP system itself organizes the values of its internal units based on its initial input patterns. It then has the ability to organize within the system entirely new patterns based on the existing values of the current patterns. It has the capacity to create new patterns by reorganizing the existing values based on its past experience and from this analysis create a better response pattern. In this way it is able to create new patterns in order to provide the most optimum response in meeting its output objectives. It has the ability to continue this process of reorganizing the values of its units to form new patterns in order to respond to any new variations called for by its output objectives. When we refer to "output objectives" we are referring to situations in life which require some kind of response from us. Whether that response is something we think, say, or do, it is first organized on the subconscious level. It reflects the value of the underlying units of the patterns we subconsciously choose in order to respond to the many challenges we face each day.

In this respect the PDP model exhibits a type of machine knowledge for it can recognize and organize the value of its units to form meaningful patterns in response to new output objectives. Further, the PDP also exhibits a type of machine intelligence for it can evaluate its output objectives and then internally organize the values of its units to form patterns which allow it to select the most optimum pattern response to meet those objectives. This type of PDP machine learning or intelligence and this type of machine acquisition and consolidation of knowledge provides invaluable information concerning the processes involved in the formation of human thought itself at the level of the subconscious mind. PDPs are still in their infancy; nevertheless, a considerable body of knowledge has already been gathered from the experimental laboratory models in existence today.[2] In this and the following chapters we will consider how these PDP models organize patterns at the level of the subconscious analog system of the mind and how these patterns are interpreted in organizing thought by the conscious mind.

Basic Functions of the Analog System of the Subconscious Mind

The most basic feature of the analog system of the subconscious mind is the value of its units making up the structure of its analog system. A PDP unit corresponds to a neuron cell which makes up the brain's basins of attraction. It is these underlying basins that determine the values of the units of the analog system of the subconscious mind. The value of the units consists of its energy and matrix configuration. Variations in the unit's value is determined by the variations of the electrochemical composition of the neurotransmitters of the neuron cells of the brain's basins of attraction which subserves the units of the analog system. Units group together to form clusters of units. Clusters form the basic imagery patterns of the subconscious mind which are referred to as the modalities and submodalities of the mind. Units forming a cluster are based on the similarities between the energy and matrix configuration of related units in the surrounding neighborhoods in the landscape of the analog system. Clusters, in turn, form simple networks of patterns based upon the attraction and repelling effect of the values of the clusters. In similar fashion simple networks join together to form complex networks. The subconscious mind interprets simple networks as simple belief systems and complex networks as complex belief systems.

At this point you might be thinking, does the subconscious mind really interpret these analog patterns formed by the brain's basins of attraction or is the subconscious mind simply interpreting the underlying electrochemical configurations of the basins of attraction themselves? Very good question. Reductionist scientific theory holds that the later statement is absolutely correct, the subconscious mind simply interprets the electrochemical configurations of the brain's basins of attraction. However, thoughts and feelings are not simply electrochemical processes, although they are based on electrochemical processes. They are more accurately described as analog pattern configurations which entails a more complex level of organization than simply being interpreted as electrochemical config-

2 For an excellent analysis of how the PDP connectionist models operate refer to the book by William Bechtel & Adele Abrahamsen, *Connectionism and the Mind.*

urations. Simply interpreting them by their underlying electrochemical compositions overlooks the hierarchical structural arrangement of the cognitive structure of the subconscious and conscious mind involved in arranging the primary personality and operational programs of ego personality as well as all of our secondary beliefs. Beliefs form endless patterns of thoughts, ideas, attitudes, opinions, feelings and emotions and are not easily organized or understood by simply describing or interpreting them in terms of their underlying electrochemical configurations. The reason for this is that human consciousness can interpret the exact same mental patterns in various different ways depending upon the entire mental field in which they arose. By strictly interpreting thoughts by their underlying electrochemical patterns of the brain's basins of attraction does not allow for differing interpretations in thought patterns because of the different relationships of the thoughts' analog patterns to the entire mental field in which they arose. Accordingly, our subconscious analog model is a more comprehensive model in explaining and understanding the complexity of the cognitive structures of ego personality and its belief systems in creating realities. It is a very useful model when organizing and applying the MetaCreation processes in creating realities in life.

Now back to our discussion on the subconscious analog system. It is imagery patterns which the subconscious mind acts upon early in life to organize the initial parameters of the subconscious primary personality and operational programs of the early ego personality. It is also these subconscious imagery patterns that the conscious mind uses to form the conscious secondary thoughts about itself and its environment. Consequently, it can readily be seen that it is these pattern interpretations described as level 8 in Chart 3-3 that set forth the description of the most critical internal processes of the subconscious mind. They are crucial for the conscious mind in organizing beliefs about itself and its world. Consequently, it is these level 8 pattern interpretations and the level 6 and 7 modalities and submodalities that we will deal with in creating and dissolving inner and outer realities in order to create a better, more enjoyable life.

We have seen that clusters of units group together to form both simple and complex analog networks in the subconscious mind. Analog networks form belief systems. Complex networks form complex belief systems. Belief systems are represented as being modular in structure for two reasons. First, there is a syntax structure to human thought. We all use language to formulate our thoughts and to express ourselves. We use a common native language that, more or less, follows the rules of grammatical structure, depending on how linguistically correct we are. This grammatical structure is what forms the syntax of human language. Secondly, we have also represented belief systems as modular in structure for purposes of studying the mind's cognitive structures themselves. For we will see that the formation of ego personality itself is a mental structure composed of these patterns. Further, we will see that our secondary thoughts are also composed of these basic patterns of the subconscious mind.

All such aspects of the human mind being composed of combinations of these patterns have an energy structure to them. It is this structure that forms our inner and outer realities of ourself and of our world. Structure, you will remember from Chapter One, is what gives form to the subatomic particles composed of energy and matrix configurations. The structure of the subatomic particle is what gives form to life itself, this world of ours and to the entire universe. We will be saying a lot about cognitive structuring of our thoughts and beliefs in Chapter 9 and 10. We will also explore how to change the structure of our thoughts forming our realities if we are not happy with them by cognitively restructuring our mind.

Keep in mind that these modular structures of language of thought are representational only. In actuality their original formation and organization at the subconscious level are processed in analog fashion by the subconscious mind. These modular structures of belief systems all arise out of and conform to their underlying analog networks. They are therefore really energy processes rather than structures. However, we use modular structures of thought for analytical purposes only for they help in understanding how thoughts represented by words combine together to form phrases and sentences. Phrases and sentences are the basic format in which we communicate thoughts and feelings and, accordingly, they, more or less, follow grammatical rules which, in turn, can easily be represented as being modular in structure for discussion purposes only. Also by visualizing the organization of thoughts in modular fashion helps in that it closely resembles the relationships of the underlying analog patterns giving rise to our thoughts.

Literally millions upon millions of these analog networks form a single cognitive level. Cognitive levels represent levels of conscious awareness of an individual's involvement in various aspects of his life. A single cognitive level forms the landscape of one aspect of a person's life in the time and space of his present lifetime. The cognitive level at which a person is presently operating represents the level at which he views everything in his life and makes personal decisions concerning them. All thoughts arise within the existing level of a person's conscious awareness and form the person's present inner and outer realities in life. Realities which represent the person's physical existence in the current time and space of his world all take place at the cognitive level at which the individual is presently organizing thoughts and feelings concerning a specific aspect of his life.

Numerous cognitive levels form the architecture of the cognitive structures of the conscious and subconscious mind. Thoughts originating at the level of the cognitive structure at which a person is currently operating do not require any significant shifts in conscious awareness as to the realities a person is currently experiencing at that level. Thoughts arising within a given level are generally characterized as logical rational thought. Logical rational thought represents the type of thought processing we normally use to organize our daily affairs in life.[3] When a person's organizing synthesis is not adequate to resolve a problem he faces in life at the level of self awareness at which he is presently operating, the problem will appear to be insolvable. Such seemingly insolvable problems are not readily resolved by logical rational thought at the cognitive level at which a person is presently operating. Typically such insolvable problems in life often involve such major issues as a death in the family, bankruptcy, divorce, terminal illness, and other such tragedies precipitating an existential crises. However, you will see in Chapter 9 a person operating from a closed, rigid, self limiting cognitive level often encounters such insolvable life style problems in ordinary daily living. Such problems do not necessarily involve major existential issues but involve less significant problems we are all faced with at one time or another. Ricky, not being able to make a commitment to Betty even though he loved her very much, is a prime example of a systemic insolvable problem encountered at the level at which he is experiencing and organizing his life. Logical rational thought alone could not resolve the problem for him.

In order to move to a higher level of organizing synthesis a person needs to move to a higher level of awareness. Moves upward between cognitive levels require changes in conscious awareness and not just changes represented by variations in our logical rational thought processes alone at the level at which a person is operating. Changes in our awareness normally require Creative Integrative Thought™ which affects the components of ego personality. It is the structure of ego personality that determines our level of self awareness in addressing problems and in living life effectively. Creative integrative thought™ is defined to mean the process by which a person transcends cognitive structures. It is the process by which a person moves upward from one cognitive structure to the next higher

3 Charts 35 and 36 in Chapter 9 set forth the organization of the cognitive levels of the mind.

cognitive structure in the hierarchy of cognitive structures. As a person moves up to a higher cognitive structure a person redefines ego personality. How this process actually takes place is explained in Chapter 9. Conscious analog thinking is a characteristic of creative integrative thought and, accordingly, analog thinking assists in making vertical shifts between levels of the cognitive structure.

Not to complicate this picture any more than is necessary, we do need to note that people can and do operate at different levels of awareness concerning different aspects of their life. For instance, a man may be operating at a rather high level of awareness in his business life where he exhibits a high level of integrity, openness, respect for others and flexibility in his business affairs. At the same time he can be very closed, authoritarian, disrespectful and demanding with his own family. How all of this plays out in real life is addressed in Chapters 9 and 10.

Chapters 9 and 10 also address how we shift between cognitive levels by using creative integrative thought in order to attain higher states of consciousness and expanded states of awareness. In this Chapter 6 we only address the structure, function and processes of the subconscious mind. We will also explore how the analog system of the subconscious mind processes mental imagery into patterns which are then interpreted by the conscious mind into components of thought. Chapter 7 deals with the origin of thought and belief systems. Chapter 8 deals with our subconscious primary personality and operational programs of the mind. Chapter 8 also deals with what we refer to as our conscious secondary thoughts which give form to the specific life style we follow. Chapter 9 addresses the levels of the cognitive structure and how these levels form the hierarchy of the cognitive structure of the mind. As I said Chapter 9 also deals with the nature of creative integrative thought in moving from one cognitive level to the next level up as we expand our awareness of ourself and of our world. Chapter 10 deals with optimum levels of beliefs forming our inner and outer realities. Both Chapters 9 and 10 also deal with the constraints on the cognitive structure of the conscious and subconscious mind that hinder us from creating expanded, more enjoyable realities for ourself.

Similarities Between a PDP System and Human Cognition

The reason we devote most of this chapter to a discussion of the functions and processes of the PDP system is that these PDP connectionist models simulate the functions of the human brain and mind. They do so by organizing the mental patterns upon which the conscious mind forms interpretations, using linguistic symbols, in order to organize and give meaning to inner and outer realities. Eight of the main features of a PDP system[4] used in human cognition are described as follows:

1. Neurolike System

The first of these features is the basis upon which PDP systems were first constructed. They resemble neurolike systems. It is often stated that they were neurologically inspired in their basic functional design. The activation of the units within the system resembles the activation of neuron cells in the brain in that the neuron cell is either in an activated or in a nonactived mode. As seen in Chapter 5, bundles of neuron cells through the processes of their neurotransmitters either attract or repel one another depending upon the composition of their neurotransmitters. In this manner they form basins of attraction and attractor pathways. Refer back to charts 27-1 through 27-4, at pages 180 for examples of how the brain's neurotransmitters form positive or negative basins and pathways.

We have seen that these basins of attraction and attractor pathways through their associations form mental patterns. It is these basins of attraction and attractor pathways which subserve the analog system of the subconscious mind. Consequently, you can begin to see how it is that the sensory stimuli coming into the human or PDP system are converted, in the case of the brain system, into electrochemical stimuli which are propagated throughout the system via the neurotransmitters. In turn, the neurotransmitters form the connections based on their ability to attract or repel surrounding neuron cells. In this way the cells organize themselves into these basins of attraction and attractor pathways. In the case of the PDP system it organizes into patterns formed by the cluster of units based on the attraction and repelling effect arising from the polarity or matrix configuration of the energy of each unit. Either way you end up with a neurologically inspired analog system. The conscious mind can avail itself of this subconscious analog system by first scanning or reading the patterns, then interpreting them, next labeling them linguistically, then placing values upon them, and finally interpreting these values in order to organize meaningful realities for itself.

2. Rule of Soft Constraints

If a unit's value in a parallel processing system is positive it activates a neutral neighboring unit to a positive mode. In this way it acts as a constraint on the value of the neighboring unit. However, if two or more units have a strong negative energy mode they, in turn, act as a constraint on a neighboring positive unit. The system searches for the optimum response among its units in order to meet its output objectives. In this last example the output would be nega-

4 Bechtel & Abrahamsen, *Connectionism and the Mind,* pp. 56-65.

tive. However, this is not arrived at as an average range of the values of the sum totals of all of the units. In fact, the optimum response pattern may not reflect any of the specific values of any of the units. The response pattern itself has its own unique composite value. This then is what is referred to as the rule of soft constraints. What is important about this property of the PDP system is that the human brain and mind function in a similar fashion in following this rule of soft constraints.

For instance, if the patterns of an ego personality prime directive is mostly negative this will negatively affect the creation of all other mental patterns arising within the affected ego personality prime directive. The ego personality primary pattern existing within the subconscious mind's analog system will influence the configurations of the patterns of the secondary thoughts and feelings arising within this primary pattern. Therefore, the negative value of the prime directive will act as a constraint upon the creation of our secondary beliefs. No amount of rearranging of the system's secondary thoughts through logical rational thought processes alone will rearrange the values within the hidden units of the subconscious mind's prime directives. Until the value of the primary pattern is changed at the level giving rise to the negative thoughts and feelings, conscious thoughts, feelings and behavior will not significantly change. Thus there will be no significant change in the inner and outer realities. What I am describing is a process constraint in the functions of the system. Many of the other major process and content constraints on the system of the mind are addressed in Chapter 9.

3. Graceful Degradation

The mind/brain is generally a very durable functioning system. However, when the mind/brain is overloaded with too many tasks to perform at one time, or overloaded with too much information to assimilate, it will start to ignore some of the information resulting in operating at less than optimum levels. Also when it is physically impaired it also exhibits this characteristic of graceful degradation. When overloaded or partially impaired it will not just crash all at once, but will continue to operate but not at peak performance levels. Sleep deficits, drugs or alcohol, high anxiety, fear or severe depression all will affect performance levels. However, in these instances, again the mind/brain's operating system does not just crash, but its level of performance is impaired. Sleep deficits, alcohol and the like impair function for they distort the energy of the neuron cells of the brain which, in turn, distort the patterns of the analog system of the subconscious mind. Substance abuse adversely affects the chemical component of the underlying neurotransmitters and anxiety, fear and depression adversely affect the energy component.

Binary serial systems use nonexceptional hard rules to make selections by the elements within the system. If a basic operational rule is defective quite often the entire system will crash. In the case of a parallel processing system, such as the mind, all units are activated locally and the combined functioning of all of the units within the system determines the weighted outcome of the system. Therefore even if quite a few units are incapacitated in a large scale system the performance of the system will not be noticeably degraded. It would take a considerable number of units to fail before the system starts to show any degradation in its level of performance. Even in this case the system continues to function, but its error rate starts to increase and it starts to exhibit slow or graceful degradation. Alzheimer's disease is a classic example of slow degradation in the performance of the mind's system. When a person first contacts Alzheimer disease the mind doesn't go all at once. As the mind starts to lose large numbers of neuron cells the mind begins to experience slow degradation, first in short term memory, then in long term memory, next in its cognitive processing centers and finally in its bodily autonomic systems.

4. Content Addressable Memory

Content addressable memory is of considerable importance to human cognition in that it is essential to memory retrieval. By recalling one or more of the properties to be remembered we are able to recall the entire thing or event to be remembered. The system is capable of searching for similarities among those characteristics that match the properties of the thing to be remembered. For instance, if you forgot the name of an old friend you happened to run into at a restaurant you will automatically try to remember the person's name by recalling the things you do remember about him. You will try to recall his name by using retrieval cues such as thinking back to the time you last remember seeing him. You will then try to remember things about him, how he dressed, his job, his family, anything that might trigger a retrieval cue which would bring to conscious memory his name. However, binary serial systems retrieval is accomplished through the use of a key word to access the particular file you are looking for. If you do not have the file's name you cannot access the information simply through a search of the contents of the file itself.

This feature of the connectionist models is very important to the MetaCreation process. We do not have to consciously recall the source of a negative event in our earlier life. We need to only create what we refer to as a "Personal Power Analog"™ composed of similar energy and matrix configurations but with an opposite polarity.

Through the intentionality of our thought we allow it to search out the source of the negative program and dissolve it. We do not have to know the details of the original negative event in order to dissolve it. The subconscious mind will supply this information in a natural way when the timing is right. At the appropriate time you will continue the process of conscious cognitive integration concerning how the event negatively impacted your life. You do not have to intentionally force cognitive integration since it will naturally occur on its own. A prime operating principle of the subconscious mind is that if the subconscious data or information is important enough to you it will be revealed to conscious mind.

5. Knowledge Acquisition

We have already seen that the brain's basins of attraction function in organizing data by arranging data based on the similarities between the composition of its neurotransmitters. The brain is able to organize elaborate neuro networks representing slightly different variations in the basin's patterns through the creation of slightly different trajectories and slightly different vector spaces in its neuro networks. In this fashion the brain is able to both quickly and efficiently account for new basin patterns without creating entirely new basins every time it is faced with new input sensory stimuli. The system is able to train its neuro networks, that is, it remembers earlier basin patterns, and by using its existing knowledge derived from earlier patterns it is able to assign incoming stimuli directly to the existing basins of attraction within similar neuro networks. By using the existing basin patterns in this way it can easily build upon what it already knows and needs only to account for any new variations in the incoming stimuli by creating new trajectories and new vector spaces within existing network patterns. This is what we mean when we say that the brain has the capacity for knowledge acquisition and knowledge consolidation of incoming data.

6. Learning by Experience

Learning by experience is the hallmark of human cognition. Connectionist models accomplish this task of learning new responses to meet output objectives by arranging and rearranging the value of its units in order to organize entirely new pattern responses to meet new objectives. In the past it was very difficult to develop this type of system for machine learning using binary serial computers. The reason for this, as we have seen, is that the binary serial systems relied on rule dominated programs to process relevant information. Rule dominated systems operated in a brittle fashion. The rules didn't naturally or gracefully provide for exceptions. In order to make an exception new rules needed to be programmed into the system.

Quite often these new rules would result in unwanted results in complicated binary processing systems. Further they could not provide for the gracefulness and subtleties exhibited in human learning.

Consequently, this is a very important feature in human learning and has its application in the the MetaCreation processes. For the MetaCreation process takes into account that conscious cognition is a form of logical processing whereas the subconscious system processes information in analog fashion. Accordingly, the MetaCreation process provides for the conscious mind to avail itself of the techniques of analog processing by adopting conscious analog thinking in order to effectively change the values of its subconscious analog system. Remember, I mentioned previously that the subconscious analog system can naturally and easily process cognitive logical thought of the conscious mind. However, the conscious logical mind does not easily process analog patterns. But once the techniques are made available and explained conscious mind can be trained to process data in analog fashion in order to both discreate negative imagery and to create positive imagery for the purpose of creating positive inner and outer realities.

7. Pattern Creation

The subconscious mind can easily and naturally create mental patterns based on the values of its units. It is these patterns as we have already seen that the conscious mind interprets in creating the many realities in life. Pattern creation is what the PDP system, as well as the subconscious mind, excel at. The PDP system is not only good at pattern creation, but it is also very good at pattern recognition, pattern completion, pattern generalization, pattern integration, pattern interpretation, pattern transformation and pattern association.[5] This feature is very important in human cognition since these mental patterns constitute the essential building blocks of the conscious mind in organizing patterns of inner and outer realities. These mental patterns are primarily defined in terms of modalities and submodalities of the subconscious mind. The conscious mind, by rearranging the modalities and submodalities, is able to dissolve old negative patterns and create new positive patterns for the purpose of developing new and expanded realities in life.

8. The Analog System

Finally the last feature of the connectionist model which is important to human cognition is the functions of the analog system itself. As conscious beings we believe that when we organize our thoughts and feelings we are in fact somehow actually consciously processing the formation of those thoughts and feelings. This is the

---

5 Bechtel & Abrahamsen op. cit., pp.106 - 146.

case for you become consciously aware of the end results of the organization of thoughts and feelings at almost the instant that they are organized at the level of the subconscious mind. We have a sense that they are processed instantaneously since the human brain and subconscious mind process data at the incredibly fast rate of a few milliseconds.[6] Therefore, we experience everything as though it is conscious awareness that is organizing our realities in life and not the subconscious mind. However, it is the subconscious mind which processes realities in life. Conscious mind doesn't process anything and therefore doesn't create anything. It merely initiates and organizes thoughts and feelings. While it directs the process it is not the faculty which actually processes realities. Subconscious mind acts upon our thoughts and feelings and creates realities in accordance with the directions of conscious thought.

Unfortunately conscious mind is not aware that it is not the one creating the realities in its life. Therefore, when we are feeling bad or down conscious mind immediately assumes that someone or something in its environment is the cause of our negative feelings. However, the negative feelings originate in the analog system of the subconscious mind. Therefore, no one can actually make you feel bad. Only you yourself can make you feel bad. Another person's actions may be the cause of you feeling bad, but whether or not you will actually feel bad is within your control.

Conscious mind merely provides you with an awareness of the results of what's going on at the level of the subconscious mind. It is the subconscious mind that carries out all of the processes involved in organizing conscious thought. The reason why this distinction is important is that you cannot, through conscious thought alone, actually create or process the patterns making up your life. Only the analog system of the subconscious mind can process realities in life. No amount of conscious logical thought, in and of itself, is ever going to rearrange the values of the units of the subconscious analog system. Another way of saying this is that you will never destabilize negative energy and images of the subconscious patterns by just talking about it. Just talking about it is merely an exercise in intellectualization. By quieting the mind and directing the analog energy of the subconscious mind to destabilize negative energy and organize new positive energies you create and discreate realities in life.

How the Subconscious Mind Creates Realities

With this background look at Chart 29 at page 235, *Schematic Diagram of the Human Mind Functioning as a Parallel Processing System.* On the left side of the chart is the input stimuli from our environ-

6 Ibid page 57.

ment that we receive every moment of every day. This stimuli is received primarily through our five sensory receptors of touch, sight, hearing, smell, and taste. The circles at the intersection of the horizontal and vertical lines of the matrix represent the clusters making up our subconscious analog networks. The horizontal and vertical lines represent the attractor pathways or trajectories leading to and from the analog clusters. Each cluster is weighted by the intensity of its energy and the vividness of its matrix configuration. On the right hand side we list in summary fashion, factors which influence the values of the clusters in the system. These factors are already reflected in the values of the system. These factors include past long and short term memories, past emotions and feelings, present thoughts and beliefs, present experiences and feelings, and future wants, needs and desires and the feelings related to these future wants, needs and desires. The output responses at the bottom of the chart represent the landscape of our inner realities in life. These inner realities give rise to our thoughts, behaviors and actions about our outer physical realities. It is these thoughts, behaviors and actions which, in turn, create our many outer physical realities in life.

New signals entering the system therefore do not change the existing system as such but rather influence the values of the existing system. For instance, if the existing system was basically a negatively weighted system and was this way for many years a new positive thought will not have much effect on such a basically negatively weighted system. This represents the action of the rule of soft constraints. New Year's resolutions are a good example of this phenomena. The resolution is well intentioned when made but often doesn't last very long unless your resolve is strong and you have a clear plan to implement the resolution. Even if the system isn't that negatively weighted, an individual still encounters difficultly since much of the negativity in the human system exists, for the most part, at the subconscious level of the system. This results in the individual not even being aware of such negativity existing in his life except when it manifests as pain, lack and limitation. This pain, lack and limitation reflects, of course, the negative weights within the system. The old saying "As within so without, as without so within" is certainly true concerning the analog system of the subconscious mind.

The new incoming signals of the mind's parallel processing system just influences the existing values of the system's clusters. It does not change them entirely. If the values in the existing system are heavily weighted in a particular direction it would stand to reason that it is going to take the application of a considerable weighted signal in the opposite direction, over time, to have a significant effect

CHART 29

Schematic Diagram of the Human Mind Functioning as a Parallel Processing System

Simplified Matrix Grid of the Subconscious Mind's Analog Internal Processing System

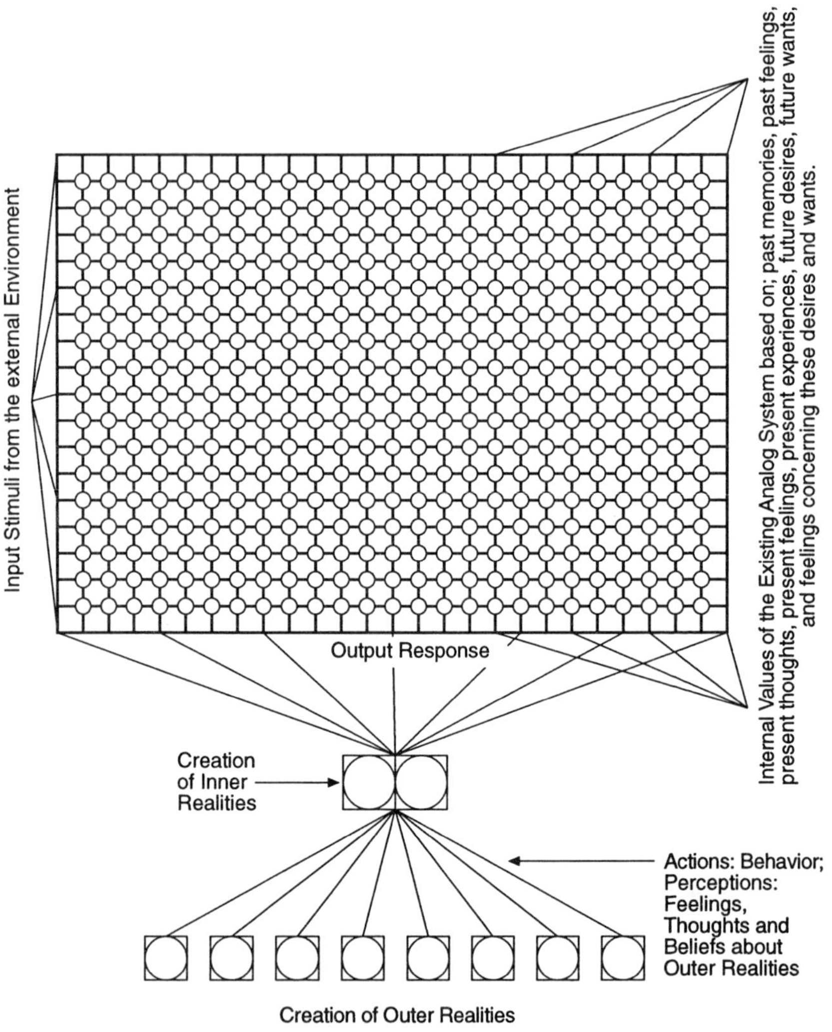

on the system. This operational bias is both good and bad. Bad in that, as we have seen, simple affirmations showing a positive intent introduced into a system which is heavily weighted in a negative direction will probably not have much effect on the overall performance of the system. For this reason daily repetitions of positive affirmations seldom work effectively to significantly change mental patterns of thought and behavior. However, this same operational bias can be used to work in our favor. Because of the mind's incredible sorting capacity the induction of a suggestion for the subconscious mind to form a superior energy and image analog symbol programmed to go on a search and destroy mission within the system can be used to eliminate negative energy and images in the primary programs of the mind. The subconscious mind will form the new positive analog symbol from the clusters closely resembling the existing negative pattern for this is the way the system works. The energy and image of this new positive analog symbol is what is used by the subconscious mind in searching out the negative energy in order to destabilize and dissolve it. This is the process used in the MetaCreation processes. In forming the power symbol the subconscious mind has an incredible array of possibilities available to it. It will pick the most appropriate symbol which holds personal meaning to the individual using the symbol. It might have universal meaning as a symbol similar to the king, warrior, or magician symbol. However, to the person creating the symbol it is very personal and is created to accomplish a specific task. In this case to remove negative energy. For a list of the more widely recognized symbols used by the subconscious mind refer to Chart 3-B at page 70.

It is the units, clusters, networks and levels which would be roughly analogous to the inner or hidden units of the PDP system. The big difference, and it is a very big one, is that the PDP models are still in their infant developmental stages. They only consist of one, two or more hidden layers of parallel processing units, as opposed to human cognitive levels of awareness in organizing realities in life. The processing units within the human mind are composed of literally billions upon billions of units and millions upon millions of analog patterns. The organizing system of the human mind is both, quantitatively and qualitatively, exponentially superior in both complexity and organizational capacity than these early PDP models. Nevertheless, with respect to the theory of mind processing, what we can learn from these elementary prototype PDP machines is of critical importance and extremely invaluable in understanding the inner processes of the human mind in creating realities in life.

How the Analog System of the Mind is Booted Up at Birth

From the time the child is in the womb the system is already fired up receiving stimuli in the process of creating the content of his life. The content of his life is represented by the billions of weighted units and the intensity and polarity of their energy creating the patterns of his life. The earliest units in the system forming the basic personality and operational programs of the mind consist of the initial weights of the original units which are in the system when the child is born. Subsequent secondary beliefs about the child's life style are influenced by the weights of these very earliest units in the primary programs of the mind. As we have seen the units weighted configurations are already set by the earliest history of the child prior to the time subsequent incoming signals enter the system. Consequently, subsequent sensory stimuli entering the system are not coming into a blank system but are coming into a system which holds considerable personality and operational content in the form of the existing weighted units within the system. Even before birth, signals, primarily from the activities of the mother, are constantly coming into the system. Our mind isn't blank at birth, but comes with quite a few basic processing programs to get us started plus whatever content patterns we received from our mother.

Like any data processing system the child's mind/brain system requires processing programs so as to process data in an orderly fashion. Otherwise how would it know to process stimuli into thoughts? No machine and no mind, when taken out of the carton, or the womb, and when plugged in, or fed infant formula, self programs its own processing functions of its hard drive. It needs basic processing programs installed by somebody, or somehow, in order to process stimuli which operates its hardware which in the case of the newborn child consists of its brain and physical body to create realities in life. A system without any process or application programs isn't an operating system at all but simply a piece of hardware. As indicated in Chapter 4, when we are born the components of consciousness are translated into process functions of the mind. As we have seen these process functions consist of self awareness represented as, "this is me". The knowledge function is represented as "I know, I organize data". The intelligence function is represented as "I think, I interpret information". Lastly, there is the volitional will function which translates into "I decide, I make my own decisions".

With a few other functions, in addition to these process functions, the system is now booted up and ready to do its work of processing stimuli into reality. But what boots it up? The answer is a simple one. All things being equal, if the child is born he is booted up.

The life force of the child is what sustains the whole mental apparatus. More precisely, we have seen that the neurotransmitters are what feed the proper nutrients to the neuron synapses which organize the electrochemical and electromagnetic forces throughout the basin of attractions and attractor pathways. These, in turn, are acted upon by the process patterns of the mind in arranging energy and images into reality. Obviously without the sustaining life force of the child there would be no system in which data is processed. This booting up process by the life force of the child is similar to the electrical current feeding power to the PDP model. If the electrical current is weak or fluctuates it could adversely affect the internal operation of the system. Likewise, if the child does not receive proper nurturing, nutrition, clothing, housing and education this also can adversely affect optimum performance of the system.

Now concerning those other additional things needed to get the system started, evolution has endowed the human mind/brain with the operational parameters of time and space in which to operate effectively in a three dimensional world. With respect to space, we have seen that any object in the space of our environment is measured by six degrees of freedom of position, three degrees of location and three degrees of orientation.[7] Any object in your life such as a desk or chair, occupies three degrees of space where it is located in your life and has three degrees of orientation where it can be relocated in your life. Similarly, the sequencing of time, if not a primordial process function, is learned from the time of birth. But even if learned at the time of birth the analog process function needed to process time is already in place at birth. After birth the child soon realizes the sequential steps in how events take place in his physical environment. These six basic process functions, namely, four of consciousness and of time and of space, act as the basic process functions of the human mind and are essential in the process of how we create thoughts which form our realities in life.

<u>The Relationship of the Brain to the Subconscious Mind</u>

Refer again to Chart 29 at page 235. On the left side of the chart are the input stimuli from our environment. These stimuli are received primarily through our five sensory receptors. The stimuli received by these five sensory receptors transfer their signals to the brain through the brain's large network of neuron cells. There are specialized networks of neuron cells within the brain which are dedicated to the task of transmitting these signals to the appropriate area of our brain where the stimuli are converted to electrochemical signals. These signals, as we have seen, are transmitted throughout the

7 Bechtel & Abrahamsen, op. cit., p. 105.

brain's network of neuron cells by the neurotransmitters. The electrochemical signals of the brain serve as the input stimuli for the human mind. In this respect the human brain subserves the functions of the human mind. That is, the mind itself is not hard wired directly to the five sensory receptors and it does not process these signals directly from the source of the sensory stimuli. Rather, the mind needs the intermediate stage in the process whereby the brain converts sensory stimuli into electrochemical signals forming the basins of attraction. The subconscious mind acts upon these basins and translates them into the energy and images patterns that serve as networks of the parallel processing system of the subconscious mind. In this fashion the subconscious mind bridges the gap between the brain's electrochemical stimuli and the highly evolved thought processes of the conscious mind. While the brain's basins of attraction subserve the subconscious analog system, they are not the analog system. The brain's electrochemical neurotransmitters do not serve the functions of the networks of the analog system. Quite often you will see in applied mind science literature how the brain thinks and processes thought. The brain doesn't think nor does it process thought. It is from the cognitive structures of the conscious mind that thinking is derived. The brain simply translates sensory stimuli into electrochemical substances and then arranges these substances into basins of attraction. At this point the brain's job is essentially done. It is up to the functions of the mind to process these basins into patterns which are interpreted as conscious thought.

A good analogy between the relationship of brain and the mind would be the functions of the computer hard drive and its software programs. The hard drive processes the software programs. However, seldom do you think in terms of the software programs as being an integral part of the hard drive itself. The reason for this is that you can purchase many different software application programs which your computer hard drive can operate. It can effectively operate these software programs for the reason that the hard drive itself is programmed with internal processing programs which operate the system. If would be more accurate to say that these internal hard drive processing programs are integral aspects of the hard drive similar to how neurotransmitters and basins are integral to the operation of the brain. The hard drive system of the human brain can operate literally thousands upon thousands of human application programs. All application programs however are run on the identical brain hard drive system which was initially booted up with its inherited processing programs. It is only when an application program isn't operating properly do we suffer conscious pain, lack, and limitation in life.

A very good example of what we are talking about here is the MetaCreation processes. The MetaCreation processes could be described as a set of mind application programs quite similar to a PC home computer floppy disc application program. The only difference is that human beings do not have a little slot at the base of their skull in which to insert a new floppy disc program. However, this doesn't pose any real problem for us for the MetaCreation processes come with a user manual similar to a "Mac For Dummies" manual. Instead of inserting a floppy disc program we simply debug the existing personality and operational programs of our minds and then install, by down loading, new software programs. Some of the floppy disc-like application programs that come with the MetaCreation user manual include: a Whole New Personality program; a Whole, Perfect and Complete program which expands your consciousness; a new Spacetime program to take effective control over linear time and fixed space; an Analog Thinking program to expand your cognitive structures; a Past Time Line program to access past histories and correct them; a Future Time Line program to create new realities; a Content Application program to create the contents of the mind for a better life; a Housekeeping Desktop program to organize your modular structures of thought; a MetaCreation program to create new inner and outer realities, and many other programs to enhance life and make it more enjoyable. If you consistently apply these MetaCreation programs on a daily basis for at least six months you will begin to experience significant positive changes in your life.

You may wonder how the mind is able to convert the electrochemical signals of the brain into energy and images forming the weights of the system's units. The answer is that the electrochemical signals of the brain themselves are weighted. However, in this case their weights are determined by both the polarity of their electrical charge and by the chemical composition of their chemistry. Variations in the weights of the thousands upon thousands of the brain's neuron cells forming neuro networks is what determines the values of the mind's units in its analog system. For example, the thousands of pictals forming a T.V. picture are made up of various hue, tone and brightness of one of the three primary colors. However, the T.V. picture itself when viewed as a whole forms a complex pattern interpreted by our vision as a moving series of action pictures taking place on the screen. The electrochemical stimuli of the brain neuron cells and the energy and images formed by the brain's basins of attraction act in the same fashion. The composite thousands of electrochemical stimuli of the brain's basins form the energy and images of the mind's analog patterns. In turn the energy and images of thousands of these analog patterns form the mental imagery patterns of the mind. The conscious mind, in turn, converts these mental imagery patterns into meaningful thoughts.

Organization Of Thought Patterns within The Subconscious Mind

We now address how the subconscious mind creates our thoughts and beliefs about ourself and our life. Remember in Chapter 5 we addressed how the brain organizes speech patterns. However, we did not address how the subconscious mind originates our thoughts expressed as speech patterns of the brain. You may recall that we described the brains's ability to create speech patterns by referring to the simple sentence "Dog who chases cat who chases mouse". In organizing this speech pattern the brain's neuro networks double up in the usage of its basins of attraction by using almost the identical neuro networks to form and track slightly different sentence structures. In this fashion the brain not only organizes word patterns but also recognizes words in the context of the sentence in which they are used. In this section we now deal with how the mind actually originates thoughts at the level of the subconscious analog system subserved by these speech pattern basins of attractions. We have seen that the subconscious mind's system for organizing and storing information is different from the manner in which logical thought is organized by the cognitive structures of the conscious mind. It is also interesting to note that while there is considerable overlapping between the brain's word basins of attraction organizing speech patterns and the subconscious mind's analog system originating patterns of thought based on these speech patterns, there is also a distinct difference between these two systems as you will see in just a moment.

We have seen in Chapter 2 that since childhood we labeled everything in our environment and by doing so developed an attitude of what served our best interest and what did not. We have seen the subconscious mind matches up representations of external objects with their internal analog patterns. If there is no inner pattern to match up with the outer object the subconscious mind will create a new analog pattern based on the most similar patterns within the system. The conscious mind then interprets these patterns and labels them so that we can consciously form meaningful thoughts, ideas and beliefs about ourself and our world. If the new incoming signal is incompatible (different energy and matrix configuration) with the existing basins subserving the analog thought patterns already present in the system, either the subconscious or conscious mind may immediately reject the incoming signal since it is perceived to be detrimental to the organization of the existing system. The conscious mind rejects it through logical thought processes. The subconscious mind rejects it by screening out the incompatible incoming stimuli. The reason for the rejection by conscious mind may be rational or irrational, it doesn't make any difference for the purpose of this discussion. The point is that if the incoming signal is perceived to be in opposition to the analog patterns representing existing beliefs the system needs to be modified or the incoming signal needs to be rejected. The subconscious mind and brain do not tolerate for very long pattern signals which create cognitive dissonance within the system. The brain and subconscious mind being analogical in nature will always seek resolution to cognitive dissonance and a return to hemostatic balances within its systems. Conscious mind, however, can live with a certain amount of cognitive dissonance if it can endure the pain of conflict. That is why a married man can juggle his time between his wife and kids and his mistress. He feels that the emotional and physical pleasure he receives from this life style is well worth the pain of conflict existing between the two conflicting relationships in his life.

In Chapter 5 we indicated that the inner somatic system has a tendency to scan the exterior landscape for negative elements that reinforce a negative system and positive elements to reinforce a positive system. In similar fashion the system will also automatically scan the exterior landscape to pick up on ideas which reinforce existing belief systems. The system also has a built-in bias to reject ideas which are in opposition to the thoughts of its existing belief systems. Consequently, what we believe determines to a large extent how we interpret the landscape of our life. When the energy and matrix configurations of the external representation is totally in opposition to the energy of the existing internal analog thought patterns, then conscious mind has a choice. It can accept the external representation as a part of its personal reality and modify the energy of its internal analog thought patterns in opposition to the new image, or it can simply reject the external representation as being either wrong or false. Also, as we indicated, the conscious mind as well as the subconscious mind can simply ignore it as not being a part of its personal reality. An example of what we are talking about here is that people who believe in UFOs often see UFOs. People who do not believe in UFOs seldom, if ever, see UFOs.

What is happening is that we don't stop to reflect on the many opposing representations appearing in our outer reality expressed as thoughts, feelings, attitudes, judgments, perceptions and opinions before they are unconsciously rejected simply for the reason that they are in opposition to the energy of our existing analog thought patterns. They are rejected without consideration of whether or not a new outer representation could serve any beneficial purpose in our life. What is going on is that the weights of the clusters representing existing thoughts and beliefs held by the system in the form of both simple and complex belief systems will act as a significant constraint concerning the formation of entirely new and novel thoughts and beliefs. This and other mental constraints are addressed in Chapter 9. Sometimes we are too rigid in our logical thinking and consequently miss out in life experiences simply because we do not understand logically all aspects of a uniquely new experience at the moment it is received by the system. We are often not consciously open and flexible enough to comfortably experiment with new and novel ideas which might hold considerable benefits for us. People are often very rigid and closed when it comes to entertaining a new idea, particularly when it is in opposition to the person's current beliefs consisting of contrary analog patterns already existing within the system. They will feel that the new idea is in some way a threat to their self image or their way of life. We will see in the next chapter how beliefs can become objectified and take on a life all their own. However, all beliefs and all life realities at the quantum level simply consist of energy and images.

When the weights of the new thoughts are totally incompatible with the weights of the existing system you have to consciously and deliberately decide to override the tension created in the system because of the incompatible energies of the weights forming the new thought. However, since the processes involved in the formation of thought take place on the subconscious level of the mind an individual is seldom aware of how these internal dynamics affect how he thinks as well as what he thinks about each day. Often he will consciously conclude that certain new ideas presented to his conscious mind just don't seem to logically make any sense. Also he may just have an uneasy gut feeling about them without quite knowing why. He will then consciously reject them before he really has a chance to evaluate the new idea to determine if it might be beneficial to him. This condition gives rise to the statement we all have used at one time or another, "I have a really bad feeling about this", without really knowing why we feel the way we do.

We have seen instances when a person in an argument needs a rationale to explain his position he will almost always find it. Take the example of a person who has been a lower income blue collar worker all of his life and has always believed that the Democratic Party is the party of the working class. Because of this long held political belief reinforced by his deeply embedded thought analog patterns formed over many years, he might always vote a straight Democratic ticket. He does so without ever analyzing the issues or looking very closely at the records of the individual candidates or their personal experience and character. What has happened, and this is true of all of us and not just blue collar workers, is that after a time patterns of thought like habits become well formed within the subconscious thought patterns of the cognitive structure. As circumstances in our life change we do not always periodically review whether older beliefs are serving us well in our new circumstances in life. The weights of older beliefs are not only well established but also their pathways are well established. However, it is through the conscious intentionality of our thoughts that we are able to effectively change the weights of the trajectories and the vector space of old established beliefs and in this way organize entirely new thoughts forming new life realities for ourself.

Previously we referred to a distinct difference between how the brain organizes speech patterns and how the subconscious mind organizes thoughts and beliefs. The brain and mind double up in the use of our basins of attraction and analog patterns in creating speech patterns and thought patterns. However the way they use these patterns is distinctly different. The system readily assimilates new speech patterns by creating new basins to reflect the new speech patterns. Except maybe for curse words, the system automatically endeavors to accommodate all incoming new speech patterns without discrimination.[8] However, arranging speech patterns into patterns of thoughts and beliefs requires judgment discrimination by the conscious mind. The reason for this is that thoughts express the content of our life and as such they create the realities we experience in life. Words are essentially tools providing us the means by which we create the content of our thoughts. Words themselves, in this context, act as process functions of the subconscious mind in creating the content of our life. Accordingly, the mind/brain will readily assimilate new words in order to expand our vocabulary, but at the same time the mind/brain will not so readily accept new thoughts and beliefs made up of our words.

The reason for the differences between logical organization of thoughts by conscious mind and analog organization of speech and thought patterns by subconscious mind is that historically the analog processes of the subconscious mind were first developed many hundreds of thousands of years ago before mankind ever started to think logically and communicate his thoughts through speech patterns. The primate ancestors to modern man survived

8 With respect to forbidden words such as curse words the child's analog system readily assimilates them before the child is admonished that they are forbidden words which should not be used.

mostly by instinct using the basins of attraction processing functions of their rudimentary brain and mind in order to just survive. They didn't use thoughts and speech to communicate what was going on in their lives. Subsequently, prehistoric man started to use more and more of the neocortex portion of his brain which subconsciously processed higher order cognitive thoughts of the conscious mind by using more complex analog patterns. Once this prehistoric man started to use a type of simple language to identify things in his environment he was able to use these well-established subconscious complex analog patterns to organize his thoughts. At the same time he was able to use the underlying speech basins of attraction to communicate those thoughts in the form of spoken words.

In summary, it is the configuration of the weights of the units forming clusters and the intensity and configuration of the trajectories and vectors that influence analog thought patterns in the creation of new thoughts and the modification of old thoughts originating within the system. We have seen that the mind/brain system doubles up in the use of its trajectories and the vector spaces of its neuro networks to organize speech by way of its basins of attraction. It also doubles up in the manner in which it creates thought patterns by organizing its thought patterns from these underlying speech basins of attraction. In this way we have the ability to learn from past experience. As we acquire new knowledge of how the conscious and subconscious mind operate to create the many analog thought patterns in life, and as we learn to effectively interpret these patterns, we increase our level of consciousness and expand our awareness. In doing so we are able to take effective control over our environment. Taking effective control over our environment also entails taking effective control over the time and space of that environment which is addressed in the next section.

Cognitive Architecture of the Mind In Time and Space

As we have seen the established application programs of the mind developed over hundred of thousands of years of enculturation are based on objective realities existing in the time and space of a fixed three dimensional objective world. The mind's subconscious application programs largely process logical rational thoughts and ideas within the program parameters of this objective world reality.[9] It is difficult for the logical mind to envision a world as being other than this objective three dimensional world which exists separate and apart from us. Because this representation of our world reality is such a very basic program of the mind it has become a very deeply

9 Obviously the mind also processes fantasy and other types of thoughts and beliefs outside physical reality but here we are only dealing with the creation of our outer physical realities in life.

entrenched individual and cultural program around which we unconsciously arrange all activities in our life. We have seen that it was formalized in the Cartesian way of thinking adopted by Western Civilization over three hundred years ago. Since that time it has never again been critically reevaluated as to whether it represents the mind's only or best way to view the coordinates of time and space of our world. This program establishes the time and space patterns for all of the other application programs of the mind. Accordingly, all of the other programs of the conscious and subconscious mind are contained, confined and restricted by this framework.

For instance, this program was in place at the time of birth and therefore defines the outer parameters for the development of our ego personality program, the personal and operational primary programs, and the secondary belief systems of the mind. To cause a quantum shift in the cognitive structure coordinating time and space would cause a quantum shift in all of these other programs and, consequently, a quantum shift in our experience of inner and outer realities of life. Such a shift in the perception of our world involves creative integrative thought which entails a shift to another level of organizing synthesis where self awareness perceives its world from an entirely new and different viewpoint in life. Seldom is logic able to bootstrap itself to such an extent that it is able to expand the parameters of its own subconscious time and space analog patterns. Conscious Analog Thinking™ and the MetaCreation™ processes assist you in making such a transition.

Therefore, logical rational thoughts rising within these patterns of time and space can not easily envision a world other than that which presently exists. Of all of the analog patterns of the subconscious mind the time and space analogs are the most well established patterns for they stabilize our sense of self existing in a fixed three dimensional world. Theoretically if you were to immediately remove these time and space analog patterns from the subconscious mind you would lose all sense of time and space. What happened yesterday and what happened last year would become so jumbled up that you would have difficulty separating the two. Certain physical impairments due to physical injury sometimes result in exactly this type of pattern loss where the individual loses any sense of whether a thing happened yesterday or five years ago. People coming out of a sustained deep coma often exhibit this type of time/space disorientation. Coma victims need to be retrained all over again in the time and space sequencing of their personal lives. All of us are familiar at one time or another with the phenomena of waking up early in the morning and momentarily not knowing what time it is or even what day it is or where we are. It is not until we reorient ourselves in a second or two as to the time and space analog frames of our subconscious mind that we recognize what time and day it is and where we are located physically.

Thoughts and feelings rising within our time/space patterns based on a fixed objective three dimensional world reality perpetuate such a world reality. What is meant by this is that our world reality is a three dimensional objective reality in fixed time and space, because we perceive it, act upon it, and bring it into cognitive realization, as being that way. The very basins of attraction of the brain and the analog patterns of the subconscious mind and the logic of conscious thought all are organized along these lines of a fixed exterior objective three dimensional world existing separate and apart from us and largely outside our control. Over hundreds of thousand of years we have enculturated ourselves, and everyone else for that matter, into believing that this is the proper and only way of viewing life in the context of our personal existence in our physical environment. *We created our own world reality in this manner by programming our brains and minds to conceive of it and bring it into conscious reality in exactly this way!*

Difficulties arise in any attempt to logically shift our existing time and space analog patterns because such shifts would also entail, not only shifts in the person's perception of how this external world is conceived and brought into reality, but shifts in the way the individual views himself and his relationships to that world reality. This then would entail shifts in his very thought processes themselves as they affect ego personality and not just shifts in the content of his thoughts in the way he perceives his world. The processes of human cognitive thought has not evolved to this level of organizational synthesis. At least not yet. But civilization is heading in this direction. Einstein through his formula $e=mc^2$ and relativity theory saw his world differently at the quantum level and at the universe level. Mankind has not fully exploited these initial findings of Einstein and expanded upon them to apply them to how the mind experiences its physical world. Consequently, logic can only arrange and rearrange information in the form of thoughts and beliefs into different belief patterns based on logical choices at the level of the cognitive structure at which we are presently organizing and experiencing life. The level at which we are presently experiencing life is an objective world of fixed time and space existing separate and apart from us.

Literature concerning applied mind science almost universally holds this viewpoint that we live in a fixed objective world that physically exists separate and apart from us and largely outside of our control. Further, it holds that we cannot change the past, only our experience of it and that we cannot mold the future simply by the processes of our mind alone. This position holds that the reason for this view of our world is that while we live out our life within the physical environment we exist separate and apart from that environment. Accordingly, the only way we can take control over this environment is largely through our physical efforts alone in taking control of it. Mankind believes that we do not control our environment through the intentionality of our thoughts alone since the energy of thought as presently organized on the subconscious level has no direct connection to this objective world reality in which we live. Therefore, the only really direct connection we have to this world is through our physical efforts in physically rearranging the landscape of that world.

Consequently today, some 80 years after Einstein first set forth his quantum and relativity theories, mankind still struggles with the concepts of spacetime of the fourth dimension and the idea that all matter exists as quantum energy organized into an endless variety of material representations. The fact of the matter is that our world wasn't initially created nor is it fundamentally constituted today as physical matter but rather the origin of its organization is represented as pure kinetic energy which is the basic building blocks of everything that exists throughout the universe. Physical matter, as we perceive it to be, at the quantum level of its organizational structure, is made up of pure energy and not quantum solid masses. This last point that all matter existing at the quantum level is energy and not quantum solid masses is extremely important in understanding how the mind creates realities and is overlooked by almost everyone in the field of applied mind science.

What is proposed is an entirely new way of viewing our world and our relationship to it. A view that recognizes that it is the energy and images of our thoughts which mold our outer physical world, where our outer physical world no longer exists outside of us and outside of our thought patterns. In fact, we have seen that the outer physical world is an inherent part of the analog processes of the subconscious mind itself in creating realities in life. It is the stimuli of our world through the cybernetic loops of our mind that provide us with the initial signals upon which we build our life and our world. Therefore, this world is as much a part of us as our knowledge and language expressing who and what we are in life, are a part of us. As we learn to take control of our world by redirecting our attention to how we create our realities manifesting as our world we are able to mold the world more to our liking. Exactly how that process works and how we can exercise control over that world through the intentionality of our thoughts impressed upon the analog system of our subconscious mind is the subject matter of the remaining chapters.

What we need to do is to change our thinking about how our world is conceived and brought into existence and our relationship to it. Once we understand and experience the effects of the cybernetic loops that format its existence in three dimensional frames we will be able to change our perceptions of our world. We will be able to reformat entirely new images in order to rearrange the signals of these cybernetic loops to create different and entirely new and unique patterns responses to life. In this way we are able to influence the component parts making up that life and change them in whatever fashion we desire for ourself. The result of all this is that we do not presently have very much control over our world because, out of ignorance, we did not realize that it is we who are influencing the type of world in which we live. Had we understood the dynamics of the mind a long time ago we would have taken far greater control of our world at a much earlier age. Had we done so we would not have endured the many disastrous effects of the current social conditions we are faced with today. The secret then is not to try to take control of your outer world reality through your physical labor alone but to also take control of it through your internal images of its physical expression in spacetime.

However in mentally conceiving a world and bringing it into manifestation keep in mind that we also follow physical laws governing the operation of physical things in our world. I do not intend to imply or suggest in any way that we actually move things around in our physical environment by some sort of mental telekinesis. Or that some other metaphysical or magical technique is operating outside the known physical laws governing the operation and existence of our physical world. Therefore, once you conceive your world reality and mentally create the realities you want for yourself you take appropriate action and do those things which support your physical creation of that reality. You do not just think about your reality and do nothing at all to support its materialization on the physical plane of existence. You might conclude from all this, "Well isn't that what we are doing, more or less, anyway, so what is the big difference"? The answer is yes, that is absolutely true. However the difference, and it is a big difference, is that we are doing it on a hit or miss basis. By not thoroughly understanding the dynamic processes of the mind in creating realities in life we are not maximizing our fullest potential. In not understanding the nature of creation, and the nature of time and space in the formation of realities in life, we do not have the knowledge nor methodology by which we can, in any systematic way, order our life and create more beneficial realities. Such constraints, and they are formidable ones, vastly limit our ability to create the type of live we desire for ourself and collectively limits the creation of the type of world realities we desire for all mankind.

Human Consciousness And Artificial Intelligence

In our review on consciousness in Chapter 4, two aspects of human consciousness were knowledge and intelligence. Knowledge is our ability to organize data by labeling everything so that we can make sense out of it. Intelligence is our ability to interpret as being either good or bad those things that we labeled. This ability of a parallel processing system to organize and then interpret the units' weights in order to obtain the most optimum response fits these basic definitions of knowledge acquisition and intelligent interpretation. The reason a sophisticated PDP system has not been able to achieve such a high level of organizing synthesis as the human brain and mind is that it lacks the other two essential elements of consciousness, namely, that of self awareness and volitional will.

If at some future date, the PDP models could be trained to make their own volitional decisions notwithstanding the internal values of their units, and were able to acknowledge and experience their own existence, they would then exhibit a type of machine volitional will and a type of machine self awareness. This would constitute a type of machine consciousness. Such a computer would then embody morphological programming capacity. The super computer "Hal" in the movie "A Space Odyssey, 2001" was just such a self aware super computer. Hal certainly seemed to be self aware for he recognized himself apart from the other occupants and apart from the spaceship itself. However, Hal's volitional will exhibited a really mean streak by locking Dave out of the spaceship and killing the other occupants who were being held in a state of suspended animation. It is obvious that the type of machine knowledge and machine intelligence exhibited by the current PDP models are primitive in the extreme when compared to the machine consciousness of Hal.

It is self awareness and volitional will then that sets mankind apart from the most sophisticated PDP system. The element of self awareness is essential to the evolutionary path of mankind leading to yet even higher states of consciousness and expanded levels of awareness. It is these characteristics of self awareness and volitional will that provides us with morphological capability in our operating programs of the brain and mind. Because mankind is self aware he has been able to get outside of himself and view things from a broader prospective in order to be able to process even more evolved thoughts and feelings about himself and his world. This organizational synthesizing ability is a characteristic of morphological programming. To date a parallel processing system has not been able to be programmed to synthesize machine intelligence to such a high level as human consciousness.

A machine lacking in self awareness has, to date, never be able to evolve its internal programs to the point that it is able to become both the "knower" and the "known". The machine lacks the capacity to know itself. It lacks a sense of a personal self which is needed to be able to take control of and change both the external input signals and the internal programming capacity. It needs to take control over the input signals and internal programs in order to take control of its environment as well as control over its own internal operating systems. The machine would have to be able to become both its own computer programmer and its own organizer of its external environment in order to be able to develop a process allowing it to evolve to a higher level of organizing synthesis, an organizing synthesis where it would be able to create and expand consciousness of self awareness of a type exhibited by humans. In the many discussions in journals and books on the processing capabilities of PDP systems there is very little theoretical discussion about the role of the programmer and the role of the environment in relationship to how such factors might affect the operating capacity of the machine itself.

PDP systems also lack volitional will even though the machine can exhibit a type of decision making capacity exhibited by volitional will. Regardless of how sophisticated its decision making capabilities are it is still governed by the rules of soft constraint. The machine doesn't have the internal capacity to volitionally override the weights of its internal units and suspend the rules of soft constraint and soft reinforcement. To date the machine has not been able to become the initiating internal programmer who creates and programs its systems nor has it been able to take control over its external environment in order to change the landscape of that environment.[10] Mankind through self awareness and volitional will has such capabilities to modify both the reality of its environment, the time and space of that environment, and the inner program parameters of its mind, which determine internally who and what he is in relationship to that environment.

Mankind and his Environment

While mankind has been able to take control over his environment, he has not been able to develop self awareness to the point that he is able to personally experience that, in fact, he is one with that environment! Mankind and his external environment are really different aspects of one and the same thing. Mankind cannot be considered as existing simply within the human mind and body separate and apart from his environment which nourishes and sustains the whole individual. For it is the world environment from which mankind is able to develop a sense of self in relationship to that environment. Implicit in the definition of self awareness is the fact of being aware in relation to someone or something else. That someone else is you and me and that something else is your personal environment. The mind of man is reflected in his environment and his environment is reflected in the mind of man. Self awareness of all mankind operating in human form has not yet fully grasped this essential fact. The primary purpose of the human mind/brain system, developed in the manner that it has, is to not only just survive but to grow and prosper within and as part of its world environment, the environment which sustains its life. Its life, in turn, sustains the processes of its mind which in turn should be able to sustain the environment which sustains its life. However, an awareness of the basic-to-life cybernetic loops between mankind and his environment has not yet been fully

10 Researchers in the field of artificial intelligence should take heart for consciousness is not something metaphysical or existing in some spiritual dimension outside the internal processes of the complexity of the dynamical self organizing systems that they are presently working on. The mechanism for achieving this recognition process is similar to the components of consciousness reflected in the process functions utilized in the formation of ego personality. The first process function of ego personality is "this is me". Accordingly, the initial programming of the analog system of the PDP needs to include the "this is me" program parameter. Theoretically this program parameter would consist of the energy and matrix configurations of clusters of units creating patterns recognitions of a localized sense of self embodied within the operational system of the PDP. Rudimentary machine consciousness would arise out of the energy configurations forming these analog patterns within the system itself. It would manifest as a result of the internal analogs patterns being able to recognize that it is the one organizing these patterns.

developed or realized in mankind's involvement with his environment. So long as mankind continues to perceive himself as separate and apart from his environment mankind will continue to suffer the consequences of his limited thoughts and actions in his futile efforts to take control and dominate that environment.

Because of a lack of such awareness mankind has not been able to learn how to fully cooperate with that world environment in a mutually beneficial way. Chart 30 at page 254 demonstrates how mankind affects, either positively or negatively, the world ecosystem and how the world ecosystem, in turn, affects the health and welfare of mankind. American Indians and the ancient Hawaiian Kahunas had a sense of this connection between themselves and their environment which they referred to as "Mother Earth". This concept of Mother Earth picks up on some of the aspects of what we are referring to when we say that mankind and the environment are two aspects of one and the same thing. Mankind and his environment are a part of a much greater reality in life. Consequently, when we destroy the earth we not only destroy our physical environment but we also destroy an inherent part of our very own internal mental processing system, a system which derives its existence via external sensory signals of how to best respond, cooperate with, and live in harmony, with its environment. An environment that mankind has been, both intentionally and unintentionally, destroying at an alarming rate over the last few decades.

The same holds true for the violence we do to one another. By introducing negativity into the internal analog processes we are unintentionally destroying the peace and harmony of our way of life and of our world. When I say that we destroy a part of our own processing system I mean that we are adversely affecting the weights and trajectories on the mind's analog system in creating positive realities in life. When promoters of violent movies and excessively explicit sexual scenes say that such material has little to no effect upon us they are clearly wrong. Such material is taken into the system as visual, auditory and kinesthetic signals and forms submodalities and modality patterns which are processed as negative influence in the weights of the system. When we see adults and children using high powered semi automatic firearms to kill one another we do violence to the internal operating analogs of the mind. Young children, teenagers, and young adults, are more susceptible to such negativity than older adults for the simple reason that the inner processing systems of such younger people are not yet as well defined or established concerning the content of their internal processing system. Society has not yet learned that all violence is a form of self violence, that all

CHART 30

The Pathology of the Human System and the World Ecosystem

Demonstrating the Interconnection and the Interrelationship of the Human System and the World Ecosystem

World Ecosystem

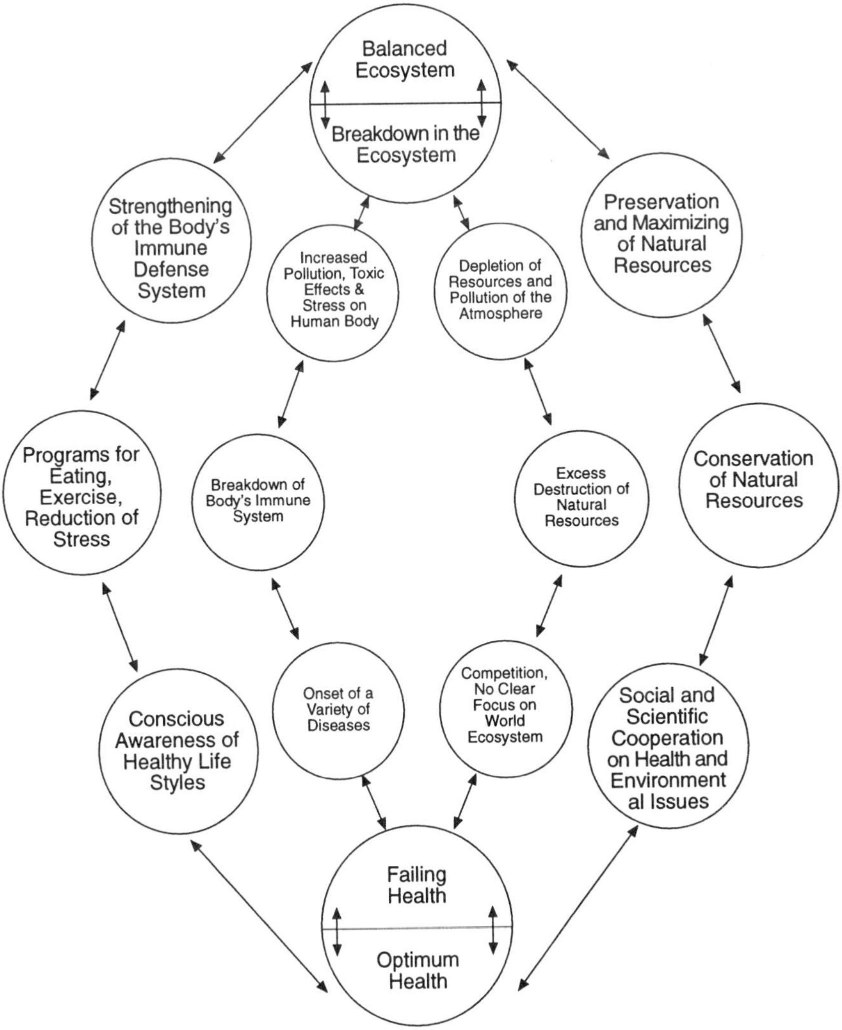

Interrelationship of the World and the Individual Self

abuse and neglect is a form of self abuse and neglect, and that all exploitation of others is a form of self exploitation. We need to become more aware and realize that by helping one we help all, by encouraging one we encourage all, and that by healing one we are beginning to heal the whole world.

Conclusions On Parallel Processing Systems

As we have seen the two characteristics of consciousness namely, self awareness and volitional will, more than anything else have become the driving force behind the evolution of all mankind. Whether you argue that consciousness is an inherent aspect of mind arising out of the operating processes of the mind itself, or whether you take the position that consciousness exists as something separate and apart from the organizing processes of the mind is totally immaterial. The fact of the matter is that mankind does possess these two characteristics of consciousness which a PDP and any other artificial processing system do not yet exhibit in their operating systems. It is for this reason, and this reason alone, that mankind is able to continue to evolve to higher levels of consciousness and expanded states of awareness along its evolutionary path toward its destiny in life.

CHAPTER 7

On the Origin and Nature of Thought

What Constitutes a Thought

What is a thought? In very simple terms a thought is a conscious expression of reality, more precisely a conscious expression of YOUR reality. A thought is an expression formed in linguistic, mathematical, musical, artistic or other symbolic forms which is consciously interpreted and understood by an individual. The process by which a thought is created is the process by which the mind organizes energy and images into patterns of reality. These patterns form meaningful distinctions, comparisons, relationships, associations, and identifications by the conscious mind in organizing a continual conscious stream of reality which constitutes your life.

We often think that thoughts we hold in conscious awareness somehow reflect or represent objective truths, or reflect or are supported by some kind of external physical reality. To the contrary, thoughts themselves really only represent the subjective processes of the mind. However, we come to believe that these subjective processes of the mind are real or are somehow intrinsically meaningful apart from the process. That is not to say that we do not draw from our experiences of our physical world in creating thought. However, as to the process itself, a thought arises from the subjective processes of the mind, which we then interpret as reflecting our objective world. We have seen that the mind/brain does not record experiences arising from the outside world as an exact representation of that world. Rather our subconscious mind relies on our past thoughts, beliefs, memories and feelings, as it interprets and processes incoming stimuli into thoughts and feelings.

Without knowing how the process works we are unaware that we objectify our thoughts and beliefs and then form the belief that they somehow represent objective physical reality residing in the landscape of our fixed outer world. Often we believe that they reflect some higher or greater truth in life. However, all beliefs reflect the outcome or results of the subjective mental processes of the mind. What a person subjectively believes to be true differs for that same person from time to time. What you hold to be true today might radically change tomorrow. What a person believes to be true differs from what other people hold to be true. Accordingly, our reality is different for each one of us depending upon what we each believe to be true or meaningful in our life at any particular moment in time.

The Creation of a Thought

The image of a thought is interpreted by the mind as an idea. An idea does not necessarily form a belief. We have many thoughts and ideas each day but we do not necessarily believe that all of them are true. We may believe that they are not true or we may have no opinion about them at all. How we form a belief about them is explained in just a moment. The energy of the thought provides the substance of the thought. The image of the thought provides the definition of the thought. The definition of the thought contains within itself the energy or substance of the thought. The definition of the thought has a defined shape or form which is referred to as the thought's "cognitive structure." The way we arrange thoughts in our mind is called "cognitive structuring" of the mind. Cognitive structuring of the mind is described in Chapter 9. How we might change the cognitive structure of our mind if we are not satisfied with our inner and outer realities is discussed in Chapters 9 and 10.

Notwithstanding the fact that thought creates our experiences of reality, it has its downside effects. All thought is by definition self limiting. If you have a thought that you are a millionaire, you may have just limited yourself from the possibility of attaining 10 million dollars, unless, of course, you change your thinking about owning millions of dollars. If you perceive yourself as owning a hundred million dollars, you again may have limited yourself from the possibility of owning a billion dollars. I think you get the point. Every thought, regardless of how uplifting and expansive it might be, is self limiting. The reason for this is that all thoughts exist in time and space once created and exist for a specific period of time which is limited in duration, and has distinct boundaries. Anything existing outside the thought exists outside the conscious awareness and volitional control of the person holding the thought at that moment in time.[1] What I am saying here is that you can only change things in life through the intentionality of your thoughts. As to the creation of personal realities in life nothing gets done without somebody first thinking about it and then puts it into action.

We create hundreds upon hundreds of thoughts each day. These thoughts when grouped together under common themes in our lives form our belief systems. It is these hundreds of different belief systems that we hold in conscious awareness each day that create for us a continuous stream of consciousness that we experi-

---

1 The flip side of the notion that all thoughts are self limiting is the thought "Whatever the mind of man can conceive and believe he can achieve." Quotation from Napoleon Hill's book, *Think and Grow Rich*. The reader might want to contemplate this quotation in light of what he learns concerning the nature of thought.

ence as our reality. Thoughts combine together to form the modular cognitive structure of our beliefs systems. Complex belief systems form complex cognitive structures within the mind. We have seen that thoughts which are in opposition to an existing cognitive structure are often rejected as being incompatible with the existing beliefs of ego personality and therefore rejected by the system. On the other hand, if a new thought appears to be useful to the ego self then the modular organization of the cognitive structure will be modified or expanded to accommodate the new thought. This is how we mentally grow and expand our awareness of what constitutes reality. A good example of how cognitive structures are modified and expanded was the widely held historical belief that the world was flat until Columbus sailed to America and discovered that the world was actually round.

Chart 31 at page 260, *The Anatomy of a Thought*, traces the directional energy flow in the life of a thought from its source to its terminal point. You will note from this chart that the organization of a thought starts with the incoming stimuli received by the brain. The brain's basins of attraction then organizes this stimuli into organized energy which the subconscious mind converts into mental imagery. The subconscious mind then sorts for a conscious expression of the thought in outer reality. It is these thoughts of our inner realities which prompts us to behave in certain ways and to take action which facilities the creation of our outer realities. Through the operation of cybernetic loops feeding stimuli from the environment back into the systems the mind modifies its thoughts, behavior and actions in order to make modifications in its outer realities. Hopefully as we become more aware of how this process works we will start to create more positive thoughts resulting in more enjoyable realities. Charts 32-1 and 32-2 at pages 261-262 depict the process the mind uses in initially creating realities for itself. Chart 32-1 shows that when thoughts arise within a closed ended or limited cognitive structure they result in the creation of limited realities with little or no change in our thinking and life styles and no expansion in self awareness. On the other hand, Chart 32-2 illustrates that an open ended fully actualized cognitive structure results in expanded realities and expansions in awareness.

The energy of the thought is what provides the thought with either positive, neutral or negative feelings in the form of emotions. The matrix configuration of the thought provides the thought with either positive, neutral or negative imagery. Positive thoughts composed of positive energy and positive imagery create positive inner and outer realities. On the other hand negative energy creates negative thoughts and negative emotions resulting in the creation of negative realities. If there is negativity in some of your originating primary programs on the subconscious level of the mind you will manifest this negativity in the form of thoughts arising from within these negative programs. If you have a positive image with a negative charge, or a negative image with a positive charge you have an ambiguous thought. If you are in conflict, and the energy and the images of your thoughts and beliefs are misaligned, you will lack congruency throughout your mind, body and affairs. This conflicted condition weakens your effectiveness in whatever you attempt to do in life.[2] For a list of negative and positive energy expressing human emotions refer to Chart 32-A, *Different Types of Energy Expressing Most Human Emotions*, at page 264.

Ambiguous thought patterns create a short circuit between the things you are doing presently and your goals and desires for your future. This short circuiting effect causes a drain on your mental, emotional and physical energy and results in tension, conflict, anxiety, pain and sickness. Until you get clear on what it is you want out of life and align the images and energy of your thoughts with the energy of your body and affairs, you will continually be working against yourself. Quite often the more you work to overcome the resistance the greater the resistance. A good example of this phenomena is when I ask you to not think of a bright red apple for the next minute, and you, of course, think of the apple. It's extremely hard not to do so. What you need to do is not resist the thought, temptation, event, circumstance, or whatever it is that is creating the tension and pain in your life but accept it as your present reality in life, because that is exactly what it is. Then discreate it by taking its energy down to zero and reframe it by creating a new image with positive energy more to your liking. How you accomplish this is explained in Chapter 10.

A good example of an ambiguous thought is when Ricky Mindful stated that, "he loves his mother dearly, shouldn't everyone love their parents?" Yet deep down inside he was extremely angry at his mother for a very neglectful childhood. When he visited her as a young man in college they always ended up arguing. He would leave thinking to himself "Why can't she just understand and become aware of how she is acting and the negative effect she has on other people?" Ricky didn't understand that it is he who needs to understand and become aware of his own ambiguous feelings toward her and toward himself. It is these subconscious ambiguous feelings that

2 The energy of of the subconscious mind giving rise to thoughts is described in Chart 18-1 at page 133.

CHART 31

The Anatomy of a Thought

The Energy of a Thought	The Image of a Thought
1. Sensory Stimuli	Unorganized Energy
2. Basins of Attractions	Organized Energy
3. K/V/A Energy	Sensory Submodalities
4. K/V/A Energy	Sensory Modalities
5. K/V/A Energy	Mental imagery
6. Feelings & Impressions	Meaningfully Interpreted Imagery
7. Feelings & Impressions	Experience of Inner Reality
8. Kinesthetic Energy	Actions and Behaviors
9. Kinesthetic Energy	Experience of Outer Reality
10. Feelings & Impressions	Modifications of Old Thoughts
11. Feelings & Impressions	Terminal Point of the Old Thought and Creation of New Thoughts
12. Feelings & Impressions	Creation of New Expanded Inner Realities
13. Feelings & Impressions	Creation of New Expanded Outer Realities
14. Kinesthetic Energy	Moving toward higher consciousness and expanded awareness

---

Tracing the Directional Flow in the Life of a Thought from its Source to its Terminal Point

CHART 30-1

Closed Ended Self Limiting Cognitive Structure Resulting in Limited Creations

The following chart represents the chronological steps in the process of creating limited inner and outer realities

Ego Personality

The Mind

Limited Time/Space Cognitive Structure	Limited Time/Space Cognitive Structure
↓	↓
Limited Primary Personality Program	Limited Primary Operational Program
↓	↓
Limited Secondary Personality Belief System	Limited Secondary Operational Belief System

Focus & Attention
(Formation of the Thought)

↓

The Energy of Limited Thought
(Negative Feelings, Emotions, Imagination)

↓

Structure of the Limited Thought
(Negative Imagery, Focus, Direction, Formation)

↓

Limited Behavior & Action to Facilitate the Thought's Creation

↓

Creation of Limited Realities
(Formation of Negative Inner & Outer Realities)

↓

Limited Sensory & Mental Feedback
(Defensive or status quo Monitoring and Processing)

↓

Little Adjustment in Thought
(Limited to no Adjustment in Belief Systems)

↓

Little to No Adjustment in Inner and Outer Realities

↓

Little or No Potential for Growth in Personal Realities

↓

Little or No Potential for Changes in Thinking, Life Styles and Behavior

↓

No Changes in the Basic Cognitive Structures

↓

Little Change in Thinking or Life Styles

↓

No Expansion in Awareness or in Consciousness

CHART 32-2

Open Ended Fully Actualized Cognitive Structure Resulting in Expanded Creations

The following chart represents the chronological steps in the process of creating expanded inner and outer realities

Ego Personality

The Mind

Expanded Time/Space Cognitive Structure	Expanded Time/Space Cognitive Structure
↓	↓
Expanded Primary Personality Program	Expanded Primary Operational Program
↓	↓
Expanded Secondary Personality Belief System	Expanded Secondary Operational Belief System

Focus & Attention
(Formation of a Thought)

↓

The Energy of Expanded Thought
(Positive Feelings, Emotions, Imagination)

↓

Structure of the Expanded Thought
(Positive Imagery, Focus, Direction, Formation)

↓

Positive Behavior & Action to Facilitate the Thought's Creation

↓

Creation of Expanded Realities
(Formation of Positive Inner & Outer Realities)

↓

Positive Sensory & Mental Feedback
(Positive Mental and Emotional Monitoring and Processing)

↓

Positive Adjustments in Thought
(Positive Adjustments in Secondary Belief Systems)

↓

Expanded Adjustments in Inner and Outer Realities

↓

Creation of Expanded Realities

↓

Expanded Positive Changes in Thinking, Life Styles and Behavior

↓

Expansions in both Primary and Secondary Cognitive Structures

↓

Further Expansions in Awareness and Higher States of Consciousness are affecting his behavior in his own life and in his relationship with his mother and later on with Betty. An ambiguous thought requires resolution of the underlying dynamics of its deep seated feelings. Ricky needs to reinterpret or clarify his thoughts in a positive fashion so that they are more supportive in life. The processes for dealing with the energy and images of thoughts themselves are explored later on in this Chapter.

A Simple Belief Classification System

All beliefs, I don't care what the belief is about, can be classified as falling into one of four specific categories. These four categories of beliefs are as follows:

1. Emotional Beliefs
2. Truth Embodiment Beliefs
3. Utility Beliefs
4. Ritual Beliefs based on traditions and habits

Emotional Beliefs consist of those beliefs which primarily stem from our feelings and emotions instead of from logic or reason. For instance, Ricky's resistance to making a commitment to Betty, even though he loves her very much, was based primarily on his emotional feelings towards his mother's abandonment and betrayal which carried over to his relations with all women in general. Also they were based on his feelings of being threatened whenever the relationship placed him in a position of becoming emotionally intimate with the one he loved. Like Ricky, most people who make emotional decisions usually do not know the real reasons behind their decisions and therefore tend to rationalize their decisions. In Ricky's case, he concluded that all women were untrustworthy and why make a commitment which might entail a financial responsibility towards Louanne's college education. Emotional decisions may not always be in our best interest. This is not to imply that all emotional decisions are necessarily a bad thing. When one acts out of love even when the response places additional burdens or hardships on a person this decision quite often calls forth our highest good toward our fellowmen. Charitable works and acts of unconditional kindness fall into this category.

Truth embodiment beliefs are based on pure logic or reasoned thought. This is the type of thought processing we normally use on the job and in most other daily activities. When one becomes obsessively intellectual in everything that he does and seldom makes a positive emotional decision he become emotionally stunted and this also is not a good orientation in life. You need to live a balanced

CHART 32-A

Different Types of Energy Expressing Most Human Emotions

Negative Energy	Mixed Dynamic Energy	Neutral/Static Energy	Positive Energy
Hatred	Ambivalence	Indifference	Love
Hopelessness	Conflicted	Apathy	Confidence
Greed	Exploitive	Selfish	Generosity
Fear	Frightened	Cowardice	Courage
Lust	Sensual	Non arousal	Pure Love
Dependent	Needy	Standoffish	Independent
Guilty	Suspicious	Defensive	Forgiveness
Rage	Anger	Irritable	Peaceful
Worthless	Incompetent	Hesitant	Confident
Distrustful	Doubting	Suspicious	Trustful
Undependable	Unreliable	Equivocating	Dependable
Prejudice	Intolerant	Indifferent	Accepting
Selfish	Self Centered	Nonattentive	Caring
Lethargic	Fatigued	Listless	Energetic
Closed	Defensive	Suspicious	Open
Depressed	Giving Up	Defeated	Hopeful
Upset	Frustrated	Worried	Calm
Inflexible	Adamant	Resistant	Flexible
Defeated	Quitting	Giving up	Successful
Unaware	Disconnected	Empty	Insightful
Overwhelmed	Anxious	Uneasy	In Control
Tense	Up Tight	Nervous	Relaxed
Lonely	Separated	Standoffish	Connected
Empty	Searching	Loss	Fulfilled
Lost	Aimlessness	No Direction	Purposeful
Insensitive	Ignoring	Numb	Sympathetic
Untruthful	Lying	Equivocating	Truthful
Sad	Depressed	Lonely	Happy
Rejecting	Critical	Skeptical	Accepting
Hurt	Disapproving	Painful	Forgiving

life where all types of beliefs take place in appropriate circumstances. This is what enriches life and makes life more enjoyable.

Utility beliefs refer to acts and responses we make in life solely because they are useful to us. They serve a useful purpose. We do not have to have a reason or logically think about them in order to make a reasoned judgment about whether we accept them as beneficial to us. We just do certain things simply because they are useful to us in some fashion. An example of this type of response is that we seem to know where we left our car keys, glasses, billfold or purse. We usually don't leave home without them as they say in the credit card ads. These simple acts serve a very important purpose in making things easier when moving through the obstacles of life. Yet we hardly think about them. They almost become automatic reflexes arising from the subconscious level of the mind.

Ritual beliefs may need a bit of explanation. Ritual beliefs arise from our enculturation, traditions and habits in life. By way of an example is the daughter who asks her mom why she cut about 4 to 5 inches off the shank of the ham before roasting it. The mother responded that this was the way they always roasted the ham in her family, that she learned it from her mother. The little girl was still curious and called Grandmother to ask why she cut off the ham shank before roasting the ham. The grandmother said that she cut off the ham shank because her roasting pan was too small. Beliefs based on enculturation, traditions and habits, while often initially serving a useful purpose, are the most resistant to change when the original purpose no longer exists. These habitual types of beliefs quite often persist even when we have overwhelming evidence that such beliefs are no longer serving us effectively. As example, automobile manufactures still place cigarette lighters and ashtrays in cars even though cigarette smoking is harmful to your health. Some dentists still give kids lollipops after their dental appointment. Men feel that they are not properly dressed for business unless they wear a suit and necktie.

The point of this is that if you try to understand a person's beliefs or behavior in terms of logic or reasoning alone you may never understand the reason or motive behind his belief. Therefore, you may never be able to assist the person in changing negative beliefs. Parents all too often try to reason with a child when the child's actions or behavior stem from an emotional response instead of a logical response. For instance, a child may resist going to bed at night because there might be a boogey man under his bed. If you first allay the child fears about the boogey man you can more easily reason with the child about going to bed on time. Remember, the subconscious mind will process a nonlogical emotional thought or a nonlogical ritual, traditional or habitual thought just as easily as a logical rational thought. In this context you could say that all thoughts are rational but not necessarily logical or even sensible.[3]

What Constitutes a Belief

How do we come to believe in what we believe to be true? What is the process of believing? Believing in something or someone is a rather simple process of the conscious mind. Believing merely involves consciously accepting the validity of a thought as being true for you personally, regardless of whether other people believe it to be true, or whether it is held to be true by all mankind. However, believing is a much more complex process when the component parts of the actual process creating a thought are analyzed for the purpose of understanding how the mental process actually works. As we have seen, beliefs must conform to the underlying cognitive structure of the mind which gives rise to the belief. What makes an unusual thought so difficult to accept at times is that we have all objectified the reality of our world around us and we are influenced by the truth held by the majority of mankind. We have unconsciously allowed our perception of things existing in what we believe to be an objective world, and the beliefs of others whom we hold in authority, admire or respect to determine who and what we are and what we should believe. We do so unconsciously without regard of whether such objective aspects of life or authority figures have our best interest at heart. In a word, you should think independently for yourself and not rely on others or what you think your world would want you to believe. To start thinking for yourself more effectively you need to understand precisely how the processes of thought operate to create realities in your life. Exercising the thought process of the mind just to create random aimless thoughts, fantasies or wishful thinking with no purpose or direction in mind quite often creates meaningless activities that just clutter up your life and waste valuable time.

The Creation of Belief Systems

Thoughts are often organized into common categories which form belief systems. Our belief systems assist us in making generalizations about the affairs of life. Belief systems can organize around sweeping generalizations involving various broad aspects of life such as religion, politics, racial distinctions, work ethic, differences between men and women, social affairs, business matters, family life, and leisure pursuits, to name just a few. Belief systems often involve

3 It is for these reasons that computer programmers have such a difficult time programming computers to process human thought. Human thought is not always logical, but it is always rational in some context.

elaborate customs, rituals, and traditions which make the belief system even more real to the believer or participant in the family, social, or business unit holding common beliefs. Belief systems are helpful in maintaining generalizations about life which do not have to be thought out individually, again and again, one thought at a one time. However belief systems can also be detrimental when we accept an entire belief system as a package deal. For example, when we join a particular religion, political party, social, ethnic, or cultural group you are often expected to accept as true all of the basic tenants of the particular group. Often we then believe in things and people without having a really good idea of why we believe the way we do.

What Constitutes a Useful or Beneficial Belief

What then constitutes a useful or beneficial belief? There are no absolute or definitive beliefs in life. Unless, of course, you believe there are. Nor are there any strict set of guidelines by which one can easily determine whether or not a belief is valid, positive or beneficial. Such determinations are strictly personal ones. Each person must determine for himself what is, and what is not, personally beneficial to him and not necessarily what is beneficial to the world at large. Such matters then become relative to each individual. This does not necessarily mean that all belief systems are relative to each other without some underlying basis to measure their validity. However, any such underlying basis itself is subjective and ultimately is determined by each individual. With these cautionary comments the guidelines to measure the validity of a belief are as follows:

1. Is the belief helpful to the individual? Does it have utility for the believer and/or others in his world? Does it serve a useful purpose for the believer and those around him?

2. Does the belief bring the believer a degree of happiness and fulfillment? If it is useful but results in creating painful circumstances in life then it isn't of much use to the believer regardless of how much utility it may have for him or others.

3. Even if the belief brings a degree of satisfaction to the believer and is useful to him in some fashion, if it is harmful or detrimental to others it is not a very valid or useful belief to either the believer or others in his life.

4. Finally, does the belief allow for the further expansion

of awareness, or at least not impede or prevent the growth in awareness? If it is both useful and the believer derives a degree of happiness or pleasure from it but it is part of a closed ended belief system which does not allow for personal growth in awareness then the belief is self limiting and not of much value to either the believer or others in his life.

These guidelines are not intended to be all inclusive or definitive in determining the validity or usefulness of any belief. They are only offered as guidelines and you may have other guidelines which better serve your purpose in measuring whether a particular belief is beneficial to you and, therefore, you consider it to be a valid belief in your life.

The Nature of Belief Systems

The one thing I would like to ask you to consider at this point is whether your basic beliefs about your own life allow for further expansions of awareness and for taking greater responsibility over all aspects of your life? The answer to this question would, on the surface, seems to be rather obvious. However, often we have become far too rigid in our thinking as to the nature of ourself and our realities. Often it is difficult to change our way of thinking in order to accommodate a broader perspective on life. What is needed is a perspective which allows us to entertain the thought that we hold far greater power over the realities of our life than we ever imagined possible. I would challenge you to look within and determine if your belief systems involve a considerable amount of rigid, fixed or dogmatic thoughts and ideas that result in narrow and fixed attitudes and opinions about yourself and others and do not leave much room for personal growth.

Again, your first reaction may be that this is a rather obvious statement. However, on a much closer examination you might find that some of your beliefs pose the potential for much greater obstacles to creating realities than you ever imagined. The reason for this is that our thoughts and beliefs, once accepted by the internal reality creating patterns of the mind, often become transparent or invisible to our conscious way of thinking. Once they are accepted and down loaded into our primary systems we commence to see right though them. However, they continue to color our view of ourselves, others and our world. Often the problem with belief systems is that they become fixed, then externalized, and finally they take on a life all their own. Their energy and matrix configurations, over time, become embedded in our subconscious analog system for processing reality and remain there unchallenged throughout our life. Once they take on a life of their own they become objectified and appear to exist independent of, and outside our mind, the creator of the belief in the first place! Once it is objectified it is viewed as another bit piece of your reality, joining thousands upon thousands of other bit pieces all of which make up the very fabric of the patterns of reality you experience each and every day.

You can begin to see why it sometimes becomes very hard to change ideas which have become externalized and objectified. They no longer exist as ideas subject to conscious scrutiny but are now a part of your inner and outer reality! Their truth or validity becomes much harder to question for they are no longer subject to examination of whether or not they are truthful or useful. Their deep seated configurations in the subconscious mind feed and reinforce their believability at the level of your inner and outer realities. They not only embody truth, they become truth itself. They now exist as truth embodied and embedded in the reality of your life. They constitute indisputable truths!

Examples of these types of ideas that have become objectified reality bits and have become indisputable truths, not subject to further examination and change, include the following: time is fixed and measured in clock and calender units only; space is fixed and measured by the physical distance between two points; the physical world lies outside and exists apart and independent of each one of us; the nature of all physical objects in our world environment is that they are solid and exist within fixed three dimensional space; a person cannot change his outer physical reality through the thought processes of his mind alone; mechanical physical laws governing the operation of our world prove scientifically the objective nature of physical reality; you can never change your personal past, you can only change your experience of it; an individual consists of his mind, which is made up of his conscious and subconscious thoughts and emotions, his brain, body, ego personality and status and possessions in life; life is primarily about maximizing pleasure and minimizing pain; suffering and pain are inherent conditions of human life; the human condition is to age, grow old, suffer and die; mankind will never be able to determine the true nature and origin of its own existence nor the origin of the universe; there is no way you can understand in any comprehensive way the enormous complexity of the human mind and brain; consciousness is a spiritual thing beyond the comprehension and control of mankind; the consciousness of each individual exists separate and apart from the consciousness of all other individuals; evolution of mankind is defined by the survival of the fittest; there is no way we can take control and direct the future evolution of all mankind; every major discovery that could ever exist has already been made; society's negative images of sexual aggression, physical violence and death have very little, to no effect, on our mental health; mental health is defined as being able to fully function in a complex society; both the capacity for evil as well as good are inherent conditions of human nature; poverty and disease are inherent conditions of human nature; mankind has an inherent right to dominate all other living species; humans have an inherent right to exploit earth's natural resources; material wealth, as well as all natural resources, are both finite and limited; the Cartesian mechanical view of the world is as true today as it was over three hundred years ago; Quantum Physics only applies to the micro and macro level of existence but does not apply to the zone of the middle dimension in life.

You can add to this list your own objectified ideas, beliefs and indisputable truths which have become part of the reality of society. We do not intend to imply that these objectified truth are completely incorrect, that they do not reflect some degree of reasonableness. They are listed, however, to illustrate that when a belief becomes objectified it is seldom, if ever, looked at critically. It therefore becomes resistant to change. This is not to say that all objectified ideas are wrong or bad, because they are not. Some serve a useful purpose in life. However, the point is that objectified ideas are seldom subject to further investigation, and consequently, they often act as road blocks to the further evolution of the accumulation of knowledge and the collective growth and expansion in human consciousness.

Objectified Beliefs of Ego Personality

If an objectified belief happens to reflect a part of your operational program or a facet of your personality, it would be termed "characterlogical" in nature. Often the content of beliefs of personality and operational systems are indeed objectified. In the case of Ricky Mindful he had an objectified belief in his primary personality program that because of his mother's neglect he was not acceptable as a human being and that he had to continue to prove himself. The fact that he was always in and out of relationships until he met Betty and the fact that he had to keep competing against everyone else in the sales division seemed to prove this out. Ricky was totally unaware of the effects of the cybernetic loops operating in his life which continually reinforced his low self image and his difficulty relating to women. He perpetuated this often self destructive behavior because he never was able to accept and believe in himself.

Remember, your personality constitutes the self identification part of you. It is what identifies your personhood within the existence of your reality. Anything that would change, alter, or eliminate such an objectified personality idea is often perceived as a personal attack on your character. Such changes then are very difficult to achieve through therapeutic, motivational or self help processes. People have fought wars and killed in a desperate attempt to preserve the status quo of the objectified beliefs making up their lives. The fact of the matter is that in their reality that is exactly what is actually going on. They do not perceive it as a war over thoughts and belief systems but rather a war over a way of life, of personal existence, of the very survival of themselves and their family or the particular group or race with which they identify. The perceived and felt experience is that they are being attacked, and if the attack is successful, a part or all of of them will die in the process. It's like the doctor saying that the operation was a success; however, the patient died in the process.

A further understanding of the basic personality program to which these types of characterlogical ideas attach themselves is explored in greater depth in the next chapter. Suffice it to say that an individual needs to understand that these objectified ideas may not be serving them very well and that the individual's essence is not bound up in the objectified idea of who and what he is. Rather his very essence involves the sense of self which is referred to as self awareness. Self awareness, without identification with its realities, as indicated earlier, is characterized by the metaphor timeless, perfection, completeness and wholeness. At the point that an individual comes to realize that he exists as self awareness, apart from the objectified idea of his personality, the discreation clearing process can be accelerated, and effectively applied, in discharging the energy and image of negative objectified personality traits. Once the objectified idea is removed from the primary personality program, then an individual can create an entirely new self empowering idea with self empowering energy which can be reimaged and accepted as the new personality program. Examples of good beliefs are found in the content of an optimal personality program versus the contents of a limited personality program and are described in detail in Chapter Eight.

Recognize that all along I have been saying we are dealing with processes and not with content. The content of your life is left solely up to you to define for yourself. However the content of life is dealt with almost exclusively in the comparison of optimal personality and optimal operational programs versus limited personality and limited operational programs described in Chapter 8. However, even in this case advice is not given about what the content of your life should be or look like. Only comparisons are made of what might best serve you and what acts as a hindrance in life. In this respect content is used simply as tools for change, and this would be an example where content itself becomes part of the process. Ultimately it is for you to decide what is in your best interest in creating the realities in life as you explore the content of your life in Chapter 8.

How Do Belief Systems Create Reality

We have been exploring how your thoughts and beliefs about yourself and your life create your reality. What then constitutes your realities in life? With respect to the process of creating realities in life we have seen that it is the thought that is the key ingredient and not the actual reality itself. Thought itself creates the experiences of reality and not the actual physical realities we encounter each day. When viewing life from the perspective of a physical body living in a three dimensional world you would logically conclude that it is the actual physical realities we encounter each day that give us our experiences of that reality. This observation is not only not true but is the very opposite of what is actually taking place in our life, and yet we are not even aware of it at all. It is the energy and images of our thoughts that provide us with our experiences of reality and not the composition of the outer physical realities themselves. The outer physical realities only provide ego personality with the sensory stimuli needed by the mind in order to form thoughts. Thoughts themselves are composed by the mind as a result of the stimuli it receives, its past memories concerning the subject of the thought, its present experiences, as well as its present and future intentions as to the purpose of its life.

The purpose of all thought is to serve the ego self by creating realities for itself. Thoughts are what provide us with impressions of what we think we actually see, feel, think, and perceive as our outer physical reality. Consequently, it is not physical reality itself that makes us feel, see, and think. It is our thoughts and beliefs which provide us with direction in life as well as the content of life. The components making up actual physical reality neither provide us with direction nor with the content of life. Yet we erroneously think and act as though they do each and every day of our lives. Once a person clearly understands how realities are created he is in a far better position to redirect and reinterpret the sensory stimuli he is receiving in order to rearrange his thoughts, particularly his negative thoughts into positive, more uplifting, more forward looking thoughts. As you start making these distinctions you will start to see real possibilities open up in your outer physical world. For if you do not recognize this relationship between your thoughts and your realities ultimately it will foster pain, lack and limitation in life. For this is the price you pay for failing to understand these fundamental principles governing the creation of all realities in life.

Charts 18-1, 18-2 and 18-3 at pages 134-136, specifically illustrate how the mental energy patterns of the mind create our realities in life. These series of charts are all closely related to one another in precisely tracking the process by which the subconscious mind organizes stimuli into patterns of inner and outer reality in life. Of all of the charts set forth in this book this series of charts is perhaps one of the more important ones for it shows you precisely how it is that the mind creates our life realities. Chart 18-1 illustrates how changes in the mental energy patterns of the mind changes the images in the mental field of the mind. It sets forth how the processes of the mind transforms K, V, A energy into images of the inner mental field of the subconscious mind. Chart 18-2 continues to track this process by illustrating exactly how the subconscious mind organizes this mental imagery arising in the inner mental field into patterns of conscious and subconscious inner realities. It illustrates the involvement of our conscious and subconscious mind in the process of interpreting the energy of our conscious thoughts into patterns of our inner reality. Chart 18-3 continues this process by illustrating how the unlimited capacity of the subconscious mind is then able to organize the mental imagery of the mind's inner realities in order to give expression to almost endless patterns of outer physical reality in life. We have indicated that it is in this manner that you manifest your reality and that your reality reflects you.

Chart 28 at page 187 is, in a way, a summation of charts 18-1 to 18-3. The purpose of Chart 28 was to illustrate that by changing any one of the possible thousands of elements influencing the creation of the mental and emotional imagery making up your inner and outer realities in life, you cause changes in your mental or emotional imagery, which in turn causes changes in your inner or outer reality in life. Consequently, Chart 28 dramatically illustrates how important it is to take control over the creation of your mental and emotional imagery so that you will be able to take control over your life's realities and create a better life for yourself.

The Intentionality of Thought

Throughout this book we have been referring to the intentionality of the thought that creates our realities. But just what is intentionality? Philosophy holds that intentionality is a mental state having or containing content or meaning, as opposed to the more traditional meaning that you intend to do something.[4] Connectionism modeling of the human mind has helped to explain how intentionality of thought acts in relating external objects in our world to their appropriate inner representational states. Intentionality of the thought consciously focuses the energy of the thought and directs its energy to form subconscious analog patterns that give rise to the subconscious imagery that constitutes our inner realities. What this means is that through the intention of your thoughts you are able to create inner representations of outer reality. This process of creating inner representations of outer reality is a continuous one going on all the time. It is what orients us in our world. In this way an individual is able to effectively evaluate and change when necessary his relationships to his environment. This is the way the individual adapts to his environment. *However, through the intentionality of our thoughts the self organizing system also actually creates that environment as well as creates adaptions to that environment.*

That is, the mind not only relates to its environment but also creates that environment through the intentionality of its thoughts. While this process of continuous interaction by the sensory receptors with its environment actually match up the inner representations of the mind with its environment, the mind is also simultaneously creating internal images of what that world should look like in order to create the landscape features of that world. It does so, as we have seen, by the process of the subconscious mind sorting through infinite possibilities in order to select the most appropriate people, places and things that bring the appropriate analog patterns to conscious awareness. It does so that we may take action to bring about outer realities in life. There are many constraints on this creation process which are addressed in Chapter 9. Suffice it for now to understand the theory behind creating reality through the intentionality of your thoughts. This is not some metaphysical thing based on some quantum flux in our energy field, or some other thing outside of us, creating our realities for us, but simply the intentionality of our thoughts. Intentionality, if used properly, activates the subconscious mind to bring to conscious awareness those people, places and things to match up your life with the desired exterior features of the landscape of your world. How you accomplish this is the subject matter of Chapter 10

One further comment on the intentionality of thought. The traditional definition of the phrase "intentionality of thought" refers to someone or something which is the object of the thought. The referential object of the thought could be inferred, implied or implicit in

4 Bechtel & Abrahamsen op. cit., p.123.

nature or it could be general or specific in nature. However, you cannot have a thought without a referential object of some type or other. Notwithstanding this traditional definition of intentionality the subconscious mind does not require an object for intentionality. Conscious mind may require an object for intentionality for the reason that the conscious mind, for the most part, operates on general rules of logic. Without a referential object a thought just wouldn't make much sense and therefore would not be entertained by the cognitive structure of the conscious mind. The subconscious mind operates, as we have seen, on the basis of the values of its units forming representational patterns in analog fashion and not by referring to objects rising from our conscious analytical thought processes. The subconscious mind does not require logic in order to execute a "nonlogical" instruction.

For this reason we can make use of an abstract power symbol without intentionally referring to a specific object in order to change negative states to positive states within the subconscious mind. The creation of the subconscious power symbol doesn't require an object in the sense of the object of intentionality of conscious thought since we are only dealing with negative or positive states of the subconscious mind. Conscious analog thinking, in this respect, also doesn't require intentionality of the object for the reason that conscious analog thinking doesn't always require conscious logical thought. Conscious analog thinking works more on intuition and holistic relationships appearing as inner and outer reality. There isn't much in the way of logic in the idea that by simply programming the subconscious mind to create for conscious self a personal power symbol you can eliminate negative energy in the subconscious analog system.[5] The use of abstract power symbols, however, are an excellent method for the subconscious mind to clear negative energy and images in the subconscious primary programs simply because the subconscious mind responds well to analog symbols including our abstract analog power symbols.

How Belief Itself is the Primary Factor in the Successful Use of the Creation Processes

Conscious mind usually will only entertain a logical or rational thought. For this reason it is often quite difficult to accept the concept that the subconscious mind can and will create outer physical realities in accordance with the intentionality of your thoughts. In order for the process to work effectively in your life you need to believe and have faith that the process will work effectively in creating the realities you seek in life. The problem an individual encount-

5. I recognize that once you understand how the system works it makes sense to make use of a personal power symbol in order to clear negative energy and in this respect the process is a logical one.

ers in working the creation processes is that the process at first does not seem logical and therefore not believable. If you can overcome this threshold difficulty in working the process you are then faced with the challenge of overcoming any primary programs which impede the creation process from working effectively in your life. Such a thought might be that in order to obtain greater prosperity in your life you need to work really hard for it, or worse yet, you are not really entitled or deserving of it. Both primary programs need to be destabilized and dissolved before you can effectively work the process of creating a better life for yourself.

Finally, once you overcome the proceeding two problems you are still faced with yet a third problem, which is trusting and having faith in the process sufficient for it to work within the time and space of your physical world. While it is true that the subconscious mind works unrestricted by time and space you still live in time and space and, therefore, for the object to manifest in your life it needs sufficient time and space in which to manifest. The reason for this is that while the mental process of creating works outside time and space the physical manifestation itself appears within time and space. Trust and faith in the process working in your life are essential in order to allow for the lapse of time from the formation of the thought until it manifests as the physical reality of your life. Chart 32-B at page 278 sets forth the levels a person normally follows when increasing one's belief in the process sufficient for the process to work in creating personal realities in life.

Factors which Promote Your Positive Secondary Belief Systems

Believing is what makes it all happen. You first have an image of what it is you want. You experience its energy as positive feelings of what it feels like to have or achieve the event or thing comprising the thought. You then must believe in its reality in order to bring it into physical manifestation. Belief is the catalyst which turns the thought into reality. Belief implies faith in both yourself as an empowered and enabled person and your ability to use the creative process itself in order to create your own personal realities. Self empowerment is addressed in the discussion on self identity programing in Chapter 8, and how to create your realities in the discussion of the creation process in Chapter 10. Here we are merely addressing the element of belief which is an essential component in the entire process of empowerment and creation. Since belief also implies trust and confidence in yourself and in the process, belief first starts with yourself. Are you confident within yourself?[6] Do you trust yourself enough to act on your beliefs? If you do not, then you need to turn within to find

6. The next chapter deals with the archetype prime directives in life which deal with self confidence among other things.

what it is that is blocking you from believing in your own power and the power of your creations, and use the discreation process to remove those blocks preventing you from realizing your dreams.The following list summarizes some of the factors that strengthen your belief:

1. Knowledge. Knowledge facilitates believing in a thing for the reason that if you have a road map of how to get there at least you now know what to do to achieve the desired results.

2. Logical approach. Logical approach supports believing for if it didn't make any sense there would be no reason to even believe in the thing in the first place. This is another way of expressing the concept of truth embodiment.

3. Acceptance. Acceptance of something whether based on logic, emotion, utility or ritual thought starts the subconscious process in creating realities.

4. Faith. Faith in yourself and your ability to create your future realities.

5. Trust. Trust in the process in order to allow sufficient time for it to manifest realities in your life.

6. Confidence. Confidence in one's own abilities to accomplish the intended task at hand.

7. Experience. Experiencing results reinforces the methods used to create greater realities in life. Once you start to experience the results you feel encouraged to continue the process.

8. Proof. Scientific or other objective evidence facilities belief since if you have empirical proof supported by cognitive reasoning as well as feeling or intuition to support your beliefs they will be stronger thereby facilitating your creations

9. Value conformation. If your beliefs are congruent with your core values in life it is less likely that you will encounter mental or emotional blocks to your creations.

10. Certainty. Certainty or conviction in one's beliefs facilities and reinforces the creation process.

11. Imagination. Imagination allows for the fleshing out of the belief to make it more real. It is what adds vividness and felt presence to your creative thoughts. Imagination involves visualization using K, V, A submodalities to make the inner reality more "real". This in turn facilities the creation of your outer realities in life.

The foregoing factors help reinforce your belief in yourself and in the processes of creation. This is not intended as an exhaustive list and you may certainly want to add your own factors which per-

CHART 32-B

Levels of Awareness and Belief in the Process the Mind Uses to Create Realities in Life

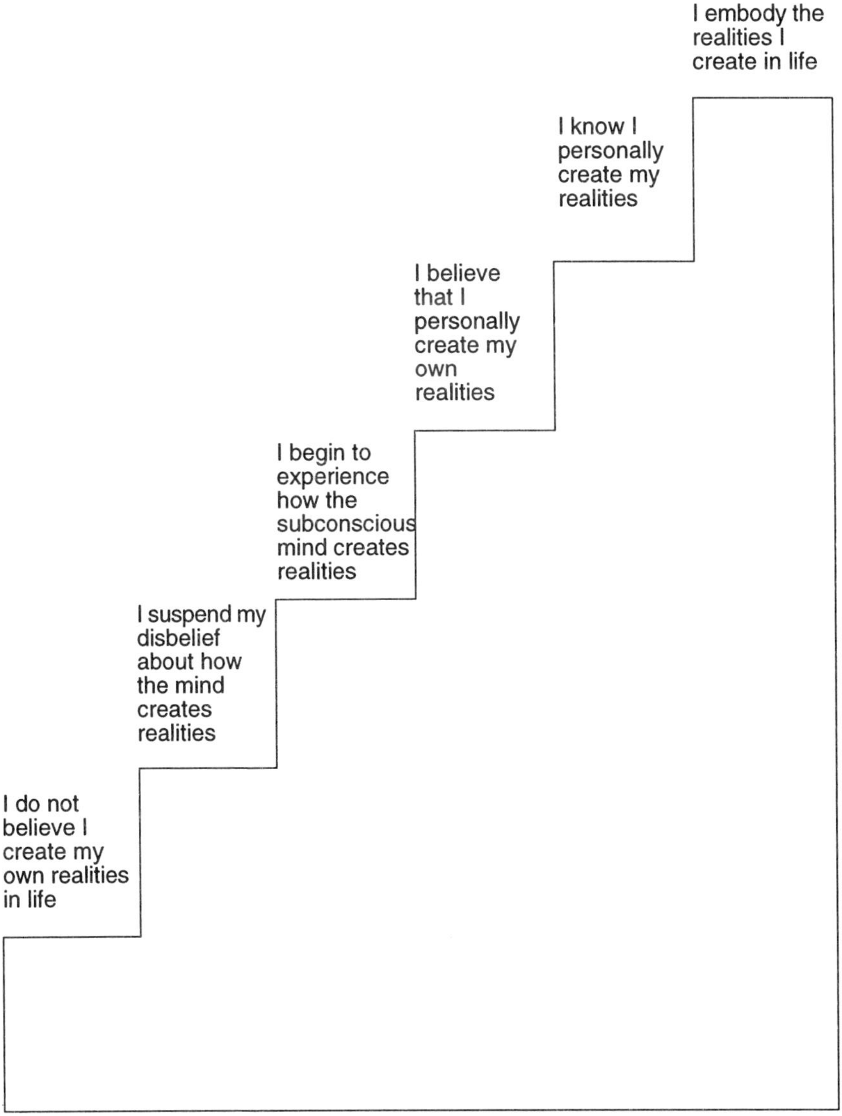

As a person raises his level of awareness of how the subconscious mind creates personal realities in life, he is able to strengthen his belief in the MetaCreation process. In this manner he is able to take greater control over the creation of the inner and outer realities he experiences in life.

sonally facilitate reinforcement of your own beliefs. However, raw belief or blind faith alone are sufficient to work the processes. However it is very difficult to muster sufficient blind faith in the process if you don't fully understand it or have any reason to believe it. If you don't understand all of the theories or technical explanations behind the processes, don't worry; the processes will work just as well whether or not you fully understand why they work. It's just that some people find it difficult to achieve very much faith in themselves and in their ability to create their realities, both in the past and in the future, without some kind of logical support to strengthen their belief. That is why we presented extensive explanations on the theory of the brain and mind, the theory of creation, the theory on time and space, and the theory of the process.

Finally, believing is a choice! It is not something outside your control. If you feel that you cannot believe in either yourself or your ability to create the life you want, then you need to look for the blocks which prevent you from believing in yourself and your creations. Remember, outside appearances or aspects of your present life are totally irrelevant when working the creation process.

The reason for this is that while outside things or circumstances may appear very real and very solid, they also just represent creations. As such they consist of the same components of images and energy as any newly created thought. They only happen to appear in the field of your reality as more condensed energy and images than the newly emerging thought. However, when it comes to their energy and matrix configurations they are on the exact same footing as the newly created thought. While they may appear as obstacles to overcome they really are not, unless of course, you believe them to be, in which case they are. It is your belief which determines whether or not they will be experienced as obstacles or opportunities.

Finally, in order to make it all happen, once you are clear on what it is that you desire for yourself, you must cooperate with your subconscious mind in order for the thing desired to manifest. Since your energy and the energy of the field of your subconscious mind are one, you need to cooperate with that energy in order for the subconscious mind to manifest whatever it is you desire. Remember, the energy of the subconscious consists of unlimited possibilities, unlimited potential and unlimited correlation. Don't you think that with unlimited possibilities, potential and correlation on your side you should be able to create whatever it is that you want? Remember, your access to infinite possibilities is through your belief in yourself and your ability to create what you want in life. It is within your conscious mind that you create the energy and image of the thought with the conviction that as imaged it must manifest as your reality. It is through the subconscious mind that you access the unlimited power to actually make it so. Claude Bristol, in his book, *The Magic of Believing,* states it this way:

> "Just as the conscious mind is the source of thought, so the subconscious is the source of power........................"
> "The subconscious mind is beyond space and time, and it is fundamentally a powerful sending and receiving station with a universal hookup whereby it can communicate with the physical, mental, psychic, and, according to many investigators, spiritual worlds, past, present, and future, as well. Such is the power of your subconscious mind. In brief, the subconscious embodies the feeling and wisdom of the past, the awareness and knowledge of the present, and the thought and vision of the future."[7]

All of us, at one time or another, have encountered in our lives situations and stories involving ourselves or someone we know who through the power of their belief in something or someone have experienced results in life not otherwise thought possible. It is that we never quite knew how to repeat these apparently superhuman feats on a systemic basis in our lives. Such events seem to be out of the ordinary and in a sense, miraculous in nature. Once you understand and experience the processes of creation and come to believe in yourself and the processes, such seemingly extraordinary events can become ordinary events on a daily basis in your life.

The Meaning of Life

One final word, on the nature of belief in self, others and your purpose in life. Because the primordial aspect of the cognitive structure includes the element of survival of the human race, the primary program structures has a built-in bias toward the positive energy and images of positive thoughts for the simple reason that such thoughts are life enhancing and life fulfilling. The primary belief system seeks at all times to preserve the life of the system. Life is evolutionary. Life seeks the greatest expression of your talents and abilities. It seeks to become congruent in all of your affairs. It seeks to be free. It seeks to become all that you desire, to become all that you are capable of achieving in life. The primary belief system structure has a type of built-in fail safe mode that will automatically endeavor to isolate any incoming data which is perceived to threaten the integrity, preserva-

7 Claude M. Bristol, *The Magic of Believing,* p. 44.

tion or performance of the system. Accordingly the evolution of the system is on your side and against any negativity blocking you from manifesting your good.

There are some who would say that life doesn't come with a meaning and that at times we might even feel that life is hopeless, that if we are to find the meaning in life we have to create our own meaning. In a sense this is true. This is why throughout history philosophers and others have always advocated the need to find your own true purpose and meaning in life. This need for meaning is best explained in Viktor Frankl's book, *Man's Search for Meaning.* Without meaning prisoners quickly perished in the Nazi concentration camps during World War II. Viktor Frankl states in his book:

> "We needed to stop asking about the meaning of life, and instead to think of ourselves as those who were being questioned by life -- daily and hourly. Our answer must consist, not in talk and meditation, but in right action and in right conduct. Life ultimately means taking the responsibility to find the right answer to its problems and to fulfill the tasks which it constantly sets for each individual."
> "These tasks, and therefore the meaning of life, differ from man to man, and from moment to moment. Thus it is impossible to define the meaning of life in a general way. Questions about the meaning of life can never be answered by sweeping statements. 'Life' does not mean something vague, but something very real and concrete, just as life's tasks are also very real and concrete. They form man's destiny, which is different and unique for each individual....."[8]

What then is the meaning of life? The meaning of life can be found in seeking your own truth. Life expects that each one of us seek, even struggle if we must, in order to find our own truth which lies within each one of us. All of the answers to life's mysteries lie within. It is up to each one of us to go within and find them for they are unique and different for each person. When we find our own truth we have also finally found ourselves. We have finally become aware of who we are and what our life is really all about. We will then know without a doubt what life truly expects from each one of us. For only through finding our own truth do we find our liberation and freedom, freedom to express who and what we are, freedom to pursue whatever career we choose in life in order to express our talents and our abilities to the fullest. To paraphrase a quote which I believe is attrib-

---

8 Viktor Frankl *Man's Search for Meaning,* p. 98.

uted to Winston Churchill, "The destiny of mankind is not decided by material computations. We learn that we are spirits not animals, and that something is going on in time and space, and beyond time and space, and whether we like it or not, each one of us is called to duty"

Life then does not expect that we remain on the sidelines but that we become involved. We are called upon to share our own truth through love by first embracing it within ourselves and by living our truth through love in all of our affairs and with all persons. We are called upon to reach out and to share that truth which we have become, first with our families, then with our community, and finally our world at large. Truth through love is what integrity is really all about. Integrity embodies the highest values in life. Integrity at this level is represented as integrity of spirit or soul, mental and emotional integrity, and finally integrity within our body and in all of our affairs. As each one of us attains our own truth through love with integrity in all of our affairs we all will then be able to experience together a new dawn, an era where all mankind will be able to experience freedom, harmony among all races and nations, and cooperation among all men. On that day, which is soon to come, the world will truly reflect our highest good!

CHAPTER - 8

Primary Programs And Secondary Belief Systems Of The Human Mind

The Primary Programs of the Mind

In order to develop a comprehensive understanding of the primary programs of the mind we will again refer back to Chart 3-3 at page 61. Chart 3-3, you will recall, gives a breakdown of level 8 of the processing functions of the mind referred to as Pattern Interpretations. Pattern interpretations is broken down into three sublevels. The first, sublevel 8-1, is described as the Personality Program Filter, the second, sublevel 8-2, is described as the Operational Program Filter, and the third, sublevel 8-3, is described as the Secondary Belief Systems Filter. The first two of these sublevels of the mind are subconscious or hidden cognitive structures of the conscious mind. They serve as the containment field from which we form all of our conscious thoughts about everything in our life. No conscious thought about anything in your life is ever formed outside of these primary program filters.

At the same time these same primary program filters constitute the subconscious analog processing systems of the subconscious mind. We have seen in Chapter 6 that the analytical cognitive structure of the mind in which conscious thought is formed and the analog processing system of the subconscious mind are interconnected and interrelated in organizing sensory imagery into conscious thought. Remember that while the conscious analytical mind has some difficulty in organizing thoughts in analog fashion, the analog system of the subconscious mind has no problem whatsoever in processing analytical thought. In fact, the subconscious mind was constituted hundreds of thousands of years ago to easily process the formation of conscious analytical thought through the operation of its subconscious analog systems.

The primary personality and operational programs provide us with our sense of our physical existence residing in a physical body in three dimensional time and space. They provide us with a sense of a personal self, of who we are. They provide us with the sense of how we see ourself and others, and how we perceive that others see us. We have seen that the basic parameters of these primary personality and operational programs were laid down very early in life and thereafter became the most significant influences throughout our life. We were seldom aware of them at the time of their installation. We were too young to remember those early events in life. However, they acted upon the environmental stimulus causing us to adopt either positive or negative, primary and secondary programs which thereafter determined our life scripts. They provided us with the sense of how to perform within our world in order to maximize our own self expression. *These personality and operational programs together with the secondary belief systems programs cannot entertain a thought, idea, belief, opinion, attitude, observation or impression outside of the processing parameters of these primary belief system structures themselves.* Consequently, everything about yourself and your life is governed and controlled by these basic primary program systems of the mind.

The consequence of this is that you cannot effectively entertain a belief that you alone can create, direct, and control your outer physical realities unless the program parameters of your primary program system allows for such a belief to exist within your mind. Analog thinking provides the means for expanding these primary programs to allow you to effectively take greater control over your outer realities in life. However, you must make this decision for yourself and trust and have faith in yourself and your ability to create your realities for the process to work effectively in your life. For a list of ways for expanding your cognitive structures by using conscious analog thinking in order to create more meaningful realities in life refer to Appendix One.

The primary program system determines how strong an individual's self identification is, and how effective his operating programs are, for functioning in his world environment. The primary programs, at their early formation in childhood, do not make value judgments or distinctions of whether or not the incoming data is positive or negative. Whatever data it receives it processes, and if the data is negative the result will be the formation of negative processing parameters in your primary programs. Value discrimination is developed later on in life, at preadolescence or early adolescence stages of development. This is about the time the individual has been able to develop a strong conscious ego identification program for himself. However, by this time the negative data in the primary systems have been set for life and, consequently inhibit the development of a positive value system in the life of the young teenager. The consequence of this failure in the formation of a strong value system early in life results in poor choices and decisions throughout his adult years.

The reason for this is that all secondary thoughts and beliefs which form the specific life style of the individual arise from within these early primary programs. If the primary systems are limited or defective in any way, they will ultimately have a detrimental effect on one's ability to carry through on his conscious secondary beliefs of what he wants out of life. Until the individual finally deals with the limitations created within his primary systems of who he is and how he relates to others he will continue to struggle to attain his goals in life. When he fails he will often blame others for his bad luck. Or he might continually keep on trying to revise and change his conscious secondary thoughts and beliefs in order to improve his life.[1] All of these efforts will be to little avail until he deals with the limitations and defects within his primary programs. The key then to the removal of the negativity in the primary programs is to access the negative program parameters themselves which reside in the primary belief systems structure, and take their energy down to zero. Then reimage and reframe them, in the case of the personality program, to a more positive self image or self esteem level, and with respect to the operational program, to a more positive self enabling level.

Once the primary program systems are in place they are seldom changed except in those rare cases where the individual encounters a crisis in his life which cannot effectively be dealt with within his secondary belief system structure. Examples of such crises would include: a death of a close loved one, loss of a long time career position, critical life threatening illness, difficult divorce after a long time marriage, or other factors which may precipitate what is commonly referred to as an "existential or mid-life crises". All such crisis events do not have to be negative in nature. A person may encounter a deep emotional or religious experience where he momentarily experiences a transcendent state of consciousness. He might even experience what is referred to as an "out of the body" or "near death experience" where he experiences states of illumination which, in an instant, forever transform his entire life in a very profound way. What has happened is that almost instantaneously the transforming event has had such a profound effect upon the person that it literally changed the polarity of the energy and matrix configurations of his primary personality and operational program analogs from negative modes to positive modes. We have seen this happen in both Ricky's and Tommy's lives.

The Formation of the Secondary Belief System Structure

As we have seen the primary system forms the structure from which the ego personality creates a secondary level of conscious

1 The reason why creative visualizations, treasure mapping, positive self talk, and positive affirmations often fail to attain the desired results is the fact that they all have to do with changes in the person's secondary belief systems. If the person's primary program system structures are blocked or impeded, dealing solely with the secondary beliefs about your life will never solve the underlying deeper problems.

thoughts and beliefs which provide the form and content of an individual's life style. The secondary beliefs set forth the specific manner in which the ego personality consciously presents itself to its world. The secondary belief system deals with all areas of an individual's life including his self image, short and long term goals in life, where he will live, career choices, spiritual and moral values, types of friends, philosophy and meaning in life, special skills, level of intelligence, relationships, personal beliefs about himself, family choices, social and ethnic beliefs, political preferences, health attitudes, society, government, and world views, all as they relate to self, others and his world environment. The secondary belief system is necessary in life for it provides the energy and images of thoughts which are needed to create an experience of the desired outer realities we seek in life. Our outer realities are imaged by the secondary programs based on the parameters set down by the primary program. We then take action based on our secondary thoughts and ideas of how best to live. Thoughts and ideas, consisting of images of the desired outer physical manifestation of a particular life style, all conform to the programs of the primary analog system of the subconscious mind. The subconscious mind then sorts through unlimited possibilities in order to bring to conscious awareness those people, places, or things in order to create the desired results in life .

This creation process works whether the images and energy of the thought are negative in nature or are positive in nature. Of course, if it is negative we have seen that it results in negativity manifesting in our physical creations. If it is positive it results in positive manifestations. Either way it works in accordance with the type of energy and images we create within our primary programs. This same process takes place also with respect to the creations of our inner realities. In this instance the primary programs create the energy and images for our thoughts, ideas, emotions and feelings which are directed inward in creating mental imagery instead of outward in planning daily activities. These inner thought patterns reinforce our attitudes about ourselves and our world and create the basis for the mental chatter or mental self talk when the mind is not involved in the activities of our daily lives.

We have looked at problems appearing at the primary programs levels. However problems also appear at the conscious level of the secondary belief system processes themselves. At this secondary level problems often arise as a result of the following: (1) the person does not even consciously realize that he is creating the realities he is experiencing in his life, (2) he has no concept or idea of how to go about creating better outer realities, (3) he is simultaneously creating thoughts and images, either consciously or unconsciously, whose energy is in opposition to an intended reality, (4) he never follows through on the creation of the intended reality, (5) he lacks focus and clarity in formulating exactly what it is that he wants to create, (6) the energy of his secondary intentions in life are inconsistent with the energy needed for the particular creation he desires, (7) he lacks perseverance in pursuing his long term goals in life, (8) he misdirects his behavior and actions in an attempt to create his desired realities in life, (9) his concept of time and space and objectified realities operationally can not accommodate an idea or thought of personal creation outside these constraints, (10) he consciously perceives that the outer circumstances of his life are inconsistent with, or would somehow not tolerate, the desired creation, (11) the possibility of creating expanded realities in life is incompatible with the existing cognitive structure of his conscious thoughts, and (12) the energy of his primary programs is incompatible with the energy and images of his secondary thoughts. This list, while not exhaustive, covers most of the reasons that can effectively block or impede the creation of the desired realities at the conscious secondary belief system level of the mind. Many of these issues impacting your creation of your outer realities in life are addressed in Chapter 10.

It can easily be seen how important these secondary beliefs are to the enjoyment of life. When the person is confronted with a situation which triggers negative primary patterns they will often react by creating what they believe to be remedies to their problems by creating additional secondary beliefs. They will then take action based on the newly formed secondary beliefs in an effort to solve their problems in life. They fail to understand that the newly created secondary belief pattern themselves will sooner or later be altered or eliminated by the energy and images of the negative primary pattern. For instance, before meeting Betty and even though he has experienced many failed relationships in the past, Ricky kept dating women in the hope of finally meeting the "perfect woman" of his dreams. No permanent relief is ever attained when an individual creates an opposing secondary belief over an opposite primary pattern. This behavioral pattern is refer to as "creating over" or "creating against" self and seldom works effectively.

Comparisons Between the Primary Programs and the Secondary Belief Systems

The following will give you a brief comparison in summary form between the processes of the primary programs of the subconscious mind and the secondary beliefs systems of the conscious cognitive mind. It is helpful to keep these differences in mind when using conscious analog thinking in spacetime:

1. Primary Programs operate on the subconscious level of the mind. Secondary Beliefs Systems operate on the conscious level of the mind.

2. Primary Programs process information in analog fashion. Secondary Belief Systems process information primarily in analytical fashion.

3.The Primary Programs set the parameters for the creation of the Secondary Systems. The Secondary Systems all arise within the parameters of the Primary System.

4. The Primary Programs give definition to the personality and operational programs. The Secondary Systems give definition to the manner, style and mode of daily life.

5. The Primary Programs can operate outside of linear time and fixed space. The Secondary Systems operate within linear time and fixed space.

6. The Primary Programs provide the general form and substance of the Secondary System. The Secondary Systems seldom alter the form or content of the Primary Programs.

7. The effects of the Primary Programs are transparent. We seldom see the effects they have on the operation of our Secondary Systems. The Secondary Systems seldom have any effect on the content of the Primary Programs.

8. The basic ego personality traits are formed by the Primary Program and are seldom changed. The Secondary Systems can change the conscious exterior traits of ego personality.

9. The Primary Programs primarily serve as process functions in the formation of the Secondary Systems. The Secondary Systems primarily serves to provide us with the content of our life.

10. The Primary Programs do not serve to form conscious thought. The Secondary Systems do not serve as process functions in the subconscious formation of thought. They do, however, serve in organizing conscious thought using linguistic patterns and forms.

11. The Primary Programs are primarily established in the early childhood years. The Secondary Systems continue to change throughout our entire adult life.

12. The Primary Programs mediate thoughts forming our inner realities as well as the thoughts forming our outer realities. The Secondary Systems provide the form and content of our inner and outer thoughts.

13. Changes in the Primary Programs quite often result in significant changes in our life styles. Changes in the Secondary Systems usually result in more gradual, less significant, changes in our life styles.

14. Analog thinking holds the potential for significant changes in the Primary Programs. Rational, logical, analytical thinking of the Secondary Systems usually results in changes in the Secondary System only and seldom results in changes in the Primary Programs.

15. Analog thinking in spacetime holds the potential for changing our world views from one where we have little control over events and things in life to one where we have almost total control over the events and things in life. Rational, logical thought of the Secondary Systems has little potential for changing the parameters of our world realities. They primarily provide the content of those realities.

16. Analog thinking using power symbols holds the potential for easily changing significant negative deficits in the primary personality and operational programs. Rational cognitive thought often acts slowly and often painfully in changing subconscious negative blocks in the primary programs.

17. The Primary Programs hold the potential for organizing more fulfilling and meaningful inner and outer realities in life. Rational thought of the Secondary Systems is limited in directly creating significant changes in our inner and outer realities in life through the power of analytical thought alone.

This sets forth a brief comparison of the differences between the primary programs of the mind and the secondary belief systems.

<u>The Organization of the Subconscious Mind in Relation to the Primary Programs</u>

The act of creating realities was defined to include anything that you could feel, think, touch, believe, learn, consider, invent, see, taste, or smell, including anything inside your body, including the body itself, and everything outside the body including all physical objects and all intangible things in your world. The subconscious mind organizes images of our inner realities of how we see ourselves and how we think we should perform in our world. The subconscious mind also organizes images of our outer realities, of how we perceive ourselves in our world, and how we want our world to appear to us. Again, this is the primary purpose of the mind and it bears repeating. *The primary purpose of the subconscious mind is to create and discreate realities in space and time in accordance with the conscious intentions of your thoughts.*

The subconscious mind does not decide what is best for you or determine what is your purpose or meaning in life. You, as ego self, decide what is best for you and what ought to be the meaning of your life. The subconscious mind simply organizes a reality to fit that meaning in order to make that meaning more real in your daily affairs. The subconscious mind is not ego personality nor your conscious cognitive structure, nor does it communicate to self awareness what the subconscious mind might think is in your best interest. It does not do so for it doesn't communicate in rational, logical terms. It has no conscious awareness of its own. Of itself it has no volitional will. In this respect the subconscious mind is like a huge parallel processing computer. Regardless of its immense capacity to create realities it still needs someone to operate it. It needs the conscious mind to communicate with it and tell it what to do. In a way it is unfortunate that the subconscious mind can sort for almost unlimited information but cannot use this facility to make volitional decisions in life. On the other hand, the conscious mind can make volitional decisions in life but does not have direct access to this immense storehouse of information in order to make more informed decisions. *The purpose of conscious analog thinking is to bridge this gap. In a way it joins conscious analytical thinking with subconscious analog thinking for optimum success in life!*

The foregoing holds the key to accessing and using substantially more of the subconscious mind's capacity in order to create expanded realities in life. You can use the awesome power of the mind to create a better life for yourself and others, or you can use the power of your mind to create pain and suffering for yourself. The choice is yours and yours alone to make. No one nor anything else can make those choices for you. I can assure you that your current problems ultimately do not directly stem from your childhood, society, parents, circumstances in life, or anything else you might have identified as the cause of your problem. I can assure you that you will find all of the causes of your problems as well as all of their solutions within the operational programs of your mind. That is not to say that an abusive childhood, unloving parents, serious illness, cruelty in school, an uncaring spouse, disrespectful children, or any other external impediments to a better life may have been a prime factor contributing to the limitations and distortions in your personality and operational programs which created the pain and suffering in your life. Regardless of the initial cause of your pain, however, the ultimate fact of life is that the source of the problem is currently right now in your internal personality and operational programs. It is your responsibility alone to deal with these limitations and distortions in your internal reality programs if you are ever going to find peace, happiness and a better life for yourself. If after taking responsibility for your problems and correcting them you still want to beat up on your parents or others in your life who were the initial cause of your pain that is O.K. by me. But I want to warn you, I don't think that's a very good idea. Since it is counterproductive and oh, by the way, it also creates negative energy and negative images, something you just don't want or need in your internal operating programs. It's like throwing sand into the gas tank of your car. It's just not helpful to the internal operations of the engine.

You have always been whole, perfect, complete, accepted and validated right where you are, right now! You don't need to prove yourself or create a future which reflects your value or self worth. You don't have to get others to like you. You don't have to please others to affirm your own self worth. You don't have to prove your self worth by overperforming, or overachieving, or in any other fashion whatsoever. For you are the essence of self worth and self esteem right here, right now. You are perfect just the way you are; you need only to accept it, believe it and personally experience it. There is absolutely nothing and no one in the whole wide world who can decide for you who and what you are. You, and you alone, have the ultimate decision to decide whether you will be rich or poor, capable or incapable, beautiful or ugly, smart or dumb, confident or fearful. You alone decide what or how you think about yourself and your environment. You alone have the capacity to change the very foundation from which you derive your sense of self worth. You alone set the basic parameters from which you will formulate your thoughts and beliefs about yourself, your world and your fellowmen. You alone create your own reality! Therefore, create it and experience it as your own positive reality today. You have the unlimited capacity to do so. I hope this book will give you some tools to bring reality to your present awareness as you experience your own positive power each and every day of your life.

Your Primary Programs - The Four Archetype Prime Directives in Life

As indicated back in Chapter 4 on consciousness, the formative ego avails itself of the originating process functions of the mind in order to activate the operation of the application programs of the mind. The very first programs which form the newborn baby's primary personality and operational programs are the four application programs referred to as the "Archetype Prime Directives"™ in life. They are called archetype prime directives, or just "prime directive" because they are the very first application programs developed by the ego personality and are fundamental to the earliest development of the child. These prime directives are the very first developmental application software programs of the newborn child as distinguished from the earlier process functions represented by the consciousness of the child.

The four Archetype Prime Directives are described as follows:

1. I like who I am
2. I have a right to be here and live my own life
3. I express myself fully
4. I am capable

The archetype prime directives as well as the other eight originating primary programs of the mind are stated in linguistic terms because we use linguistic expressions to communicate our thoughts and desires about life. However, they essentially organize energy into analogs in order to operate and express themselves in life as functional application programs of the subconscious mind. Their energy forms the basic structure of our ego personality and operational style in life. We have seen that the subconscious mind processes these primary programs in analog fashion, for the subconscious mind organizes our conscious linguistic symbols in analog fashion. Anything we think, feel or do will conform to the these basic primary application programs. We never consciously or unconsciously think a thought or plan any type of action outside of these basic primary programs.

The first two prime directives "I like who I am'" and "I have a right to be here and live my own life" form the initial characteristics of the ego's emerging personality. The second two prime directives, "I express myself fully" and "I am capable", form the initial characteristics of the ego's emerging operational life style. These prime directives arose out of the earliest primordial development of the human brain and mind. They are what self awareness first selects to develop the content of ego personality. These prime directives are the fundamental application programs from which the child develops all other primary and secondary personality and operational systems. Such systems provide the child with his self identity and operational characteristics for functioning productively throughout his entire life. You will notice that these archetype prime directives are stated in positive terms for the reason that they are intended to create a positive ego and a meaningful life for the child. Every child conceived and brought into this world has an inherent right to be able to like and accept who and what he is. If the primary caregiver creates a negative or hostile environment these prime directives will be accepted by the child in negative terms, in which case they can have devastating effects during the entire life of the child!

Consequently, we look a little more closely at each one of these prime directives in life. First, "I like who I am" is essential to the wellbeing of the child. This prime directive forms the basis for the child's self identity, self worth, and self esteem. If the child's environment consists of negative stimuli it has the effect of programing the child to not feel very good about himself. The child will create a poor self image and low self esteem. Yet, far too often parents and others in the child's life, whether out of ignorance or out of their own dysfunction, give negative feedback to the child resulting in a negative self image. Such feedback includes such things as physical, mental and emotional abuse and neglect in the form of lack of love and nurturing. Probably the one most critical factor in the development of the child's self image is whether or not the child is loved and wanted by the parents or other care giver. Even today far too many parents do not understand and apply these fundamental principles in raising healthy, happy and productive children. Far too often parents are preoccupied with their own needs and wants to provide a nurturing environment for their children. Children just don't grow up to be good. They are programmed by their parents and environment to either live a productive life or a dysfunctional life. Yet a dysfunctional life can, in almost every case, be avoided by the proper love and attention given the child in the earliest years of development.

With respect to the primary program, "I have a right to be here and live my own life", the same admonitions hold true. Besides being accepted and loved the child should also be made to feel that he is wanted and appreciated by his parents, even in the case of an unwanted pregnancy. Teenage pregnancy is probably one of society's greatest illnesses for the reason that we have children having children. Often the mother was unwanted and unloved herself and, consequently, developed low self esteem and difficulty in developing mature responsible relationships with men. The mother, who is still herself a child, interprets physical pleasure as love and affection. By bringing an unwanted child into this world she continues the cycle of mental and emotional deprivation in the formation of the newborn child's prime directives.

The prime directive "I express myself fully" has to do with the fundamental right of each individual to be given the opportunity to fully express himself in life. Again, far too often the old adage, "children should be seen but not heard" is still all too alive and well in today's society. Children are often made to feel as though they do not have the right to develop their own unique personal life styles, or the right to make their own choices in life including career choices. They often constantly seek the approval of others to ratify who and what they are. The needs of others often come before the fulfillment of their own legitimate needs and desires leaving them with an anxious feeling that life is somehow passing them by. They feel that while their peers are achieving what they want in life they are on the sidelines forever waiting for their light to shine until it's too late. Finally toward the end of their life their light fades without them ever realizing that the small tiny flicker within was always there waiting to be nourished and fed in order to blossom and brighten their way.

If parents only knew the devastating effects a negative environment has on their children they would think twice about having children in the first place. If parents are not fully prepared to provide a positive nurturing environment essential to the child's development they should not contemplate having children. The proper development of these basic prime directives are critical in the early childhood years. If these prime directives are not properly developed and are significantly distorted as a result of neglect or abuse their negative energy in the form of shame, quilt, fear, anger, hatred, remorse and hopelessness all work to significantly inhibit all of the child's best efforts to attain a happy and productive life. While this book is not geared to providing effective child rearing techniques the theory and practices set forth should be of some help to parents and practitioners in the child development field in understanding the crucial role the environment plays in the mental and emotional development of children.

The last prime directive "I am capable" is an essential ingredient in the emerging operational program of the child. The reason for this is that if the child believes as a result of early negative feedback that he is not competent he will constantly struggle against such a negative unconscious prime directive in an effort to succeed in life.

This negative prime directive will work against all of his best efforts to achieve success. An environment which indicates to the child in so many ways that he is a failure is sure to result in an unconscious belief that he is no good and that he will never succeed in life regardless of what he tries to do. Often such a child becomes dependent on others to give direction and meaning to his life. As a result he grows up with ambivalence towards those who sincerely want to help. He develops a love/hate relationship toward such individuals. This response to life is often referred to as co-dependency or borderline personality disorder. As the child grows up he is not consciously aware that the fault lies within his subconscious defective prime directives in life which continually operates against all of his best efforts to succeed.

The reason for this tension between what the child wants out of life and what he is able to actually achieve is due to the opposing energy forces within his operating systems. The more fundamental of the two forces is the one that will always prevail notwithstanding the best intentions of the child. For this is the way the subconscious analog system works. Remember, the subconscious mind does not make value judgments. In this case the subconscious does not communicate to the conscious aware mind that the defective prime directive is operating to create failure and needs to be reprogrammed for success. If life were so easy we all would be healthy, happy, millionaires Consequently, the young adult, after already encountering several failures early in life, such as poor grades in school, failed relationships, and loss of job opportunities, will logically conclude that either people are conspiring against him, that he is accident prone, that he is just an unlucky person, or that the world itself is a hostile and foreboding place in which to live, or a combination of all of these. Little does he know that neither people nor the world are his problems. Rather, it is simply his very own subconscious childhood programing which is sabotaging all of his best efforts at obtaining happiness. Just exactly how these prime directives influence the child's self identity, his ability to function, and the creation of the physical realities of his life are described below.

The Twelve Originating Primaries™ of the Mind

These four archetype prime directives form the subconscious portion of the primary cognitive structure which, together with the next eight originating primaries, are referred to as the "Twelve Originating Primaries™ of the Mind". The essential twelve originating primaries, which include the four archetype prime directives, are divided into two groups of six each. The group on the left side of the page below constitutes the formative primary personality programs of the child and the group listed on the right side constitutes the child's formative primary operational programs. These twelve originating primary programs are as follows:

<u>Primary Personality Program</u>	<u>Primary Operational Program</u>
1. I like who I am	7. I express myself fully
2. I have a right to be here and live my own life	8. I am capable
3. I like my physical body	9. I am intelligent
4. I control my thoughts and emotions	10. I am responsible
5. I like my sexuality	11. I make my own decisions
6. I am loving and lovable	12. I create my own reality

You will begin to see how these essential twelve originating primaries of the mind start to arrange themselves into the hierarchical form of the cognitive structure of the subconscious mind. How the mind organizes its cognitive structures for relating to life is the topic of the next chapter. The remaining eight originating primaries were chosen because they deal with the most essential characteristics of ego personality and a functioning operational program. For instance, the two described as "I like my physical body" and "I like my sexuality" obviously describe essential attributes of the human body as ego personality. A person who is comfortable with his or her physical body and his or her sexual identity will probably develop normal sexual relationships and always feel physically comfortable with himself or herself in normal social settings with others of the same or opposite sex.

While the foregoing is such an obvious statement you would hardly think it is worth stating. However, far too many children, as a result of early childhood sexual and physical abuse, grow up not being comfortable with their own physical bodies nor with their sexual feelings. Far too often because of the repressive upbringing by their parents they are made to feel ashamed of their bodies and guilty over their sexuality. Since these elements of the ego personality are developed at such an early age they are very near the top of the hierarchy of the factors making up ego identity. The negative events giving rise to these cognitive structures are often far removed from conscious awareness yet they have a pervasive effect on the character and self identity of the child. For this reason they are often referred to as "characterlogical" in nature and are very difficult to reverse by traditional therapy, or other spiritual, mental, emotional or physical self help workshops, motivational groups or by other means.

Similarly, the primary operational program traits of "I am responsible'" and "I create my own reality" are traits developed at a very early age. The child doesn't necessarily think in these terms. However, the child develops at a very early age a personal sense that he can succeed in life by taking responsibility for his own acts, or that he is a failure and he is not responsible for what happens to himself or others. Responsible nurturing develops responsible children. Neglectful and abusive development creates irresponsible children, children who fail in life simply because they have not been given the means to develop a confident self image which supports the belief that they can achieve whatever they want in life.

The originating programs "I control my thoughts and emotions" and "I am loving and lovable" are also essential to the development of a strong self identity and for intimate relationships. If a person does not feel loving about himself that person will not be able to give and receive love which is necessary to sustain committed intimate relationships throughout life. This was a factor in Ricky Mindful's relationship with Betty. The originating programs "I am intelligent" and "I make my own decisions" are also included as originating primaries for they are basic to the development of a sense that life does hold meaning and is not hopeless, and that one has sufficient intelligence and decision making capacity to pursue an education, career or work one finds fulfilling. Here again a neglectful or abusive environment will often make the child feel that he is stupid and cannot make his own decisions in life. He believes that he will never be able to graduate from school or hold down the type of job or career he desires for himself.

The Secondary Belief System Forms the Details of Life

These essential twelve primary programs form the child's basic subconscious cognitive structure which gives rise to the child's entire secondary belief system for the development of the child's specific life style. The twelve originating primaries have been chosen for they, in summary form, represent the originating personality and operational structure of the human subconscious analog system.[2] Chart 33 beginning at page 298 list 14 examples of secondary personality beliefs and 11 examples of secondary operational beliefs a person may hold about himself. There is no magic in this particular list of examples of secondary beliefs arising from within the primary pro-

2. These twelve primaries are simply generic models which are intended to fit almost every personality and operational style of the subconscious mind. However, we recognize tailoring the descriptions of these primaries to suit your own needs may assist you in better understanding your own personality and operational style in life.

CHART 33

Examples of a Positive Primary Personality Program and its Corresponding Secondary Belief Systems

Subconscious Primary Belief System:	Conscious Secondary Belief System:
Relationship to Self	**Relationship to Self**
1. I like who I am	I am comfortable with myself
2. I like my physical body	I am a beautiful/handsome person
3. I have a right to be here	I am entitled to a life
4. I can easily express myself	I am persuasive with others
5. I believe in myself	I believe I can earn a college education
6. Sex is good	I view sex as healthy and I enjoy a good sex life
Relationship toward Others	**Relationship toward Others**
7. I am loving	I have close loving relationships
8. I release the past	I am not stuck in the past, I have no regrets
9. I learn from the past	I view all events as learning opportunities
10. I create my own future	I create the lifestyle I desire for myself
Relationship toward Your World	**Relationships toward Your World**
11. My life is meaningful	I find purpose and fulfillment in life
12. I have a positive outlook on life	I always try to see something good in everything
13. I live in the present moment	I make every moment count
14. I am prosperous in all things	I can create the material comforts I desire in life

Examples of a Positive Primary Operational Program and its Secondary Belief Systems

Subconscious Primary Belief System:	Conscious Secondary Belief System:
Relationship to Self	Relationship to Self
1. I am capable	I can accomplish whatever I really set out to do
2. I am responsible	I take responsibility for my own actions
3. I enjoy life	I live a balanced life of work and play
4. I grow mentally and emotionally	I view events in life as learning experiences
Relationship toward Others	Relationship toward Others
5. I cooperate with others	I work easily with the people in my office
6. I communicate effectively	I can easily make myself clearly understood
Relationship toward Your World	Relationship toward Your World
7. I know what I want	I have specific plans and goals in life
8. I have special talents	I develop and use my talents to enjoy my life
9. I am satisfied	I am content with what I have accomplished
10. I am unlimited	I do not let other people or things limit me
11. I make sound decisions	I can critically evaluate a situation

Relationship Between The Internal Mental Reality And External Mental Reality Of a Positive Personality Secondary Belief System

Internal Secondary Belief:	External Secondary Belief:
Relationship to Self	**Relationship to Self**
1. I am comfortable with myself	I am comfortable with people in different situations
2. I am a beautiful/handsome person	I respect myself and keep myself attractive at all times
3. I am entitled to a life	I function effectively in group situations
4. I am persuasive with others	I am effective in retail sales
5. I believe I can earn a college education	I will earn a PhD degree in Philosophy
6. I view sex as healthy and I enjoy a good sex life	I am committed to the one I love and we sexually express our love for one another
Relationship toward Others	**Relationship toward Others**
7. I have close loving relationships	I have loving relationships with significant others in my life
8. I am not stuck in the past, I have no regrets	I can easily forgive and forget and move on with my life.
9. I view all events as learning opportunities	If I fail I try to learn what went wrong, I don't take it personally
10. I create the lifestyle I desire	I live, work and play in the fashion that I choose for myself
Relationship toward Your World	**Relationship toward Your World**
11. I find purpose and fulfillment in life	My career is purposeful, I don't just work for a paycheck
12. I always try to see something good in everything	By staying positive, I open myself to greater possibilities
13. I make every moment count	I don't procrastinate and work diligently to get a job done
14. I create whatever material possessions I desire	I can create the career and things in life I want for myself

Relationship Between The Internal Mental Reality And External Mental Reality Of a Positive Operational Secondary Belief System

Internal Belief:

Relationship to Self

1. I can accomplish whatever I really set out to do
2. I take responsibility for my own actions
3. I live a balanced life of work and play
4. I view events in life as learning experiences

Relationship toward Others

5. I work easily with the people in my office
6. I can easily make myself clearly understood

Relationship toward Your World

7. I have specific plans and goals in life
8. I develop and use my talents to enjoy life
9. I am content with what I have accomplished
10. I do not let other people or things limit me
11. I can critically evaluate a situation

External Belief:

Relationship to Self

I can past the test; get the job; and get the promotion
People trust me because they know I am responsible
I enjoy sports and leisure activities
As I learn I become more expert in the things I do

Relationship toward Others

I motivate and encourage others to do their best
I can express complicated ideas so others understand

Relationship toward Your World

I plan to have sufficient income to retire by age 40
I use my talents to achieve those things I desire in life
I do not compete to show superiority
I don't listen to negative people tell me I can't do it
I have the ability to make sound investments in life

Relationship Between External Mental Reality And External Physical Reality Of a Positive Personality Secondary Belief System

External Mental Belief:	External Physical Reality:
Relationship to Self	**Relationship to Self**
1. I am comfortable with people in different situations	I am comfortable in my new career position
2. I respect myself and keep myself attractive at all times	We are attracted to each other and plan to get married
3. I function effectively in group situations	I handle myself effectively in my new corporate position
4. I am effective in retail sales	I just got promoted to V.P. Sales
5. I will earn a PhD degree in Philosophy	I now have my PhD degree from State University
6. I am committed to the one I love and we can sexually express our love to one another	My life is fulfilled as I am able to express my love for my partner
Relationship toward Others	**Relationship toward Others**
7. I have loving relationships with others in my life	I have loving relationships with spouse, children, parents, others
8. I can easily forgive and forget and move on in life	I have forgiven my parents for a harsh and neglectful childhood
9. If I fail I try to learn, I don't take it personally	Learning from my mistakes has advanced my career in sales
10. I live, work and play in the fashion that I choose	Creating independent wealth allows me to travel extensively
Relationship toward Your World	**Relationship toward Your World**
11. My job is purposeful, I don't just work for a paycheck	I created a nursing career which gives me meaning in life
12. By staying positive, I open myself to all possibilities	A positive attitude has advanced me rapidly in my sales career
13. I don't procrastinate, I work diligently on the job	Diligence has made me the head of my department
14. I can create the material comforts I desire in life	I created the penthouse, luxury car and other things I desire

Relationship Between External Mental Reality And External Physical Reality Of a Positive Operational Secondary Belief System

External Belief:	External Physical Reality:
Relationship to Self	**Relationship to Self**
1. I can pass the test; get the job; get the promotion	I passed the test; I got the job; and I got the promotion
2.People trust me because they know I am reliable	Taking greater responsibility earned me the promotion I wanted
3. I enjoy sports and leisure activities	I play a good game of tennis & golf, and I enjoy sailing and skiing
4. As I learn I become more expert in the things I do	I have become an expert quantum physicist
Relationship toward Others	**Relationship toward Others**
5. I motivate and encourage others to do their best	Because I effectively work with others I get regular promotions
6. I express complicated ideas so others understand	By effectively communicating I obtain leadership positions
Relationship toward Your World	**Relationship toward Your World**
7. I plan to have sufficient income to retire by age 40	By age 40 I retired and moved to a farm in Vermont
8. I use my talents to achieve those things I desire	By developing my talents I have become an expert musician
9. I do not compete to show superiority	I have achieve what I want and take pride in other's success
10. I do not let other people or things limit me	I disregarded negative advice in achieving my business goals
11. I have the ability to make sound investments	My 401-K Retirement Plan and real estate investments have greatly increased in value this last year

grams of the subconscious mind. They were chosen as our model for being fairly typical of secondary beliefs representing the underlying primary systems since they represent secondary beliefs held by most people, or with which most people can easily identify.

These examples of 14 personality primaries and 11 operational primaries represent variations on the basic twelve originating primaries. These charts are organized by first referring to one of the primaries, or a variation of a primary program, with its corresponding secondary belief described opposite it. Next, I describe an expression of its corresponding internal mental belief and opposite this description is an expression of its external mental belief. These examples of internal and external beliefs represent examples of a person's internal held reality and a person's external held reality. Finally, opposite the external mental belief is a description of the actual physical outer reality of the person holding these primary and secondary beliefs. The purpose of this organization is to track the close correlation between the subconscious primaries and their conscious inner secondary beliefs. I then outline the correlation between the inner and outer expressions of these secondary beliefs and how they affect the actual creation of the person's outer physical realities. Chart 34 starting at page 306 follows the exact same process in creating realities in life except that they state in negative terms what Chart 33 sets forth in positive terms. In this manner you get a comparison of how a positive system creates a rewarding and fulfilling life and a negative system creates a painful self-limiting life.

When reviewing these series of charts keep in mind that the definition of a primary program is that it is a subconscious primary program for the development of ego personality and an operational system for living. It is largely unconscious and defines who you are. It defines your self identity and operational style in life. This is contrasted to the secondary beliefs you hold in life which, for the most part, are consciously held beliefs and define not so much who you are but rather what you are and what you do in life. They define how you present yourself to others and what type of life you lead. They define your life style, your personal appearance, career choices, relationship choices, and all of the others aspects of your personal life style.

All of the various primary and secondary beliefs set forth in this series of charts relate to one another horizontally within groups of related secondary belief systems forming the cognitive structure of the conscious mind. They illustrate just how the subconscious primary programs in life influence the internal reality about our conscious secondary beliefs about life. The internal secondary beliefs, in turn, influence our external beliefs about our life. These external beliefs about our life influence the actual physical realities encountered in life. When we refer to the internal secondary beliefs about life these would include the inner mental imagery we have, including our hopes, plans and goals about the future. It would also include our fantasies and daydreams about life.

Since our dreams arise during periods of sleep when we are not consciously aware of our thoughts, dreams are not included in the internal secondary belief system structure. However, analog patterns arising within our dreams may hold significant information which often is helpful in making conscious decisions in our awake state. When we refer to external beliefs we are referring to secondary beliefs about the external images we create in our mind of the way we perceive, feel and think about how our outer physical realities should exist in our life. It is these beliefs, perceptions and feelings about our external affairs which in turn actually influence, either positively or negatively, the creation of our actual physical realities. We are seldom aware of how these primaries influence our secondary beliefs and how our internal and external secondary beliefs actually affect the creation of our outer physical realities.

You may wonder how is it that beliefs can influence, to such an extent, our actual physical realities in life. Or stated another way, isn't physical reality a fixed objective reality outside of ourselves and certainly outside our thoughts and beliefs about life. The simple answer is that the subconscious mind will always follow the dictates of conscious thought. The subconscious does not second-guess the conscious mind. Once a thought is accepted by the subconscious mind, whether it is a good thought or a bad thought, the subconscious will sort through the infinite possibilities for people, places and things which could manifest an awareness of a particular reality conforming to the thought created by conscious mind. It can easily accomplish this type of sorting for realities for the reason that analog systems are programmed to operate in this fashion. Remember, the subconscious primarily works to create realities in life. The subconscious does not like tensions in its systems set up by a conscious thought which remains unexpressed in physical reality. It is the subconscious mind's job to find physical expression of the conscious thought generated by the ego personality. Of course, if you fantasize a lot about a certain type of lifestyle and you do not really intend to create the fantasy for yourself, the subconscious mind will only give expression to the internal imagery of the fantasy and not give it an actual physical expression.

This analog system of creation of realities is the way the

CHART 34

Example of a Negative Primary Personality Program and its Corresponding Secondary Belief Systems

Subconscious Primary Belief System	Conscious Secondary Belief System
Relationship to Self	**Relationship to Self**
1.I do not like myself	I do not respect myself or others
2.I do not like my body	I am an ugly and never look good in my clothes
3.I am incomplete and unfulfilled	I crave food, drugs, and sex for fulfillment
4.I don't know my boundaries	I don't know how to say "no"
5.I am obsessive and compulsive	I am a compulsive person and I can't control my cravings
6.I am dumb and stupid	I will never amount to anything in life; I fail at whatever I try to do
7.I am unhappy	Nothing ever pleases me; I am dissatisfied with myself and others
8.Sex is bad	Sex is dirty , shameful and base
9.I am insecure	I worry and get depressed when I think about the future
Relationship toward Others	**Relationship toward Others**
10.I am not lovable	I think only about myself and never really care about other people
11.I am a negative person	I always seem to end up in abusive negative relationships
12.I don't care about other's welfare	I am often intolerant, hateful and prejudiced towards others
13.I hate those who hurt me	I keep going over the same hurtful past
14.Life is difficult	I am fearful of the future

Relationships toward Your World	Relationships toward Your World
15.Life has no meaning or purpose	I find that I fill my days with meaningless activities
16.I have a negative outlook on life	I always seem to see the bad in everything
17.I am untrustworthy	You just can't trust people because they lie, steal, & cheat on you
19.I am poor	I never seem to hold onto money
20.I lack choices	I never have any choices in life; my life is dull and will always be so
21.I am always chronically ill	I always seem to be ill; I never take good care of myself

Example of a Negative Primary Operational Program and its Secondary Belief System

Subconscious Primary Belief System	Conscious Secondary Belief System
Relationship to Self	Relationship to Self
1.I am not capable	I have trouble accomplishing anything in life
2.I am incompetent	I never succeed at anything I try to do
3.I am not resourceful	I don't know how to get things done
4.I am not responsible	Others are to blame for my mistakes and failings
5.I can never relax and enjoy myself	I always seem to be up tight about everything going on
6.I lack confidence	I never seem able to do anything right
7.Life is survival of the fittest	Others will take advantage of you, if you don't take advantage of them
Relationship toward Others	Relationship toward Others
8.I do not know how to cooperate	I always seem to be fighting with others for no good reasons at all; they always seem to pick on me
9.I am a loner	No one wants to work with me, I just don't fit in

10.I don't know how to communicate	No one seems to listen to me
11.There is no decency in the world	We live in a dog eat dog society
Relationship toward Your World	Relationship toward Your World
12.I will never achieving my goals in life	I never do anything right regardless of how I hard I try
13.I have no special talents or abilities	I don't have anything special going for me at this time
14.I am dissatisfied with my life	Nothing seems to turn me on; I'm bored all the time
15.I was born into the wrong family	Because of my background I'll never amount to anything

Relationship Between The Internal Mental Reality And External Mental Reality Of a Negative Personality Secondary Belief System

Internal Belief	External Belief
Relationship to Self	Relationship to Self
1.I do not respect myself or others	I let others use and abuse me and, in turn, I victimize others
2.I am ugly and never look good in my clothes	I always seem to be overweight and dress unattractively
3.I crave food, drugs, and sex for fulfillment	I am promiscuous; overindulgent; and out of control
4.I don't know how to say "no";	I let others take advantage of me
5.I am a compulsive person	I crave tobacco, alcohol, drugs and food
6.I will never amount to anything in life	I fail at whatever I set out to do; others see me as a failure
7.Nothing ever pleases me	I am dissatisfied with everything in life
8.Sex is dirty, shameful and base	Men use me to satisfy their own lustful desires
9.I worry and get depressed about the future	I seem to self sabotage myself when I am on the verge of success

Relationship toward Others	Relationship toward Others
10.I only think of myself and not others	I only care about what others can do for me
11.I end up in abusive negative relationships	I don't know why I always get involved with the wrong men/women
12.I am prejudiced and intolerant of others	I seem to be drawn to hate groups, they understand me
13.I keep going over the same hurtful past	I never seem to get any relief regardless of how hard I try
14.I hate those who hurt me in the past	I like to get back at people who hurt me in the past
15.I am fearful of the future	I always seem to fail whenever I try to better myself
Relationships toward Your World	Relationships toward Your World
16.I fill my days with meaningless activities	I am always bored, the excitement and zest for life are gone
17.I always see the bad in everything	I see the hate in people and the bad in the world
	I just seem to be drawn to it, I don't know why
18.You just can't trust people	I might be a little paranoid; but you just can't trust anyone these days
19.I never seem to hold onto money	I compulsively gamble and spend beyond my means
20.I never have any real choices in life	I am in a rut and I just can't seem to ever get out of it
21.I always seem to be ill	I have trouble following a weight loss and exercise plan

Relationship Between the Internal Mental Reality And External Mental Reality Of a Negative Operational Secondary Belief System

Internal Belief	External Belief
Relationship to Self	Relationship to Self
1.I have trouble accomplishing anything in life	Anything I ever try to do turns out badly
2.I never succeed at anything I do	When I fail it is usually other people's fault
3.I don't know how to get things done	Nothing ever goes right; I don't know why

4.Others are to blame for my mistakes	I believe that others are at fault for my shortcomings
5.I always seem to be uptight about everything	I often drink, use drugs or sex to relieve tensions and anxiety
6.I never seem able to do anything right	My mother always said that I would never amount to anything, I guess she knew what she was talking about
7.Others will take advantage of you	No one better stand in the way of me getting what I want
<u>Relationship toward Others</u>	<u>Relationship toward Others</u>
8.I always seem to be fighting	People always seem to want to cause me trouble
9.No one seems to want to work with me	No one is ever going to get ahead of me so help me God
10.No one seems to listen in me	No one cares about me, so I really don't care about them
11.I believe that we live in a dog eat dog world	You really have to fight hard to get what you want in life
<u>Relationships toward Your World</u>	<u>Relationships toward Your World</u>
12.I can never do anything right	Dad always said that I couldn't do anything right; I guess I proved him right after all
13.I don't have anything special going for me	If I just got that college degree; job promotion; won the lottery; got that beautiful boy/girl I could be somebody!
14.Nothing seems to turn me on anymore	Life just doesn't seem to hold any excitement anymore, guess what I need is a new relationship, clothes, car, etc.
15.I'll never amount to anything in life	I could excel if I had: a better education; wife/husband; more money; car; house; clothes; opportunities

Relationship Between External Mental Reality And External Physical Reality Of a Negative Personality Secondary Belief System

External Belief	External Physical Reality
Relationship to Self	Relationship to Self
1.I let others abuse me and I victimize others	I just had another fight with my woman and she left me like the last one
2.I always seen to be overweight	I never have any dates, I don't know why
3.I am promiscuous and out of control	Men just use me, they are all the same, just dirty old men
4.I let others take advantage of me	I really don't like him, why do I keep going to bed with him
5.I crave tobacco, alcohol, drugs and food	I try to quit, but my friends won't let me
6.I fail at whatever I set out to do	I failed at my marriage and job; will things ever change
7.I am dissatisfied with everything in life	I just bought a new car and clothes, yet I'm dissatisfied
8.Men use me to satisfy their lustful desires	All men ever want is dirty sex, then they just dump me
9.I seem to self sabotage myself	This is the third job I lost this year, I don't know why
Relationship toward Others	Relationship toward Others
10. I only care about what others can do for me	I divorced my wife/husband; they just weren't good enough for me
11.I get involved with the wrong men/women	This is my third abusive marriage I don't know what's wrong
12.I seem to be drawn to hate groups	The militia is right; someone should take over by force
13.I never seem to get any relief on the past	It's been 45 years ago that I left that abusive household but it feels like yesterday. I wish I could get them out of my life
14.I like to get back at people who hurt me	I don't know why I keep hurting the ones I love
15.I always seem to fail	I don't know why I opened my big mouth when I almost had that job offer. I always seem to shoot myself in the foot

Relationships toward Your World	Relationships toward Your World
16.I am always bored	Nothing gives me satisfaction; I often feel depressed and suicidal
17.I see the hate in people I seem to be drawn to it, I don't know why	There is so much violence in the country; what we need is more guns, jails, stricter laws and lock the bastards up
18.I might be a little paranoid	I don't trust my wife, I think she is running around behind my back, just like the last one
19.I compulsively gamble & spend too much	I can't pay my bills again this month; they repossessed my car
20.I am in a rut and can't seem to get out of it	If it weren't for the kids I would have left him a long time ago
21.I have trouble following a weight loss plan	I'm overweight, my back hurts, my ulcers are acting up, I am in bad health, I am at my wits end

Relationship Between The External Mental Reality And External Physical Reality Of a Negative Operational Secondary Belief System

External Belief Relationship to Self	External Physical Reality Relationship to Self
1.Anything I ever try to do turns out badly	I just got laid off and I have seniority over the younger men
2.When I fail it is usually other people's fault	Its the bank's fault that they foreclosed on my house
3.Nothing ever goes right; I don't know why	I just smashed up the car and got a DUI, life isn't fair
4.I believe that others are at fault	It's the fault of my parents and police that I am in jail
5.I often drink too much or use drugs or sex to relieve tensions and anxiety	My girl friend is pregnant and wants me to marry her, Hell, it is not my fault she didn't use protection
6.My mother said I wouldn't amount to anything in life, I guess she was right all along	I called my mom after I lost my job. She said that I was a no good bum, God how I hate that women

17. No one better stand in my way of what I want	I joined the militia. We are going to show those bastards in Washington a thing or two
<u>Relationship toward Others</u>	<u>Relationship toward Others</u>
8. People always seem to cause me trouble	They got me up on manslaughter, I didn't mean to hit him that hard; he just really pissed me off
	He tried to cut me off so I rammed him, that will teach him a thing or two
9. No one is ever going to get ahead of me	
10. No one cares about me,	No I'm not going to Dad's funeral. He never cared for me
11. You really have to fight hard in life	OK, so I took advantage of her and used the grocery money for booze; its her fault for making me so mad
<u>Relationships toward Your World</u>	<u>Relationships toward Your World</u>
12. Dad said that I couldn't do anything right	OK, so I flunked out of school what does the old man care anyway
13. If I just got that college degree; job promotion; won the lottery, I'd be somebody	I got dealt a bad hand in life, I'll just stay at this dead end job and hope that maybe I'll win the lottery someday
14. Life just doesn't hold any excitement	Maybe this new girl/boy friend will add some excitement, the last one didn't after just three months
15. I could excel if I had: a better education; wife/husband;more money; car; house; etc.	Once they recognize my talents and give me that big promotion then I'll really be somebody important

mind/brain works and has been organized in this fashion from time immemorial. It reflects the evolutionary processes of mankind. It is a basic primordial characterization of the DNA of our genes. It has served mankind well in its evolutionary path leading to the creation of expanded more complex realities in life. We all can accelerate this evolutionary path of mankind by becoming more aware of how these systems work and then make them consciously work for our good. This evolutionary path of mankind is illustrated in the earlier Flow Chart 16 at page 130 entitled *Interrelationships Between the Flow of Pure Energy and Human Consciousness.* The purpose of that and related charts was to trace the consciousness of mankind. Consciousness first expresses itself through the creation of the cognitive structures of the mind, and than to a level of awareness described as self actualization where mankind directs the energy flow of his mind in organizing realities in his life. The next level is enlightenment where mankind is able to experience self as the source of his realities. The dynamics of this highest level of awareness is not directly addressed in this book. However, as you use the MetaCreation processes primarily to create expanded realities for yourself you will automatically be led in the direction of these higher stages of consciousness referred to as enlightenment. The evolutionary path of mankind is towards wholeness, perfection and completion. Each of us ultimately determines what these metaphors mean personally in the context of our own life. In fact these metaphors set up power analogs in the subconscious mind which provide meaning and direction in life.

So much for theory. Now back to the specifics of how exactly thoughts create reality. Look at Chart 33. The very first primary program "I like who I am" for the most part remains an unconscious program about yourself, at least up until you start consciously thinking about it. The corresponding secondary belief, of which you are consciously aware, is something like "I am comfortable with myself". Recognize that this and the other primary programs and secondary beliefs set forth here could be stated in a hundred different ways. These are simply models of generalized responses in life to illustrate the points I am making. They are not intended to act as a specific response elicited from any particular person. In this respect they represent internal states of mind. They represent perceptual positions in life rather than specific expressions of intent. In this manner they represent a type of generic or generalized states of mind with which most people will readily identify personally. This internal reality "I am comfortable with myself" gives rise to the belief expressing the person's perception of his external reality "I am comfortable with people in different situations". This external belief in turn gives rise to the person's actual external physical reality "I am comfortable in my new career position". It is this series of beliefs stemming from the originating program "I like who I am" upon which the subconscious mind sorts unlimited possibilities to find the people, places and things which gives expression to the thought. It is the intentionality of a thought that forms the realities defined by this series of subconscious and conscious thought patterns in life.

Now refer to Chart 34 which show how a negative secondary belief has the effect of creating a negative outer physical reality. The negative primary belief about self would be "I do not like myself". This negative belief represents a state of mind which in this case is a negative perception of oneself based on a negative primary program. Seldom does a person actually consciously think to themselves "I really don't like myself". Although sometimes they do. More often such an unexpressed thought simply represents a subconscious state of mind. Of course there might be times when everything is going badly in your life and you temporally think at that moment "I don't like myself" or "I am really stupid for doing what I just did" in the context of the stresses you are encountering at that particular moment. This simply represents a situational expression of a bad episode in life and nothing more. This is not what we are talking about in connection with this negative prime directive. This situational example is simply an expression of a secondary belief which is easily removed when things get better. However, if you feel like you do not like yourself most of the time then this becomes a characterlogical trait making up your ego personality indicating the existence of a negative prime directive in life. This negative trait should be dealt with since it has the effect of impeding your actions leading toward a happy and productive life.

In continuing with our analysis, the negative primary "I really don't like myself" creates a negative internal secondary belief "I do not respect myself or others." This secondary belief gives rise to the external belief "I let others use and abuse me and, in turn, I victimize others" This external belief in turn gives rise to the negative reality "I just had another fight with my women and she left me like the last one." Now again remember these responses are conceptual generic responses intended to merely illustrate a given perceptual position representing an internal state of mind in response to the negative primary "I am dissatisfied with myself". As indicated a moment ago, there are a hundred and one different ways of expressing these positions. Therefore, watch what you say and what you think for you might end up with what you don't want.

Here I am only trying to convey the substance of the

thought, not a specific intent for any particular person. Now take a minute to look over many of the other examples set forth on Chart 34. They illustrate how the negative primary programs give rise to the secondary beliefs which, in turn, affect all areas of your life in literally hundreds of different ways each and every day. Remember, the subconscious mind doesn't second guess the conscious mind. It only scans its memory banks and the experiences of our outer physical realities by way of its cybernetic loops. By the way the subconscious mind remembers everything even if some things can't consciously be readily recalled. In this manner it is able to come up with the most appropriate response to express a conscious thought as an awareness of our outer physical reality.

For another example of how the positive secondary beliefs in life affect our outer physical realities look at item 6 under the positive primary personality program in Chart 33, "Sex is good". The internal secondary belief forming the corollary to this positive primary is "I view sex as healthy and I enjoy a good sex life". The external secondary belief would be "I am committed to the one I love and we sexually express our love for one another ." The external outer physical reality then becomes "My life is fulfilled as I am able to express my love for my partner." However, when the unconscious primary is "Sex is bad" as shown in item 8 in Chart 34 the internal secondary belief might be something like "Sex is dirty, shameful and base." The external thought would then become "Men use me to satisfy their own lustful desires," which gives rise to the actual physical reality of "All men ever want is dirty sex, then they just dump me!"

The point of this is that our outer realities track with the deepest unconscious primaries originating in our formative personality and operational systems which arose from the earliest events in our life. In our adult years we may not have any present conscious awareness of how these early events affect our life. Therefore, there is no way the adult would even know that his problem of poor relationships stems from a hidden negative personality trait buried deep in the subconscious cognitive structures of his mind. Until the negative energy of this negative primary is removed there is little chance of total intimacy and openness in a committed relationship. This was symptomatic of Ricky Mindful's problem in making a commitment to Betty. You probably have heard the old adage "You must be able to love yourself unconditionally before you can wholly love another." You may want to again refer to all of these charts in this series. These charts illustrate how energy and images of thought express the primary processes of the mind which in turn create the inner reality secondary beliefs about life, which reflect the outer physical manifestations in life. Once you begin to understand the dynamic relationships between our inner and outer thoughts in creating realities in life you finally are able to start taking control over those realities, and if you are not satisfied with them you can start to change them.

Summary on the Primary Programs and the Secondary Belief Systems of the Mind

This is the area in which individuals, families, societies, religious,ethnic and political groups, and other socio economic groups have created sometimes complex and often very rigid beliefs systems about themselves and others. They have defended such belief systems to the death, or worse yet, have started religious, ethnic, and political wars often resulting in massive amounts of pain, suffering and death to others who opposed such belief systems. Such destructive belief systems could, and in the future can, be avoided. The participants need only expand and broaden their cognitive structures and their resulting beliefs systems in order to become more fully aware of who and what they are, and what they are doing to themselves and others. From this more enlightened cognitive structure they would be able to create for themselves and others within their community more enriching and edifying realities instead of self limiting and destructive realities. In this way they would be able to develop new belief systems based on love, understanding and cooperation among all mankind. This, in turn, would result in a more peaceful world environment where the human worth of each man and women would be honored and respected.

CHAPTER 9

Hierarchy Of The Cognitive Structures Of The Mind

Introduction to the Cognitive Structures of the Mind

In this Chapter we explain what prevents you from living the type of life you want. All of the previous chapters culminate in this Chapter and the following Chapter 10 on creating your future realities in the spacetime of your mind. We have seen how the theories of quantum physics and relativity theory of Chapter One and the complexity of self organization of Chapter Two set the stage for understanding the mental processes of the brain and mind in Chapter Three. Chapter Five dealing with the neurological structures of the brain and Chapter Six, dealing with the subconscious mental processes, set the foundation for understanding how mental processing works in creating realities. Chapter Four on consciousness, Chapter Seven on how thoughts are formed and Chapter Eight on the system of primary and secondary beliefs, all set the stage for this Chapter Nine dealing with the cognitive levels of ego personality in forming your realities in life. For it is the level of your awareness of your ego personality which determines how meaningful and fulfilling your life will be. No amount of changes in your conscious thoughts, beliefs, actions, behavior, plans or strategies for the future will ever significantly change the fundamental characteristics of your ego personality which is the mechanism forming your inner and outer realities. Such changes cannot come about unless you are capable of changing the components of ego personality making up the very fabric of your life.

Chapters 7 and 8 described how both the energy and images of our thoughts and belief systems, primarily of ourselves, our relationship to others, and to our environment, give us a continuous experience of what we call our "personal reality." It is because we cannot assimilate rapidly enough the literally thousands upon thousands of data bits of information our sensory input channels are bombarded with every minute of every day that our ego personalities create a containment field. This containment field acts to limit the amount of data it can consciously handle at any one time in any meaningful way in order to rationally function in its world environment. The cognitive structure is just such a containment field. Your cognitive structure can be defined as "the mechanism used by ego personality to order an individual's world in a meaningful fashion so that a person can make sense out of it." It is the instrumentality which gives rise to our experience of a personal reality.

We create literally hundreds upon hundreds of thoughts each and every day which give substance and structure to our life. Everything outside this immediate frame of our personal reality exists as a physical dimension consisting of physical objects in the three dimensional space of our physical world. However, it is not necessarily a part of our personal physical reality. We only define our personal reality to include the physical dimensions of our existence which we personally perceive and interact with in some fashion. It is this personal reality which defines who and what we are. Therefore, it is this personal reality we are primarily involved with in creating new and expanded realities in life.

The term cognitive structure includes not only the structure from which ego personality determines our thoughts, feelings, behavior and actions, but it constitutes the very fundamental structure giving rise to all of our realities. In this context the cognitive structure provides all of the conscious and subconscious psychological programs from which we formulate for ourself our entire range of thoughts and feelings concerning ourself and our world throughout our entire lifetime. Thoughts originating at the level of the cognitive structure at which a person is currently operating do not require any significant shifts in the conscious awareness the person is experiencing at that level. Thoughts arising within a given level are generally characterized as logical rational thought. Logical rational thought represents the type of thought processing we normally use to organize our daily affairs. We have seen that when a person's organizing synthesis is not adequate to resolve problems a person faces at the level of self awareness at which he is presently operating the problem appears to be insolvable. Indeed the person's problems are insolvable at that cognitive level of self awareness.

Ricky's reluctance to make a commitment to Betty is a good example of addressing a seemingly insolvable problem at the level at which the problem arose in his life. In order to move to a higher level of organizing synthesis a person needs to move to a higher level of awareness. Ricky was able to do so with the assistance of Tommy in the north Georgia mountains. He finally consciously understood that his resistance to making a commitment to Betty was the result of his unconscious fears of being hurt all over again. These fears stemmed from his early childhood and not as a result of his relationship with Betty. Moving upward from one cognitive level to the next higher level requires changes in conscious awareness and not just changes represented by variations in our logical rational thought processes arising within a given cognitive level. When Ricky finally became aware that his surface problem with his relationship with Betty stemmed from his subconscious deep seated fear and anger with his mother and himself he was able to then release them and make a vertical shift in conscious awareness and deal with his problems from a higher level of organizational synthesis. From this higher level of awareness it became obvious that making a commitment to Betty was in his best interest. Changes in our awareness normally require Creative Integrative Thought™ of this nature. Creative Integrative Thought arises within Conscious Analog Thinking in spacetime. Exactly how you can utilize analog thinking in spacetime to expand your awareness in order to create superior realities in life is dealt with in this Chapter, in Chapter 10 and in Appendix One.

Levels of the Cognitive Structure of Ego Personality

As we have seen the cognitive structure of the mind consists of both a conscious structure and a subconscious structure. The subconscious structure consists of the subconscious aspects of ego personality. This would consist of the primary personality program and the primary operational program. These subconscious programs form the basis for the organization of all conscious thought making up all aspects of our life. Conscious secondary beliefs make up the self image or conscious aspects of ego personality. Conscious secondary beliefs consist of our beliefs making up our inner realities, the secondary beliefs concerning our outer realities, and our experiences of our outer physical realities themselves. All of these aspects of our secondary belief system constitute aspects of the processes of our conscious cognitive structure. We referred to all conscious thought regardless of its content as the secondary belief systems of the mind as contrasted to the subconscious primary programs of the mind. We have seen that the subconscious aspects of the cognitive structure do not constitute thought themselves but rather form the basic application programs from which all conscious thought arises. The personality program contains the application programs which provide the containment field and the outer parameters within which an individual formulates his basic belief systems concerning his ego personality. The primary personality program determines who and what you are and how you perceive yourself to be in your world. The primary operational program serves essentially the same purpose and sets the foundation for all thought concerning how well you will function in life. All subconscious processes exist and function in the spacetime of the fourth dimension. All conscious processes exist and function in fixed time and space of our three dimensional world. We have seen that as you move self awareness from consciously processing realities to subconsciously processing you begin to experience the effects of the deeper levels of these processes themselves. You can

CHART 35

Components of the Cognitive Structure Level at which We Create Realities in Life

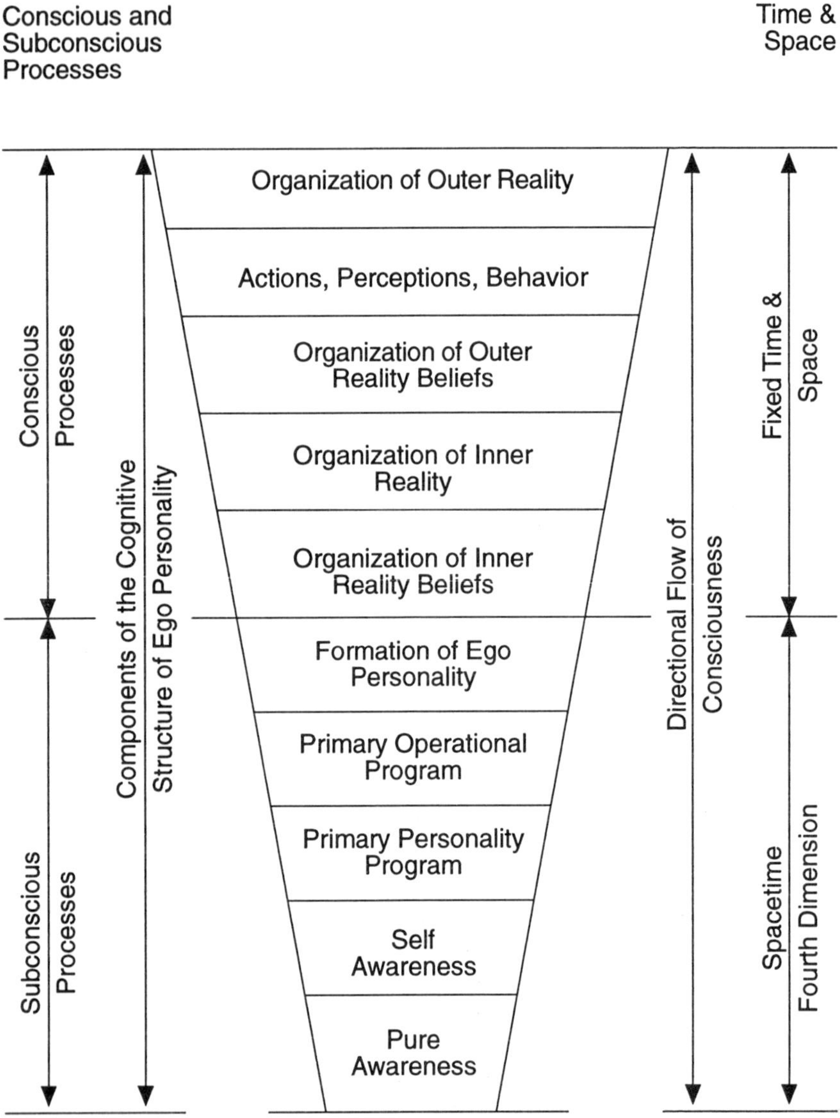

NOTE: As you move consciousness of ego personality to experience the deeper levels of the processes organizing realities in life you can more effectively take control over those processes to create expanded realities for yourself.

then effectively start to take control over these processes in order to create expanded realities for yourself.

Refer to Chart 35, page 321, *Components of the Cognitive Structure Level at which We Create Realities in Life.* The level of the cognitive structure at which a person is presently organizing his life consists of all of the conscious and subconscious components involved in organizing realities in life as shown on this chart. We have reviewed the functions of each one of these components separately in previous chapters. We will now look at how all of these separate components function as a single system referred to as "ego personality". All of these components taken together represent various aspects of the awareness of ego personality. The reason for this is that ego personality avails itself of all of these components working together in combination to create our inner and outer realities in life including our reality of ego personality itself. Ego personality exists as the resulting outcome of these combinations of components of the cognitive structure working together to create realities. In Chapter 4 ego personality was defined as "the interface mechanism by which self awareness is able to create for itself physical realities in a three dimensional world in time and space." Ego personality consists of the composite mental, emotional and physical characteristics and traits an individual develops in order to effectively function in his physical environment. "Self awareness" was defined as "the faculty by which ego personality became aware of its own self existence." "Consciousness" was defined "as the process functions by which self awareness, through the mechanism of ego personality, is able to organizes life realities for itself."

Self awareness itself does not exist as the resulting outcome of these combinations of components of the cognitive structure working together to create realities. It is ego personality which exists as the resulting outcome of these combinations of components working together to create realities. While self awareness generates the outcome of our life it is not reflected in the outcome of our life. Self awareness is what determines the characteristics of ego personality. Ego personality, both creating and being acted upon by its environment, determines the characteristics of the human mind and the characteristics of human reality. It is ego personality that reflects the content of our life. *It is ego personality that creates a sense of who you are and what your world will look like!*

Charts 36-1, 36-2 and 36-3 starting at page 323 are all related and should be viewed as a whole. Chart 36-1, *Hierarchy of the Cognitive Structures of Ego Personality,* sets forth a very simple schematic diagram of the architecture of the levels of the cognitive structures of

CHART 36-1

Heirarchy Of The Cognitive Structures Of Ego Personality

Time & Space		Cognitive Levels
Holistic Thought Spacetime Fourth Dimension	Pure Awareness Whole, Perfect, Complete	7
Holistic Thought Spacetime Fourth Dimension	Self Enlightened - Holistic - Positive Cognitive Structure	6
Logical Thought Fixed Time and Space	Self Realized - Expanded Positive Cognitive Structure	5
Logical Thought Fixed Time and Space	Self Actualized - Open Positive Cognitive Structure	4
Logical Thought Fixed Time and Space	Self Existence - Closed Positive Cognitive Structure	3
Logical Thought Fixed Time and Space	Self Existence - Closed Positive/Negative Cognitive Structure	2
Logical Thought Fixed Time and Space	Self Existence - Closed Negative Cognitive Structure	1

States of Consciousness

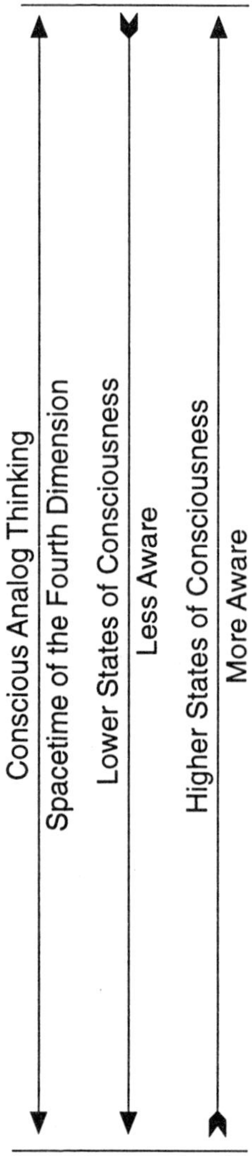

As you move to higher states of consciousness and experience expanded states of awareness you move vertically through the cognitive levels of the mind and experience higher levels of organizing synthesis in creating expanded realities in life.

CHART 36-2

The Individual Components of the Levels of the Cognitive Structure

Components of Self Awareness	Level One	Level Two	Level Three	Level Four
Primary Personality Program	Negative - Closed	Positive/ Negative - Closed	Positive - Closed	Positive - Open
Primary Operational Program	Negative - Closed	Positive/ Negative - Closed	Positive - Closed	Positive - Open
Ego Personality	Negative - Closed	Positive/ Negative - Closed	Positive - Closed	Positive - Open
Secondary Beliefs Inner Reality	Negative - Closed	Positive/ Negative - Closed	Positive - Closed	Positive - Open
Organization of Inner Reality	Negative - Closed	Positive/ Negative - Closed	Positive - Closed	Positive - Open
Secondary Beliefs Outer Reality	Negative - Closed	Positive/ Negative - Closed	Positive - Closed	Positive - Open
Actions, Perceptions Behavior	Negative - Closed	Positive/ Negative - Closed	Positive - Closed	Positive - Open
Organization of Outer Physical Realities	Negative - Closed	Positive/ Negative - Closed	Positive - Closed	Positive - Open

This chart describes each cognitive level in terms of the individual components making up the cognitive structure. As you start to move up the heirarchy of the cognitive structures of the mind you start to become more aware of who and what you are. As you become more aware of who and what you are you start to shift from negative energy and negative images in life to more positive energy and positive images in life.

CHART 36 -3

Descriptions of the Levels of Cognitive Structures of Ego Personality in terms of the Personality Traits of the Secondary Belief System Structure

Charact-eristics of Ego	Level One	Level Two	Level Three	Level Four
Self Love	Little-None	Mixed	Yes	Yes
Responsibility	Little-None	Mixed	Yes	Yes
Truthfulness	Little-None	Mixed	Yes	Yes
Morality	Little-None	Mixed	Yes	Yes
Values	Little-None	Mixed	Yes	Yes
Integrity	Little-None	Mixed	Yes	Yes
Character	Little-None	Mixed	Yes	Yes
Concern For Others	Little-None	Limited	Limited	Yes
Social Concerns	Little-None	Limited	Limited	Yes
Charitable Concerns	Little-None	Limited	Limited	Yes
Respect For Others	Little-None	Limited	Limited	Yes
Accepting of Others	Little-None	Limited	Limited	Yes
Accepting of Others Beliefs	Little-None	Limited	Limited	Yes

The above chart sets forth descriptions of the secondary belief system specific personality traits of ego personality in determining the level of awareness of ego personality. As you move your level of awareness up the cognitive structure of ego personality you move from negative closed personality traits of the secondary belief system structure to more positive open personality traits thereby manifesting in your life a more open, accepting, positive ego personality resulting in the creation of positive more edifying realities in life.

ego personality. This simplified diagram shows seven levels comprising the entire cognitive structure. In reality each level consists of many subordinate interconnected and interrelated levels existing within each major level described in this chart. For representational purposes this chart sets forth only the significant characteristics of each major level of the structure itself and not the many interconnected and interrelated subordinate levels existing within each major level. Chart 36-1 characterizes the energy and images organized by each component of the first four cognitive levels as being either negative or positive in nature and either open and flexible or closed and rigid. We start with level one, the lowest level on the chart. Level one is defined as being both closed and negative. Therefore, realities in life generated at a level one cognitive structure of ego personality will become both negative in outlook and rigid and inflexible in adapting to change.

One example of limited awareness at level one is the result of repression, denial and projection. In very simple terms repression is when you are not even aware of the source of your negative feelings.[1] Denial is simply not being in touch with the source of your negative feelings. Projection is when you justify your negative feelings and actions by projecting them onto others or events in life which you then blame as the cause of all of your problems. In Ricky's case his aggressive business style and lack of intimacy in his relationships with women was a projection of his negative feelings arising from his early childhood. Your created realty reflects back to you the very things you do not like about yourself but are unaware of their source. We stated that the effects of repression, denial and projection represented only one example of a negative closed cognitive structure for the reason that your entire created reality in life is a reflection of the level of the cognitive structure at which you are experiencing and creating a life. Your life consists of a lot more images of reality than just negative images you might project onto others.

A level two person exhibits both positive and negative aspects in life and is still somewhat rigid and inflexible in adapting to more positive changes. A contrast between a level one person and a level two person is that the level one person's four archetype prime directives are all substantially negative reflecting a negative outlook. A level two person has a combination of both negative and positive

---

1 Repression is when you bury your negative feelings so deep into your subconscious that you are not aware that they even exist. However they form feelings and images of your primary programming in life. Suppression on the other hand is when you are consciously aware of your negative feelings but you try to ignore them.

primary programming. However, both a level one and a level two person are still somewhat inflexible and rigid in adapting to positive changes. The reason for this is that the essential characteristics of their ego personality once formed in early childhood are substantially set for life unless something significant happens to make them change as we have seen is Ricky's case. A level three person is characterized as being essentially positive in nature but still pretty much fixed in his ways and is slow to change the processes by which he changes his inner and outer realities. The reasons for such resistance will be made clear in just a minute as we review both the process and content constraints to creating a superior life for yourself. A level four person, on the other hand, is both positive and open and flexible in adapting to changes, particularly changes which might better his circumstances in life. This person is characterized as being, for the most part, a self actualized individual.

While these levels appear to fit into neat little descriptions on paper, in reality these levels are overlapping and interrelated and interconnected. For instance, a person can easily be at a higher level in his family life and at a much lower level of awareness in his business and social life. Also a person can be flexible and adapting in his business life and very rigid and authoritarian in his family life. This does not necessarily create cognitive dissonance for the reason that most of us unconsciously separate different aspects of our life and can operate with different moral and ethical standards depending on the aspect we are dealing with at the moment. For instance, Ricky could be loving, kind and considerate toward Betty and Louanne and also at the same time be extremely competitive and ruthless in his business practices. He could be highly competitive and cutthroat in business and in sports, and yet at the same time he would be considerate, respectful and show considerable deference toward his brother Tommy even when he didn't always understand or even agree with him. Ricky could even love Tommy and Betty dearly while at the same time he consciously hated his mother and subconsciously even held considerable self hatred toward himself and still not create cognitive dissonance within himself. This is not to say that such conflicts did not create pain in the form of tension and stress in Ricky's life. It is just that it did not create conscious cognitive dissonance in his life. In a word Ricky was simply unaware of what was going on in his life.

You will note that we only address the first four levels of awareness of the ego personality and do not address levels five, six and seven in the charts. The reason for this is twofold. We have always said that it is not our intent to create enlightenment but rather to create realities in life. When creating realities in life the MetaCreation process, as an unstated outcome, also points you in the direction of enlightenment. It is inherent in the process for as you move upward from level to level you automatically increase the level of your consciousness and attain higher states of awareness. However, this is just not a stated goal of this book. The stated purpose of the book is to show you how to create more enjoyable realities in life.The second reason is that the vast majority of society falls into one, or a combination, of these first four levels of self awareness. How many people do you actually know personally who you would characterize as being fully self realized, self enlightened or have reached a transcendent state characterized as perfection in life. Society is starting to move in this direction at an accelerating rate. Such levels of consciousness will be reflected in the global consciousness of all mankind and in expanded realities of the world which we all will share in common at some future point in time[2].

Chart 36-2 describes these first four levels of the cognitive structure in a slightly different fashion. In Chart 36-2 each cognitive level is described in terms of the individual components making up the cognitive structure. You will note that as you move from one level to the next higher level you move from negative to positive with respect to each component. Chart 36-3 describes some of the major characteristics of the level of awareness of ego personality and describes the first four cognitive levels in relation to these characteristics. You will note that these characterizations are really descriptions of personality traits. For it is the personality traits of ego personality which drive the creation of inner and outer realities. These personality traits have been chosen for they represent the most significant aspects of ego personality. These personality traits, as with all other personality traits making up the character of ego personality, are simply descriptions of secondary belief system components of the cognitive structure of ego personality. Remember, all conscious thoughts and feelings regardless of their content arise within the secondary belief system structure of ego personality. Ego personality not only embodies these personality traits described in Chart 36-3 and everything else you might think about, but also includes the human physical body as an attribute of ego personality for purposes of creating physical realities.

You will note that those personality traits at each level relate to both the inner reality of ego personality and to people, places and

2 Notwithstanding our position in this respect we do include a brief description of levels 5 and 6 at the end of this chapter. We can only describe level seven in abstract symbolic terms since level seven transcends descriptions of cognitive thought. For a symbolic metaphor describing level seven refer to the Epilogue at the end of this book.

things lying outside of the inner personal frame of reference, or inner reality, of ego personality. Those personality traits which refer to the inner reality of ego personality at level two are referred to as mixed. However, those personality traits that relate to people and events in life lying outside the inner images of ego personality at level two are referred to as limited aspects of ego personality. The reason they are limited is that ego personality itself is, for the most part, egocentric by nature. A level one or level two ego personality seldom describes a sense of self in terms of how it can unconditionally help, assist or promote the welfare of other people, events or things in life in a purely altruistic way without first determining unconsciously how such actions will affect its own sense of self worth, self esteem or self existence. I say unconsciously for the reason that the primary personality program is on and functioning all of the time and monitors every thought and action to see that it conforms to the basic parameters of these subconscious programs. Whether these subconscious programs are limited, negative and inflexible or positive, open and flexible is reflected by what level you are functioning at in life and in the realities you are creating at that level.

If a purely selfless act appears, in the slightest way, to make a level one or two ego personality look bad it will not consciously be acted on at these lower levels of organizational synthesis. A level one or two ego personality primarily defines its own sense of self identity in terms of how other people, places, and things enhance its image of itself and not in how it might enhance the lives of others. Seldom will it act in circumstances, whether consciously or subconsciously, which appear not to be edifying or enhancing of its own self image.[3] On the other hand a level four self actualized person will, in appropriate circumstances, act unconditionally or unselfishly, to help and assist others in his life without the intention of enhancing his own self image. The reason for this is that the level four person no longer defines who and what he is solely in terms of the physical, mental and emotional aspects of ego personality. He has become more inclusive in how he views his world, a world that embodies not only an expanded sense of self identity but also a world where the person increasingly is becoming consciously aware of being connected to everyone and everything else existing in his environment.

Many people are at level three where they are positive, open, loving, kind, considerate and accepting of themselves, their loved

3. These are general descriptions only concerning the affairs of a level one and two personality. Of course, such individuals show love and concern for members of their own families and others in their life. However, these descriptions are only given as broad, overall, general descriptions of a level one and two life style.

ones, family, friends and associates. They are successful both in their business life and personal life. However, they operate out of a closed or fixed cognitive structure where they have limited acceptance, respect or concern for those who are entirely different from themselves because of their race, nationality, religion, sexual orientation, or socio-economic status. They are limited in such acceptance because they still operate from a perspective of living separate and apart from everyone else and everything else in a separate and independent objective world which lies outside themselves and largely outside their control. This level three person often has little concern for the welfare of these groups of people who are different from themselves. A perception of living in an objective world where everyone exists separate and apart from everyone else is probably the most significant mental constraint on the advancement in human consciousness of the whole human race. Process and content constraints of the mind are discussed later on in this chapter.

This is really the deeper message that was revealed to Tommy Mindful by the Mystic Shaman in the Superstition Mountains. While the organization of the inner processes of Tommy's brain and mind are exceedingly important for an understanding of how Tommy, and for that matter all of us, organize our realities, the realization of our deeper interconnection to everyone and everything else is by far the most important revelation. For it holds the key to the course of the future evolution of the consciousness of the whole world. The ancient Anasazi revealed to Tommy through metaphor that the spirits of their dearly departed ones, the consciousness of those living today, and the yet unborn future generations, are all interconnected and interrelated. The metaphor of mother earth nourishing and sustaining all of us and that we collectively walk on holy ground, so to speak, reveals this innermost secret of this interconnection to one another and to our earth environment.

In the near future, all mankind will come to this understanding, and will reach a deep appreciation and experience of this oneness and our connection to the environment which sustains and nourishes each one of us. At that point, individually and collectively, we will begin to move from a level three personality to a level four personality. The old mental representations of the Cartesian way of life will ultimately fall away and make way for a new stage in the evolution of all mankind represented by the quantum reality that we each exist as part of this greater whole, where we not only honor ourselves and our own family, community and country, but we learn to honor all mankind and the earth itself. At that point we will move from self existence to actualization, realization, enlightenment and stages which lie beyond. A level four person is included in the description at the end of this chapter when I describe level 5 and level 6 personalities. Refer to Chart 39 at page 349 which describes a limited cognitive structure of a level one and two person versus an expanded cognitive structure of a level four person.

The list of personality traits of ego personality in Chart 36-3 is not intended to be a complete or an exhaustive list of all of the characteristics of ego personality nor is this book intended to give an extended discourse on social responsibility, morality or ethical conduct. Rather, it sets forth only the most salient characteristics of ego personality in order to give you a representational sampling of what is involved in the organization of cognitive levels and what needs to be addressed when moving vertically from one level to another. You can add to this list or change this list to better suit your own circumstances or preferences. This is simply a model, and like any model it is only valid to the extent that it is useful in understanding the processes by which we create realities. I am not so much interested in the content of one's life, morally, ethically or socially, as I am in the processes used in creating the content of that life. The content is left up to you. If you are not satisfied with the content of your life by the end of this book I hope I have provided you with some models and tools in which to change the content and create for yourself a better life.

<u>The Nature of Self Awareness as You Move Up Levels of the Cognitive Structure</u>

Self awareness seldom personally experiences itself apart from its ego personality. It therefore concludes erroneously that it is in fact, ego personality. It concludes that there is nothing more to self other than what we experience as our ego personality, that ego personality is the sum total of all there is to the conscious awareness of self. This problem started early in life when we labeled ego personality as "self". Remember when we were very young we started to label everything in our life and after awhile we had accumulated a large body of knowledge and an attitude. Well one of the very first things we labeled is this personal sense of self which we refer to as "ego personality" and in doing so we, of course, unconsciously subscribe tremendous value to it. "Hey, why not, it's me and my life you are talking about here," you might be inclined to say. However, in the process of subscribing so much value to ego personality we never stop to think about the limitations we place on ourselves by creating it and then defending it under almost all circumstances. We would be better served if we were to recognize it simply as our creation and then take control over the components of that creation instead of defending it. In this way you would be able to change and expand its definitions as circumstances warrant in order to create both a better you and a better life.

However, ego personality seldom, if ever, concludes that ego personality is just another creation of self awareness. It is no better or worse than any other creation in the context of the physical realities of our existence. We are not aware of our sense of a personal self as self awareness instead of ego personality for the reason that you do not physically experience self awareness through the five sensory systems or the subconscious somatic system. You do not physically experience self awareness for self awareness exists as the process of life not as the content of life. It processes, that is, it creates life. It does not exist as the content of life. In Chapter One we stated that the universal construct in the universe as well as in life itself was energy and not quantum solid masses. The energy of the self awareness process function is dynamic energy forever organizing and dissolving mental states and organizing and dissolving realities in life. In Chapter Four we indicated that self awareness was a process function of consciousness and did not represent the content of life. As a process function it translated an undefined sense of self into the ego personality process component of "this is me" which was then translated into the prime directive "I like who I am." It is this prime directive which we experience as the embodiment of ego personality, not the undifferentiated sense of self awareness represented by the process component of consciousness.

By now you might be asking if the identity of self awareness doesn't lie in the characteristics of ego personality and in the material things of this world, then where does it lie? The answer is simple. It lies in spacetime, as process, represented by the metaphor, whole, perfect and complete, at the subconscious level of the mind. Remember the subconscious mind processes realities in spacetime. You place the content of your life in linear time and fixed space. Therefore, since self awareness is process, you can become personally aware of its true nature by implementing it in your life. Implementing it in your life is very difficult without the requisite knowledge and technology needed to make significant shifts in your understanding of the very nature of mankind. Implementing it in life is what this book is all about. We have seen that such shifts often entail shifts in ego personality, that is, shifts in moral values, integrity, character, respect and cooperation with all others in your life. They often call forth greater responsibility and compassion toward ourselves and our fellowmen. Truthfulness in all of our affairs and a tolerance toward others is often required. In a way it is a shift in the very understanding of the nature of your very own being in order to experience your true nature and the nature of life realities created in a manner which reflect and support your purposes in life. This is what ultimately gives meaning and direction to life.

Therefore, you can start to identify with self awareness by identifying with the positive effects it produces in your life. Once you realize that you are not ego personality, nor the many realities forming the content of your life, you can then start to make rapid progress in changing them for you realize that you do not exist as any of these aspects of life. Your true identity lies in the source of the process of self awareness, the dynamical energy itself which empowers you and gives life to your personal concept of self awareness. Therefore, when you start processing your life from the level of self awareness instead of from the level of ego personality you start to experience self awareness as a sense of well being, empowerment, freedom and detachment. You are then able to move up the cognitive levels of your mind. You start to get a personal experience of what the metaphor, whole, perfect and complete, truly means in a very personal way when you are able to take control of your life and fashion it in the manner you desire. The level five and six personalities described at the end of this chapter set forth the nature of self awareness as experienced at these higher levels of the cognitive structure of the mind.

Accordingly, your own personal answer to your identity as self awareness will become very clear to you by the time you have completely read this book. This is not to imply an intent on my part to put the reader off in this respect. However, the profound truth which lies behind the answer can only be understood and personally comprehended by exploring, uncovering and personally experiencing this truth for yourself, for it is revealed personally through the process! Each person must follow his own path in discovering his own answers for this truth is different for each one of us, for the process of life is different for each one of us. It is revealed to each one of us in different forms and at different times in life. Yet all will find the answer and experience their true identity if they sincerely seek to know their own truth. This is a case where "if you know the truth the truth will set you free".

The vast majority of people, however, create their thoughts about themselves and their world somewhere within the first three levels of the cognitive structure. While for the most part such people strive to better themselves they do so, more or less, by expanding and enhancing ego personality for that is the way their life is organized. That also ultimately is the way the whole world is organized. Seldom do people make dramatic shifts within their lifetime up the hierarchy of cognitive structures where they are able to achieve considerable expansion of self awareness in broader more flexible terms much less transcend the ego personality entirely.

Such shifts, when they do occur, are usually haphazardous and spontaneous and are usually characterized as being mystical, transcendent, or religious in nature. Often people who have had near death experiences report encountering a white light or beings of a higher order. The person experiencing such cosmic or mystical episodes reports that he reached a mystical or cosmic high where time and space stood still. He experienced a transcendent state where he felt that he personally experienced and understood the nature of God or the Source of all creation, if only for just a brief moment. Often after such an experience the person retains a greater degree of illumination and understanding of the nature of the world and human nature than he had before his experience. For the rest of us, however, life remains a mystery where we are born, live out a life where we endeavor to maximize our gains and minimize our loses before we die never quite knowing what it really is all about at the deepest level of human awareness.

We didn't quite know that its is the mind/brain, consciousness, and self awareness that operated to create for us an ego personality for the creation of our life realities. We were creating realities which were often unintended and unappreciated but nevertheless tolerated for we don't quite know how to live our life in terms of higher levels of knowledge and understanding. Knowledge and understanding that would show us the way of life and the means for taking control of that life in order to change things for the better. Finally, we need to come into an understanding of who we are and why we are here and most of all what we are supposed to do about it now that we are here. We need to be able to ask the right questions in order to find the right answers. At long last we need to live life to the fullest, a life that ultimately leads us to expanded levels of consciousness and higher states of awareness. Hopefully by understanding how the brain and mind work together to create realities in life you will be able to use the processes spelled out in this book to build a better life for yourself and in the process build a better you.

The Structure of Thought Within The Cognitive Level You Are Currently Experiencing Life

The cognitive structure of ego personality contains within the structure itself the content of all thoughts arising within the secondary belief systems which make up the conscious portion of the structure. We have represented belief systems as modular in structure for purposes of studying the cognitive structure of ego personality. Conceptually you could visualize the dimensions of a thought as having height, width and length much like an abstract transparent cube, a cube which holds within its six sides the energy and matrix configuration of the thought itself. Similarly related thoughts and beliefs which originate within the same underlying subconscious primary program that contain the first thought, arrange themselves in similar modular fashion in close relationship to the structure of the first thought. The reason for this arrangement is the similarities in their relationships and the attracting force of the polarity of their energy. Chart 37-1 at page 336 entitled, *The Originating Ego Personality,* conceptually sets forth the form of this structure. The outside confines of the cognitive structure forms the outer limits of the structure. All related thoughts arising in relationship to the first thought of the original program are constrained by the larger structural system of the original program. In turn these related thoughts form their own cognitive structures and act to constrain subsequent related thoughts. An example of this modular structure is set forth in Chart 37-2 at page 337 which sets forth the twelve originating primary programs of the subconscious mind. Thoughts relating to the same belief system are interrelated and interlock with each other through the affinity of their cognitive structures for one another. In this fashion many related beliefs form complex networks. Refer to Chart 37-3 at page 338 for an example of how these networks interconnect with one another to form complex networks of belief systems. We have seen in Chapter 6 that many millions of these complex networks of belief systems form the landscape of the cognitive level at which you are presently experiencing your world. It is through these complex networks of thought that the cognitive structure of ego personality creates experiences of reality.

The analogy to the structure of a cube serves the purpose for visualizing this structure. However, the mental processing systems for organizing thought are based on the values or the weights within the analog processing system of the subconscious mind. These values consist of units composed of various degrees of positive or negative energy and are not cubical or modular in shape. The matrix configuration of the units comprising thought has definition. Thought, not unlike the basic atoms and molecules we reviewed in Chapter One, has form and shape which gives the thought its definition. It is this form and shape which is represented here as being modular in form. For this reason this modular format makes a useful, though not totally accurate, representational model. Notwithstanding this distinction we still use our analogy of cubes and modular structures as a representational model to explain conceptually how ego personality organizes thought in hierarchical fashion to create

CHART 37-1

The Originating Ego Personality Cognitive Structure Described as, "This is Me". Including the First Four Archetype Prime Directives in Life

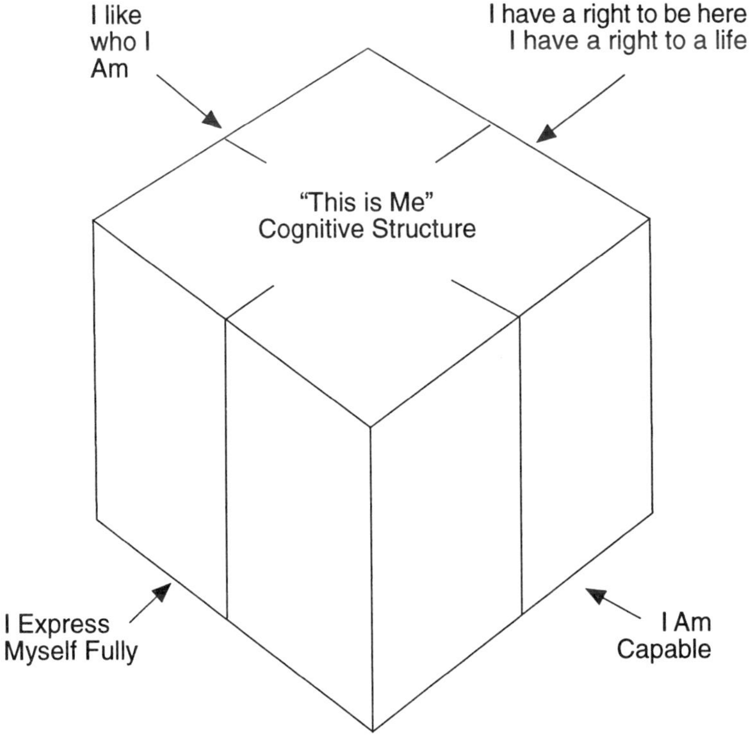

CHART 37-2

The Originating Ego Personality Cognitive Structure Including the Twelve Originating Primaries of the Mind

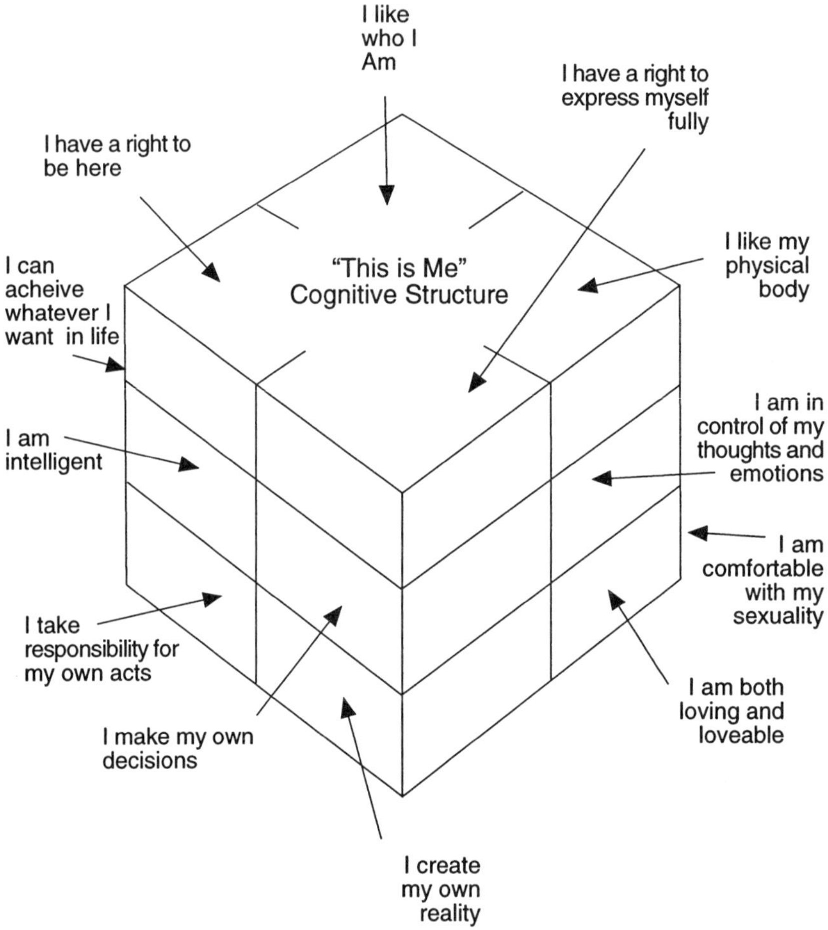

CHART 37-3

Modular Cognitive Structures of the Mind

Examples of Horizontal Shifts Within Related Groups of Secondary Belief Cognitive Structures

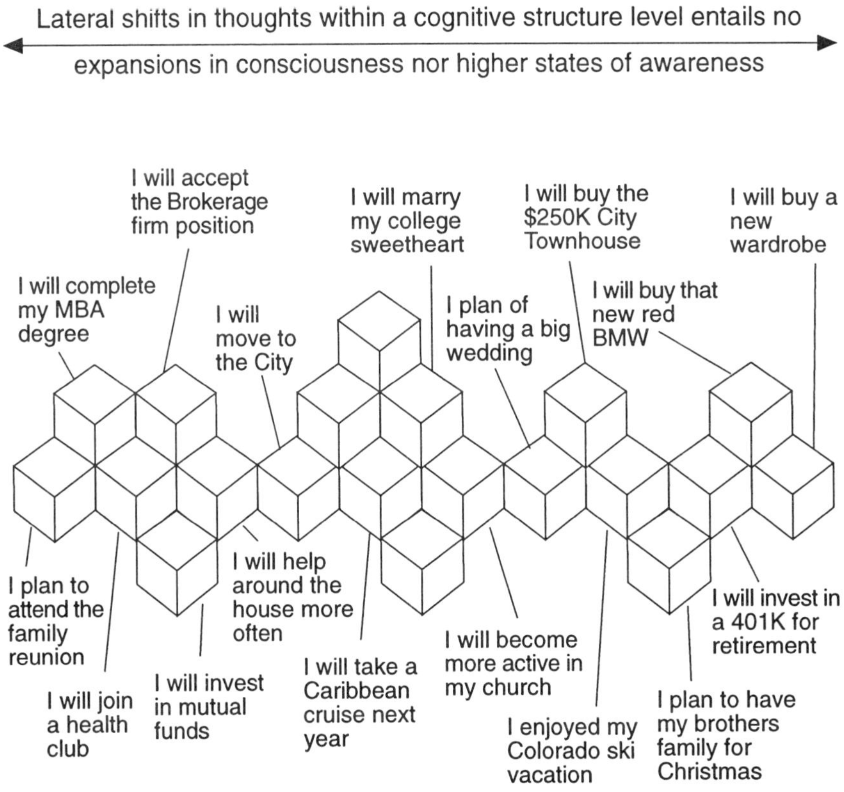

NOTE: When a person makes horizontal shifts in thought within a cognitive structure level no shifts take place in the definition of the ego personality. Such shifts only bring about additional realities that are experienced by the existing ego personality. However vertical shifts between levels of the cognitive structures entails shifts in redefining ego personality as it relates to its world and results in more inclusive views of the world and the creation of expanded realities in life.

realities. Keep in mind, however, the underlying systems are not primarily expressed as structure and function. It is the dynamical processes of the system which are by far the most significant characteristic of the cognitive structure. How they look structurally is secondary to the process features of how they work, but how they look helps in understanding how they work.

Many millions of these networks of complex belief systems in turn form the cognitive level at which ego personality, using rational logical thought, rearranges our system of beliefs within the level of the cognitive structure at which we view ourself and our world. In this manner we are able to create for ourself new and different experiences of reality. When a person makes horizontal shifts within a cognitive level no shifts take place in the definition of the ego personality. Such shifts only bring about more numerous realities that are experienced by the existing ego personality at the level at which these new thoughts and realities are created. However, vertical shifts between levels of cognitive structures entails shifts in redefining ego personality as it relates to its world. Vertical shifts upward between levels of cognitive structures can only come about through creative integrative thought since vertical shifts also involve redefinitions of ego personality as well as changes in realities.

Vertical shifts entail redefinitions of ego personality as well as expanded realities for the reason that in order to arrive at a new level of organizational synthesis in creating realities you need a shift in the mechanism which perceives and processes such reality shifts. This mechanism making these processing shifts is the ego personality itself. If the ego personality cannot or does not change its own perceived configuration of itself and its world it will never have the internal processing capacity to create different or expanded realities. On the other hand, if ego personality is only creating more numerous realities at the same level of its own awareness this does not require redefinitions of the ego personality. It only entails the creative application of logical rational thought in order to create more numerous realities at the level at which ego personality is currently operating. Broader more flexible definitions of ego personality result in more inclusive views of itself, its world and the creation of expanded realities.

For purposes of understanding how the hierarchy of cognitive structures works they are organized conceptually in a hierarchy of simple two and three dimensional diagrams. In actuality the systems are far more complex in that they involve thousands upon thousands of connections for the creation of thought and are organized multidimensionally. Their structure far exceeds what is shown in these simple charts. To give you an idea of just how complex these structures are the neuro networks of the brain involves millions upon millions of these connections, and each of these neuro networks is interrelated and interconnected to thousands of other neuro networks. All of these neuro networks support and sustain the creation of thought by the cognitive structure of ego personality. Thoughts arising within a broad belief system can be changed and modified as the modular structures of the beliefs within the system are arranged and rearranged within the cognitive structure comprising the belief system. The theory behind this organizational structure and specific examples of how cognitive structuring actually works in our daily life is explored below and in Chapter 10.

The hierarchical cognitive structures at the level of the secondary beliefs continue indefinitely organizing themselves into arrangements of complexity both during the formative years of the child and thereafter throughout his adult life. By now you should begin to see a pattern in that ego personality operates on thoughts and beliefs about itself and its relationship to its environment. If the child's primary personality or operational programs are defective the child will have a difficult time sustaining a fulfilling and meaningful life. The reason for this is that there is a direct relationship between how the child thinks and perceives himself and his world and how he will organize and experience that world. *The realities we experience are the realities we create!*

It is the vertical shifts that allow you to change the definition of the mechanism by which you create your life. Solutions rise out of the synthesis attained at the new level of organization at the higher cognitive structure. What has happened is not a resolution rising out of rational logical thought which takes place within a cognitive structure or laterally across a group of cognitive networks within a given cognitive level. A quantum shift takes place within the structure itself out of which the processes of thought themselves arise and operate to create realities. When the cognitive structure is transcended to a higher level logical rational thought is then processed within a much broader structure. Higher, more inclusive perspectives are achieved allowing for new higher levels of organizing synthesis enabling the individual to create for himself an integration of old realities with the expanded new realities in life. What this means is that as you move up the cognitive structure of the mind greater more functional realities emerge allowing a person to respond more effectively to the stimuli of his environment. In effect he is able to take greater control over his environment. What has happened is not that the older level of viewing reality has been totally eliminated but rather it has been transcended by a more comprehensive view of life, a view which allows greater flexibility and control over all of life when viewed from the larger context of the relationship of the individual to his environment. Such changes allow for much greater flexibility and greater possibilities to emerge as realities of the individual.

By continually viewing your world from this higher level you reinforce your consciousness at this higher level in all situations. Remember from the analysis of the brains's basins of attraction and attractor pathways, if the basins or pathways are not used eventually they will begin to decay, destabilized from nonuse, and they will deteriorate. Conversely, those newly formed basins and pathways rising from a higher organizational synthesis, if repeatedly used over time, will strengthen and become stronger. You will find that by using these newly formed perspectives in life you are able to create even greater, more expansive realities than you were able to do from your lower cognitive structure. You don't lose the cognitive processes used in the lower cognitive structures. You just incorporate them into a more holistic, more integrative viewpoint of yourself and your world.

Once you make these creative integrative shifts to higher order synthesis you will notice a dramatic shift in the way you experience reality. The quality of your life changes dramatically. Your world will start to look entirely different. You start to see things from a whole new perspective. Things that bothered you before no longer pose a problem. Your attitudes and behavior toward yourself and others changes dramatically. From this higher perspective you understand and empathize with those who disagree with you. However, you do not feel that you need to fight and defend every point upon which you disagree. You experience less tension and are more content and accepting of those conditions which you cannot change. Things that bothered you before no longer hold the importance that they did before changing your perspective on life. You begin to see people and things differently. You sense that you see things more clearly. You feel different about yourself and others. From this broader vantage point you can, by continually using the MetaCreation processes, create for yourself newer and more expanded realities in life.

Horizontal Arrangement of Thought Within a Cognitive Level

We have just seen that all thoughts within the level of the cognitive structure at which a person is experiencing life are organized by their relationships to their underlying energy and matrix configurations which act as constraints on the creation of opposing thoughts within a specific belief system. We will now explore this same arrangement of thought, but we will now look at this arrangement in terms of the content of the thought itself rather than its underlying energy and matrix configuration. Chart 38 at page 343 entitled The *Horizontal Arrangement of Conscious Secondary Belief System Thoughts,* organizes the content of thought on the basis of how it relates to ego self, others and its environment. Since it is thought that creates our realities within our environment, thought is organized in accordance with how it relates, in one form or another, to that environment. All thought then can be organized under three main headings: (1) Relationship to Self, (2) Relationship to Others and, (3) Relationship to Your World. Each of these categories have two internal aspects and two external aspects. For instance, the organization of thoughts under the heading "Relationship to Self" is organized under the internal aspects "How you view yourself" and "How you experience yourself'.

The cognitive structures concerning how you view yourself have to do with such things as your physical appearance, hair style, style of wardrobe, accessories, jewelry, physical mannerisms, mode and style of speech, behavioral styles, how you present yourself to others, and on and on. Whereas, "how you experience yourself" has to do with such things as how you feel about your appearance, how you feel about your body, how you feel in a business setting, in a social setting, with a few other people, in a large crowd, with someone of the opposite sex, with a group of people of the same sex, with people in authority, with family members and on and on. The external aspects would be "How you present yourself to your world" and "How you experience yourself in your world". How you present yourself to your world could be different than how you view yourself personally in that you might always be smiling when you really feel down or you might convey an air of confidence and not feel confident at all. In such situations you are obviously not congruent and such incongruence saps energy and creativity.

The purpose of this Chart is simply to set forth the horizontal arrangement of thought within a given level of a cognitive structure. All thought and all belief systems fall within one or the other of these horizontal arrangements of belief systems within the cognitive structure. Sensory data and information lying outside these belief systems structures and in opposition to them are often treated as anomalies by the cognitive structure, create cognitive dissonance, or they are simply ignored for they are generally not readily assimilated into the person's existing cognitive structure.You might want to refer to Chart 38 for organizing your thoughts as they relate to Self, Others and your Environment.This outline was used when grouping secondary beliefs in Chart 33 and 34 in Chapter 8 as they related to self, others, or to our world. It gives us the means to systematically categorize

CHART 38

The Horizontal Arrangement of Conscious Secondary Belief System Thoughts by Their Content Within a Given Cognitive Level of Ego Personality

1. Relationship to Self

a. Internal Aspect - How you view yourself
b. Internal Aspect - How you experience yourself
c. External Aspect - How you present yourself to your world
d. External Aspect - How you experience yourself in your world

2. Relationship to Others

a. Internal Aspect - How you view others
b. Internal Aspect - How you experience others
c. External Aspect - How you perceive others view you
d. External Aspect - How you experience the way others view you

3. Relationship to Your World

a. Internal Aspect - Your view of your world
b. Internal Aspect - How you experience your world
c. External Aspect - How you perceive the world views you
d. External Aspect - How you experience the way the world views you

---

The purpose of this Chart is to set forth the horizontal arrangement of thought by its content within a cognitive structure level. All thought and all belief systems fall within one or the other of these belief system categories. Sensory data and information lying outside the structure of these belief systems categories are treated as anomalies by the mind/brain, create cognitive dissonance, and are generally not readily assimilated into an individual's existing thoughts and beliefs systems at the cognitive level at which he is experiencing his world.

thoughts by related groups of thought as they relate to the content of our belief systems in life.

The difference in horizontal movements within a cognitive level and vertical moves between levels is that one organizes thought horizontally in relationship to self, others and the environment to which they relate, and the other organizes thought vertically in relationship to their importance to ego personality in creating realities. You could also characterize these two arrangements as one being an outer arrangement of thought as it relates to self, others and your world and the other is more of an inner arrangement by the order of importance of the thought as it relates or reflects your own ego personality. Finally the organization of thought horizontally within a single cognitive level creates quantitatively more things in life.The vertical arrangement of thought between levels creates qualitatively expanded realities in life.

Shifts in Cognitive Structures Redefines Ego Personality

The development of ego personality has been the greatest asset mankind has ever been endowed with and at the same time also poses the greatest hindrance to higher states of consciousness and expanded states of awareness. The ego is what keeps us focused outward continually striving for a better physical life for ourself and for those we love. At the same time it is ego personality that prevents us from looking within for the source of the pain, lack and limitations we often experience in daily living. It is ego personality more than anything else which prevents us from experiencing expanded levels of consciousness and higher states of awareness. It keeps us moving thought horizontally within networks of cognitive structures in order to continually redefine and expand ego in terms of greater power, accumulation of wealth, status and success in society. In this manner we hope to attain greater happiness and prosperity in all areas of our life. Little do we realize that more wealth, status or success do not bring higher states of consciousness and expanded states of awareness. They only bring more things and activities which we interpret as forward movement in fulfilling our life purposes.

What I have been saying is that the mechanism by which you view your world is the mechanism which determines what that world will look like. The mechanism is the cognitive structure of ego personality. Therefore, if you are satisfied with who you are and only want more of what the world has to offer by way of physical wealth and pleasure then use logical rational thought in creative ways by using the MetaCreation processes. In this way you will manifest greater wealth and prosperity in your life at the level at which you are now operating and experiencing your life. However, if you are not totally satisfied with your self image, that is you are not satisfied with how you view yourself, and how you present yourself to your world, which is reflected in your realities, then you need to change your self image. This entails changing the mechanism by which you view yourself and your world. The MetaCreation process also provides you with ways to actually change these components of your ego personality that you are not satisfied with.

By changing the mechanism of ego personality you can accomplish two things. First, you can create a better self identity for yourself, and secondly, you will be able to create expanded realities. We have seen that Ricky did this by removing negativity from his primary programs, and as a consequence he was able to redefine himself in terms of a person capable of self love and intimacy. He was able to open himself to the woman he truly loved. He was able to forgive himself and consequently was able to stop beating up on himself and others. He then was able, in the process, to build better business relationships in the office which better served his interests and the interests of his business associates as you will see shortly.

As we start to move up the hierarchy of the cognitive structures of the mind as shown on Chart 36-1 we start to become more aware of who and what we are. As we become more aware of who and what we are we start to shift from negative energy and negative realities to more positive energy and images as shown in Chart 36-2. As we start to exhibit more positive energy and imagery and correspondingly become more open and flexible to all life's circumstances we become more self actualized. As we become more self actualized the personality traits of ego personality become more positive, inclusive, accepting and expansive as shown in Chart 36-3.The question then becomes, do we first shift the energy of the subconscious programs of the mind, which then results in redefining ego personality in more inclusive terms, or do we first shift the characteristics of ego personality in order to then effectively shift the energies of the subconscious mind?

The answer lies in the fact that changes in either aspect of ego personality take place at the same time. Remember in our little physics lesson the polarity of the energy and the image of the matrix configuration of the atom is what constitutes the basic building blocks for the formation of all realities including ego personality and its many life realities. By changing energy we change ego and by changing ego we become more aware of what needs to be changed at the level of our subconscious energy. As we change the mechanism by which we create our realities we can effectively change those realities in order to change the quality of our life.The reason for this, as we have seen in the series of Charts 36-1 through 36-3, is that the energies of the subconscious mind, the system of our cognitive level, self awareness itself, and the characteristics of ego personality, are all interconnected and interrelated. We separate them for discussion purposes only, but in reality these are different aspects of the one and same thing. That one thing is self awareness, operating as a human being, with an ego personality, having a human experience in fixed space, linear time, in an objective world, in order to create a life for itself.

If the major condition in life is change, why do we often think and perceive that which we identify with as ourselves isn't subject to change? The reason is simple. Any perceived change to ego personality is perceived as a threat to our self identity. Guess what, the perceived threat is also a belief. True it is a strongly held belief, an emotionally felt belief, and an unconscious deeply buried belief, but nonetheless a belief. Like all beliefs it is made up of energy and images and as such it is also subject to change. Not only change but subject to being broadened out to become much more inclusive of life in order to become more accommodating to perceived differences between you and the problem, others, and the events in your life. The MetaCreation theory and process help clarify these concepts and assist you in making the vertical shifts needed to overcome systemic problems in ego personality which are often the root cause of all life style problems.

How do you get outside an existing cognitive structure to get a broader perspective on the problem you might ask? Often it is very difficult to make these shifts because of the unconscious demands of the ego personality in creating a material world reality for itself. As you become more aware of how we limit ourselves by viewing our world from within a self limiting cognitive structure you start to gain a greater appreciation of how you actually create your life. Creative Integrative Thought™ starts to move you up the cognitive structures of the mind. Creative Integrative Thought brings about conscious awareness of the inherent limitations of our rational, logical thought processes which only moves you horizontally through the landscape of your existing cognitive structure. Also accessing the subconscious mind in order to remove negativity as well as becoming aware of the inherent limitations of rational linear thought helps in moving up the hierarchy of cognitive structures. As you do so you reach a level that expands awareness to allow for organizational synthesis which resolves life style problems.*Absolutely nothing and no one can stop your from creating and enjoying the type of life you desire for yourself except you yourself.* But be forewarned. This exception can be a formidable one unless you fully understand the organizational structure of the mind and brain in creating realities in life. Additionally, you must be willing to work within this organizational structure to create the life you choose for yourself. As we have seen the problem is not one of rational thought alone but one of cognitive structuring in which thoughts themselves arise.The higher you move up the hierarchy of cognitive structures the greater organizing synthesis you will achieve in life and the happier you will become.

The bottom line is that your reality is not really "real" in the sense that you think it is. It is not really solid. On the quantum level of existence it is composed of pure energy. You yourself are not really solid either and on the quantum level you are composed of pure energy. Once you fully grasp the significance of these facts alone and start to put the theory of the mind into operation in your life you can start to direct your attention toward reorganizing the energy of your reality and reorganizing the energy of self, the energy forming the definition of ego personality. You can then reform realities and ego personality into ideas and images more to your liking. You do not start with reality or the image of self as a solid mass forming solid objects in fixed time and space and then disassemble them and reassemble their constituent parts into different assemblages in an effort to create different realities. Rather, by directing the power of your mind alone in the form of the intentional energy of your thought you simply need to shift the energy flows, currents, patterns and polarizations into positive patterns and polarities flowing in entirely new and different directions, forming entirely new and different more positive energy patterns and images. In doing so you will be able to create an entirely new you and an entirely new reality for yourself.

We have seen that analog thinking helps in changing ego personality and in creating expanded realities. Analytical thinking, for the most part, can only accommodate rational linear thought images arising within your existing physical reality at the cognitive level at which you are presently operating. Analog thinking can operate outside of your experiences of physical reality for the reason it can expand your cognitive structure to perceive and allow for the assimilation of much greater stimuli from your environment. It can then organize such additional data on the level of the subconscious mind in such a manner to create for itself newly expanded realities. Here again we see the morphological field effect. Remember, the subconscious mind can with proper programming sort for unlimited possibilities, unlimited potential and unlimited correlation among those potential possibilities to bring to conscious awareness newly expanded realities. Chart 39 at page 349, entitled, *"Summary Compari-*

son of a Limited Cognitive Structure versus an Expanded Cognitive Structure" illustrates how personal realities change radically when you make quantum shifts in the expansion of ego personality and in your cognitive structures from which you form all thoughts expressing your life.

Lastly, you will want to again review the components of analog thinking and how to adopt conscious analog thinking as discussed back in Chapter 3 and Appendix One. You will recall that analog thinking involves letting go and releasing control of the events and things of this life and relying on the higher energies and images of the consciousness of self awareness for direction and guidance. We have seen that analytical thinking often rationalizes existing negative energy and images arising from the subconscious mind rather than dealing with them at the time that they arise. A person using logical rational thought to order his world will often find many arguments and excuses to explain his negative or restrictive thinking, feelings, behavior and circumstances in life. He will use arguments and excuses instead of dealing with the origins of the negativity. Such negativity or restrictive thoughts will continue to manifest for they have a nasty habit of never going away nor will they be denied expression in one form or another until they are finally dealt with. Analog thinking, on the other hand, both on the conscious and subconscious levels of the mind, clears negative energy and restrictive images as it moves toward greater and more expanded states of awareness, states of equilibrium, mental homeostasis and higher order synthesis.

Summary on Moving Up the Cognitive Structures of the Mind

An expanded cognitive structure allows you to assimilate much greater incoming stimuli and you acquire the ability to organize it on a much higher level of organizational synthesis. Situations that appear in opposition to one another can now be reconciled in life. Ricky was able to effectively do this when he found self love and forgiveness for his mother and as a consequence was able to propose marriage to Betty. In effect he was able to redefine himself, his ego personality, in terms of intimacy, caring and self love instead of defensiveness, aggression, self criticism and noncommitment. As a result of redefining ego self Ricky was able to create an expanded life reality for himself by marrying Betty.

Ricky was dealing with deep seated characterlogical deficits of ego personality, deficits which are difficult to address and deal with effectively in traditional psychology. The reason for this is that traditional psychology has not developed a comprehensive model which quantifies the structure, functions and processes of the human brain

CHART 39

SUMMARY COMPARISON

Limited Cognitive Structure Versus Expanded Cognitive Structure

Limited Cognitive Structure	Expanded Cognitive Structure
1.Cognitive thinking involves rational linear thought	1.Creative thinking also involves spontaneous holistic thought
2.Thoughts mostly line up with and conform to physical reality	2.Thoughts can also be formed independently of physical reality
3.Thought processes arise mostly in Beta rhythms of the mind	3.Alpha states of mind regularly accessed for daily creative thought
4.Creations result from rational, logical reasoning consistent with the appearance of outer reality	4.Creations are spontaneous, arising from the impulse of the energy of the thought image itself independent of outer reality
5.Creations take physical work, effort, and labor to manifest in outer reality	5.Creations can easily and effortlessly manifest in outer reality without much physical labor
6.Reality is experienced as outside of and independent of oneself	6.Reality is experienced as inside and controlled by oneself
7.Your world is experienced as though you have little direct control over circumstances	7.Your world is experienced as though you have total control over all circumstances of life
8.An individual experiences little or no control over his past or future	8.An individual has total control over past and future events, since the past and future all arise in the present
9.Life is experienced on a time line from birth, youth, middle age, old age, to death	9.Life is experienced in the present here and now until you make the transition to another plane of existence
10.An Individual has less control over things and events as they progressively move further out in his space	10.An individual has total control over things and events regardless of where they are located, or at what point they occur, in his space

11. Life is often experienced as a series of problems and and a search for their solutions	11. Life is experienced as an unfoldment and advancement of awareness to completion and fulfillment
12. Life is dealt with in order to maximize pleasure and minimize pain	12. Life is experienced as an expansion of awareness as love, peace, joy, happiness and harmony with all people and things
13. Life is experienced as separate and apart from all other living things	13. Life is experienced as connected to, one with, and in harmony with all other people and the environment
14. The individual often experiences himself as out of alignment with the energy of his body and world	14. The individual experiences himself in alignment and congruent with his personal energy and the energy of his world environment
15. The individual feels separated from the energy giving rise to and sustaining all of creation.	15. The individual experiences himself as intimately connected to and one with the greater forces sustaining all of life
16. Life styles are based on competition and the accumulation of wealth, often at the expense of others	16. Life styles based on self awareness, cooperation with other people, sharing of natural resources and protecting the environment
17. Freedom and responsibility is primarily left up to the the individual	17. Life experienced as freedom and responsibility of all people both individually and collectively
18. World environment is not a primary concern of each individual	18. World environment primary concern of all people
19. Emotions arising from blocked desires result in frustration, anger, prejudice, hatred, hostility and other negative emotions	19. Emotions arising from expanded awareness are peace, harmony, love, help, support and cooperation for all people and protection of the environment
20. Values in life based on accumulation of wealth, competition and material success	20. Values lies in attaining higher states of consciousness and expanded states of awareness

21.Purpose in life is to succeed materially notwithstanding the general welfare of others	21.Purpose in life is to become more aware and support the welfare and evolution of others and the planet
23.Life itself is often considered as relatively cheap	23.Life is always considered as valuable and sacred
24.Other people and situations can cause you a lot of pain in life	24.I am solely responsible for my pain and its elimination
25.Life is experienced as an attachment to and identification with material wealth and possessions	25.Life is experienced as a connection with, but a detachment from, the material manifestations of the world
26.Life is experienced as a striving to achieve levels of success and status in a material world through hard work and competition	26.Life is experienced as an expansion of awareness, material possessions do not define the person, integrity, honesty, responsibility and cooperation do
27.Life is experienced as though the past is history and the future is yet to come	27.Life is experienced in the present here and now which embodies both the past and the future
28.You try through physical effort to maintain maximum control over your environment	28.You release all control and are guided by higher consciousness and expanded awareness to guide your involvement in the world
29.Conscious mind mostly creates from images arising from within its experience of physical reality	29.Creative mind can also create from images arising outside and independent of its experiences of physical reality
30.Experiences of physical reality mostly accommodates rational linear thought images	30.Experience of expanded awareness can accommodate unlimited possibilities, unlimited potential, and unlimited correlation of thought images
31.Happiness is derived from the experience of physical pleasure arising from the physical senses	31.Happiness is derived from the experience of pure awareness expressing as love, peace and harmony with self, the world and others

and mind in organizing ego personality and its many realities in life. The basic desk reference on clinical psychology referred to as the DSM manual describes over 400 types of mental dysfunction.[4] However it does not provide those practicing in the field with a comprehensive model of how the brain and mind work in creating a life. The MetaCreation Technology effectively fills this gap by providing this needed comprehensive model and therefore provides the foundation for an entirely new psychology we refer to as Metapsychology.™ Basically Metapsychology™ addresses the origins of characterlogical mental deficits at their very source in order to eliminate them. Metapsychology™ addresses such dysfunction at the quantum level where the atomic and molecular energy and matrix configurations create distortions in the neurological basins of attraction and in the primary analog systems of the subconscious mind. The theoretical foundation for Metapsychology™ and its corresponding clinical therapy referred to as Analog Therapy™ is set forth in Appendix Three.

However, this book is intended primarily as a self help book for the layman seeking to empower himself by enhancing both ego personality and the creation of more prosperous and meaningful life realities for himself. While it sets forth a comprehensive model providing insights into the internal dynamics of the mind and brain it is not written primarily for use by the professional in the fields of psychology, child development, mental health, neuroscience and artificial intelligence. Rather, this book was primarily written so that anyone with a high school education can easily comprehend and apply the technology. If you do not fully understand all of the theory of the dynamics of the brain and mind in creating ego personality and its many realities in life you can, nevertheless, still very easily work the MetaCreation processes to effectively create a prosperous and meaningful life for yourself.

<u>Processing Constraints on the Cognitive Structures of the Mind</u>

The preceding descriptions of how the cognitive structure of the mind is organized sets the stage for the second purpose of this chapter. This section is a brief review of the constraints on ego personality in preventing us from creating the type of life we desire. We developed, to a considerable extent, how the negative energy of the primary programs act as a constraint on ego personality in creating the type of life its seeks for itself. However, there are many other constraints on ego personality in creating realities in life. The more relevant ones will be addressed at this time. These constraints relate to

4 The most widely used diagnostic classification system used in the United States is the one published by the American Psychiatric Association entitled "The Diagnostic and Statistical Manual of Mental Disorders", which is commonly referred to as the "DSM Manual".

the expansion of thought within a cognitive level and between levels. Charts 40-1, 40-2, and 40-3 beginning at page 355 all set forth the organization of both the content and process constraints on ego personality in creating realities. By far the three most major constraints to creating a fulfilling life are: (1) perception of linear time, fixed space and an objective three dimensional world, (2) subconscious primary programming which hinders us from forming the type of life we desire for ourself, and (3) the conscious constraints of ego personality itself which hinders us from moving from one cognitive level to the next higher level in order to create changes in the quality of life. The following section of this chapter will also briefly address the minor constraints on creating realities in life.

We will not go into any further detail here concerning three dimensional reality versus spacetime of the fourth dimension since the theory concerning this topic was covered in the previous chapters. The second major constraint on creating realities is the constraints of the subconscious primary programs of the mind. This constraint has been dealt with in Chapter 8 and will only be briefly touched upon again in this chapter. The third major constraint is the limitations of ego personality. This major constraint has been addressed above in considerable detail. Keep in mind that while we address all three major constraints as separate topics for organizational purposes all three major constraints are simply three different aspects of one and the same thing. That one thing is how self awareness organizes a perception of its ego personality in spacetime in order to create and experience realities in life.

Minor Constraints in the Processes the Mind uses in Creating Realities in Life

Chart 40-1 sets forth the processing constraints affecting the processes of the mind. We refer to them as process constraints because they primarily affect the process functions of the mind whereas the content constraints primarily affect the application functions of the mind. They deal with the content of our thoughts and not so much the processes by which we create thought. The first process constraint is the knowledge processing constraint. The purpose of this book is to provide you with the knowledge you need in order to understand the processes by which you create realities for yourself. For if you don't understand these processes there is no way that you can effectively alter undesirable realities in order to create new more desirable realities.

You may believe that doing a rain dance and chanting to the rain Gods will bring rain, but no rain appears for many, many months. Apparently you need a better knowledge system for raising crops in a dry climate. Similarly, if you believe that wearing a copper bracelet or a garlic bag around your neck will ward off the common cold or flu and you come down with the Asian Flu this winter you might need a better knowledge system concerning infectious diseases. Syntax and semantic constraints have to do with the linear organizational structure of our language in expressing thought and in communicating thought to others. The form of our language reflects the way we view nature. We view ourselves as living in a fixed three dimensional frame in fixed time and space and therefore we express our realities in terms of linear thought in cause and effect structures which roughly follows the syntax of our own native language in expressing our experience of nature in a fixed three dimensional frame.[5]

This isn't surprising if you think about it for a minute for we had to create a language which is consistent with, reflects, and expresses what we perceived ourselves to be in a linear three dimensional world. As indicated in Chapter 4, we could have developed a musical mind, a mathematical mind, or an artistic mind using different patterns of linguistic forms to express ourselves. In fact Pathagoris did exactly that in creating the Western style musical scale. The 8 note musical scale is not based on logical rational thought although it follows a form of logic. It was based primarily on patterns of relationships between the sharps and flats of the notes in the scale. In this fashion it primarily follows an analog system of relationships instead of primarily following an analytical system of logical arrangement. Of even greater significance is the fact that the primary medium of organization of the subconscious mind is not linguistic patterns but rather symbolic and metaphorical patterns. By using symbols and metaphors the subconscious mind has a much broader range of organizational synthesis than does conscious mind using predominately logical linguistic patterns of expression.

Belief was dealt with in Chapter 7 and will again be dealt with in Chapter 10. Here I only point out that belief itself can pose a considerable constraint in our ability to change our perceptions of our world which would allow us to alter our perceived realities in life. Here I am referring to belief in the context of being the basis for truth embodiment, that is, what we believe to be true. What we believe to be true holds considerable influence over what we perceive our realities in life to be. For instance, I understand that you can still attend a monthly meeting of the Flat Earth Society which holds that the earth is really flat notwithstanding what Columbus proved five centuries ago. Many people still believe that more guns make for a safer society; more jails and capital punishment are the most effective way to

5 Since spacetime is holistic, conscious analog thinking is holistic instead of linear in nature.

CHART 40 - 1

Processing Constraints of the organizational processes of the Conscious, Subconscious, and Neurological, systems of the Mind/Brain

I Conscious Mind - Cognitive Processing Constraints - Secondary Belief Processes

1. Knowledge Processing Constraints - Secondary Belief Processes
    - a. Knowledge (Cultural & Emotional) Constraints
    - b. Syntax and Semantic Constraints
    - c. Belief (Truth Embodiment & Emotional) Constraints
2. Logical Processing Constraints - Secondary Belief Processes
    - a. Logical Connection Constraint
    - b. Causal Connection Constraint
    - c. Rational Connection Constraint
3. Environmental Processing Constraints - Secondary Belief Processes
    - a. Time Sequencing Constraint
    - b. Spatial Location Constraint
    - c. Objective Reality Constraint

II Subconscious Mind - Analog Processing Constraints - Primary Belief Processes

1. Conscious Mind Processing Constraints on the Subconscious Mind Processes
    - a. Knowledge Processing Constraints
    - b. Logical Processing Constraints
    - c. Environmental Processing Constraints
2. Analog Processing Constraints - Primary Belief Processes
    - a. Primary Personality Constraints
        - i. Energy (Emotional) Constraints
        - ii. Pattern (Image) Constraints
    - b. Primary Operational Constraints
        - i. Energy (Emotional) Constraints
        - ii. Pattern (Image) Constraints

III Neurological Processing Constraints - Subconscious Processing Mind/Brain

1. Neurological Stimuli Constraints
2. Electrochemical Neurotransmitter Constraints
3. Neurological Basins and Neurological Pathway Constraints

NOTE: It is these mental processing constraints on the conscious mind, subconscious mind and the neurological circuits of the brain that gives stability to the creation of inner and outer realities. Altering these process constraints alters the perceptions and experiences of inner and outer reality. If you were to distort these constraints you would distort the experiences of inner and outer reality. If you were to entirely remove these constraints you would collapse experiences of inner and outer reality.

CHART 40 - 2

Content Constraints in the Organization of the Conscious, Subconscious, and Neurological, Systems of the Mind/Brain

I Conscious Mind - Cognitive Content Constraints - Secondary Belief Formation

1. Outer Environmental Perception Constraints:
   - a. Perceptions and Experiences of Outer (Personality) Reality
   - b. Perceptions and Experiences of Outer (Operational) Reality
2. Outer Reality Belief Constraints:
   - a. Outer Secondary Personality Beliefs (Thoughts & Feelings) about Outer Reality
   - b. Outer Secondary Operational Beliefs (Thoughts & Feelings) about Outer Reality
3. Inner Reality Constraints:
   - a. Inner Secondary Personality Beliefs (Thoughts & Feelings) about Inner Realities
   - b. Inner Secondary Operational Beliefs (Thoughts & Feelings) about Inner Realities

II Subconscious Mind - Analog Content Constraints - Primary Belief Formation

1. Conscious Mind Content Constraints arising from the Content of the Subconscious Mind
   - a. Outer Environmental Perceptions Constraints
   - b. Outer Reality Belief Constraints
   - c. Inner Reality Belief Constraints
2. Analog Content Constraints - Primary Belief Formation
   - a. Ego Personality System
     - i. Archetype Prime Directives Constraint (Thoughts & Feelings)
     - ii. 12 Originating Primaries Constraint (Thought & Feelings)
   - b. Ego Operational System
     - i. Archetype Prime Directive Constraint (Thoughts & Feelings)
     - ii 12 Originating Primaries Constraint (Thoughts & Feelings)

III Subconscious Mind/Brain - Neurological Content Constraints

1. Pattern & Intensity of Incoming Stimuli
2. Electrochemical arrangements in the Basins of Attraction and Attractor Pathways
3. Pattern Configurations arising within the Basins of Attraction & Attractor Pathways

NOTE: It is these mental content constraints of the conscious mind, subconscious mind and the neurological circuits of the brain that constrain the creation of inner and outer realities. Altering these constraints alters the perceptions and experiences of inner and outer reality. If you were to distort these constraints you would distort the content of the experiences of inner and outer reality. If you were to entirely remove these constraints you would collapse experiences of inner and outer reality.

CHART 40-3

Mental Constraint Flow Chart

Any change in any of the inner or outer constraints on the mental systems of the mind potentially can alter, either positively or negatively, any of the other constraints within the system and our experiences of inner and outer reality.

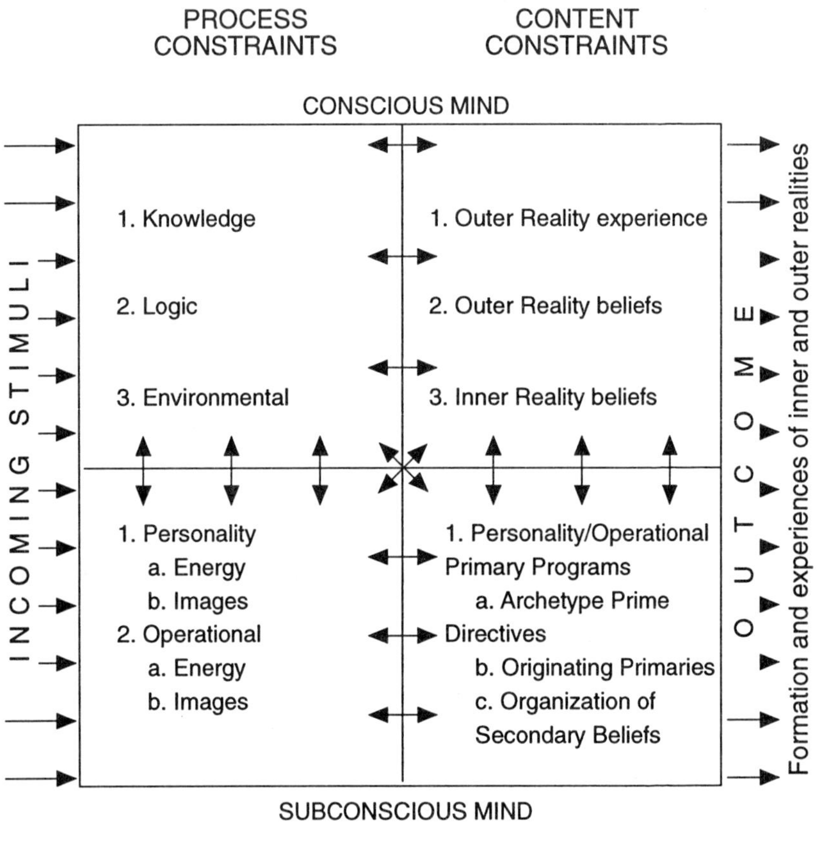

By changing any one of the above described constraints within the system potentially can change any of the other constraints within the system and thereby have the effect of changing a person's perceptions and experiences of their inner and outer reality.

deter crime; that the most effective way of solving any social problem is to throw more money at it; and that there is no direct correlation between violence in the movies, videos, and print media and the level of violence in society today; that our primary source of energy should continue to be fossil fuels, notwithstanding the fact that fossil fuel is rapidly being depleted and is almost the sole factor in global pollution. For more fixed and objectified beliefs refer back to Chapter 7 on the origin of thoughts and beliefs.

Logical processing constraints concerning the formation of conscious secondary thoughts have to do with whether the thought makes sense, that is, whether or not the thought is a logical one, causally connected, or otherwise rational. It was pointed out in Chapter 7 that thought does not have to be logical or rational to be held in consciousness as true. This categorization of thought ignores the other three categories of thought as being either emotional, utilitarian or ritual, that is, traditional or cultural in nature. Emotional and ritual thought does not have to be logical in order to make sense to the person holding the emotional or ritual thought. It is for this reason polite company at a cocktail party often will not engage in political or religious discussions because such discussions create so much emotional feeling by the participants that they end up with hurt feelings. Subconscious constraints arising from the primary personality and operational programs have been dealt with in Chapter 8. Neurological processing constraints concerning the brain's basins of attraction and attractor pathways were dealt with in Chapter 5.

It is these mental process constraints on the conscious mind, subconscious mind, and neurological circuits of the brain that give stability to the creation of inner and outer realities. Altering these process constraints alters the perceptions and experiences of inner and outer reality. If you were to distort these constraints you would distort the experiences of inner and outer reality. If you were to entirely remove these constraints you would collapse experiences of inner and outer reality.[6] An example of this is a person kept in isolation for an extended period of time and deprived of sensory stimuli would start to hallucinate for the reason that the mind would start to create images of a constructed inner and outer reality if deprived of the actual stimuli of such realities we experience each day living in a three dimensional objective world. If left in isolation long enough inner and outer reality could become so confused and distorted that the person could literally go crazy. Going crazy in this context is simply losing

6 For how changes in the energy of the mental patterns of the mind change the images in the mental field of the mind refer back to Chart 20-1 which explains how the processes of the mind transform energy in the form of K,V,A into images in the mental field of the mind.

touch with outer physical reality in a three dimension frame within fixed time and space.

Content constraints referred to in Chart 40-2 deal with our perceptions and experiences of outer reality. Our perceptions and experiences of our outer physical reality are largely framed up by the space and time in the three dimensional world in which we perceive ourselves. Changing the frame to one of spacetime in the fourth dimension gives us an entirely different perspective on the time and space frames of our life, particularly as to how we create realities within them. This time and space perspective also holds true with respect to our secondary thoughts and beliefs about our outer physical realities as well as our thoughts and beliefs about our inner realities. Subconscious analog content constraints have been dealt with in Chapter 8. Chart 40-3 simply sets forth a mental constraint flow chart which illustrates that any change in any of the inner or outer constraints on the mental systems of the mind potentially can alter, either positively or negatively, any of the other constraints within the system. Remember, the system functions as a dynamical multidimensional self organizing system where one change can propagate throughout the system and potentially affect other diverse processes within the system.They thereby can have the effect of changing a person's perceptions and experiences of his inner and outer realities.

It is the mental content constraints of the conscious mind, subconscious mind, and the neurological circuits of the brain which constrain the creation of inner and outer realities. Altering these constraints alters the perceptions and experiences in inner and outer reality. If you were to entirely remove these constraints you would collapse experiences of inner and outer reality. An example of this is a person on hallucinogenic drugs such as LSD, marijuana, cocaine, ecstasy, etc. A person on these types of drugs can alter perceptions and experiences of inner and outer reality. When you are inside such an experience there is no way of telling which is a drug induced reality and which is the normal objective reality for they both rise out of the exact same mental processes of the mind/brain. However, one is a drug induced reality and the other is a perceptually induced reality.

<u>Subconscious Primary Program Constraints on the Formation of Realities</u>

You will recall from Chapter 6 on the subconscious mind, and our discussion on the subconscious processes of the primary programs in Chapter 8, that the subconscious mind, acting as an immense analog processing system, continues to create a negative ego personality and negative realities in response to the negative programming which was installed in very early childhood. The subconscious mind doesn't think in terms of rational logical thought. It just keeps processing realities day in and day out, night in and night out, like the energizer bunny. It just keeps going and going. It just doesn't stop. If the programs are negative, the processing is negative, and the reality is negative. It is as simple as that. There is no mystery to how all of this comes about. Its not a metaphysical thing; it's not a karmic thing, it's not a mystical thing; it's not a curse that you are born with thing; it isn't a reincarnation thing; it isn't a past lives regression thing; it is not some sort of cosmic punishment for being bad thing; it's not even a conscious cognitive thing. On the deepest level of the subconscious processing by the mind it isn't even your fault, nor the fault of your father or mother or other primary caregiver early in life. They may have been the cause for downloading the negative parameters into your primary programs very early in life. But when it comes to changing the results from negative energy to positive energy fault doesn't play a role in the process for the simple reason that faultfinding is a conscious cognitive process not a subconscious analog process.

That is not to say forgiveness is not a good thing for mental and emotional growth. *It is just that fault, blame, victim, perpetrator, forgiveness and love are not essential components to the process of destabilizing and dissolving negative energy in the primary programs of the subconscious mind.* Nor is cognitive understanding essential to removing negative energy from the subconscious processes. Of course, since we are all cognitive beings forgiveness and love, as well as cognitive understanding, are central to conscious cognitive integration and emotional and mental growth on the conscious cognitive level of existence. However, they are not initially necessary in repairing the damage in the subconscious analog systems of the mind.[7] *Only the positive intention of your thought is necessary to work the process of destabilizing and dissolving negative energy in the primary programs of the subconscious mind.* Far too often we believe (truth embodiment constraint) that we need to fully understand the dynamics and content of a past traumatic episode in our life

7 Cognitive therapy often gets hung up on this point and spends far too much time in understanding the dynamics of neglectful or abusive episodes in life and in trying to sort out who is victim and who is perpetrator and who needs forgiveness and love. Such factors are paramount in healing in the inner child therapies and in abuse, sex and incest support groups. However, much of what goes on in these types of therapies is not really essential in changing the negative energy in the subconscious analog system of the mind into positive energy and images.

in order to fully discharge its negative hold on us. This becomes a transparent belief of ours. That is, it is an operative belief constraining how you think without you being consciously awareness of its effect. *You think the way you do simply because you believe that this is the proper way to think.* You could just as easily change your thinking to the point of view that it isn't necessary to understand and reexperience the content of the traumatic episode in order to discharge its negative energy.You can simply discharge its negative energy through the use of subconscious power analogs alone. Thereafter cognitive integration comes about naturally at the appropriate time and of its own accord.

The Relationship Between the Brain's Neuro Networks and the Cognitive Structures of the Mind

You can probably see a similarity between this system of cognitive structures of the mind and the neuro networks and neuro pathways of the brain. In Chapter 5 on the functions of the brain we discussed in some detail the relationships between the brain's basins of attraction and their attractor pathways. Positive thoughts and energy strengthened positive basins of attraction and their attractor pathways and negative energy and thoughts destabilizes a positive basin of attraction and attractor pathway. The reason for this neurological arrangement of the brain is that it tracks the same operating theory of the cognitive structures of the mind. That is the mind and brain represent different aspects of a single mind/brain system for the creation of human realities in life.

We have seen how thoughts arrange themselves into cognitive structures which, in turn, arrange themselves into a hierarchy of complex systems designed to handle the complexity of the many highly complicated thought processes of the mind. It is ego personality which determines the specific manner in which a person's cognitive structures are organized. No two person's cognitive structures are arranged in the exact same manner. Just as no two people's neuro networks are hard wired the same way. Therefore, no two people have the exact same personalities or functioning mannerisms for experiencing reality in the exact same way. Two children might be raised in the same neglectful family environment and yet one child turns out to have a good self image and a positive outlook on life and the other child turns out with a low self image and a negative outlook on life. How can you explain this difference? What has happened is that each child interprets the same environmental stimuli in different ways. One is able to overcome the effects of a hostile environment and develop a strong self image and the other is unable to do so.[8]

You might be wondering what all of this has to do with the neuro networks of the brain. Simply this. It is the ego personality which also determines the general direction of how the millions of neuron cells are organized into basins of attraction and attractor pathways. This is not done on a conscious logical level however. Recent brain research has determined that the environment plays a major role in the hard wiring development of the brain.[9] What this research is really saying is that it is ego personality's reaction to its environment that determines the development of the neuro networks and neuro pathways of the brain. The environment itself doesn't directly determine the development of the brain. The environment merely provides the stimuli which is interpreted by ego personality that ultimately determines the structure of the neuro network of the brain as well as the hierarchical organization of the cognitive structures of the mind. They both go hand in hand and operate together to create realities in life.

The brain's neuro network and its basins of attraction and attractor pathways and the cognitive structures of the mind are two aspects of the exact same complex human operating system. One could be viewed as the hardware and the other as the software used by self awareness to develop an ego personality and functional operating systems. They both work in tandem like a single symbiotic system. You need a brain to provide the hard drive which provides ego personality with the physical means of developing a thinking reasoning mind to function purposefully in a physical three dimensional world. The mind on the other hand is needed to provide ego personality with the means of developing thought in order to experience the many realities of that physical three dimensional world. One cannot exist without the other. Since both systems are aspects of one complex organism the activity of one greatly impacts the activity of the other. For instance, if you change the chemical composition of the brain it will directly affect the cognitive structures and thinking process of the mind. That is why alcohol impairs driving. If you rearrange the cognitive structures of the mind it will directly affect the neuro networks of the brain. That is why clinical depression deteriorates the quality of life but can be treated with psychotropic drug therapy.The reason for all of these effects is that we are talking about two different aspects of one single unitary complex system making up the mind-

8 Good examples of this are found in Viktor Frankl's book, *Man's Search for Meaning,* which describes the horrors of the Nazi Concentration Camps during WW II on the human Psychic.

9 Refer to the book by Rima Shore entitled *Rethinking the Brain, New Insights into Early Development,* 1997.

brain structure of man.

Research over the years has clearly shown that substance abuse of drugs such as cocaine, marijuana, alcohol, nicotine and other prescription and nonprescription drugs all negatively affect the electrochemical and electromagnetic processes of the brain. The detrimental effects of drugs and other substances on the brain's neuro network, in turn, detrimentally affect the thought processes of the mind.[10] Conversely mental stress, anxiety, and depression over prolonged periods of time have been shown to detrimentally affect the chemical composition of the brain. Tranquilizers and antidepressants have a physiological effect on the brain's neurotransmitters which in turn often have a pronounced positive effect on the mental processes of the mind. Clearly few people would argue the point that the organization of the mental energy of the mind directly affects the processes of the brain and that the chemical processes of the brain directly affect the mental energy and imagery of the mind.

One further point needs to be made concerning the hardware of the brain. Often in the neuroscience literature it will be said that once the brain's neurological organization is physically established by age three or four, such physical organization can seldom be changed, and therefore the subconscious mental patterns of the child are well established and become resistant to change in the adult years. This is not entirely correct for in order to change subconscious mental patterns it is not necessary to change the hard wiring of the brain's neurological circuits. As we have seen you need to only rearrange the energy patterns of the brain and mind, not the hardware itself. Negative basins of attraction can be reorganized into positive basins without rewiring the basin. Changing the polarity of the energy changes the pattern response of the basin. It's like changing channels on a T.V. set. To change a channel from a violent movie to an inspirational channel you don't rewire the T.V. set, you merely change the television frequencies of the channels themselves by using your remote control switch to tune into a new channel. Similarly with the brain, through the intentionality of thought, you change negative basins to positive basins. You do so by changing the frequency and polarity of the basin's energy and matrix configurations of its neuron cells and not the hardware itself. The reason that well-established mental patterns, particularly negative characterlogical patterns, are so difficult to change in the adult years is not because of the hardware wiring of the brain's neurological circuits but because we have not developed adequate therapies to deal with such problems. Cognitive and behavioral therapies just do not effectively

10 Ibid p. 43 - 46 refers to recent medical research on the effects drug use has on the neuro network of the brain.

address the energy and matrix configurations of the neurological basins and the analog patterns of the subconscious mind.

The Nature of Reality as You Move Up to Levels 5 and 6 on the Cognitive Structures

As you move consciousness of self awareness up to levels 5 and 6 on the hierarchy of the cognitive structure of the mind you begin to reach the highest states of consciousness and the most expanded states of awareness available to mankind. At this point you finally experience personally, on a first hand basis, what the metaphor for the universe, "whole, perfect and complete", truly means. It no longer is simply a metaphor but it becomes a reality in life, the ultimate reality of your life. You, at long last, come to know who you truly are, why you are here, and personally experience the ultimate meaning of your life.

"Whole, perfect and complete" takes on a whole new meaning for you. You personally experience a unity between your world, the universe and yourself. While you are involved in the many facets of this world you are, at the same time, detached and seem to just observe the passing events of the world. While you still have a purpose to fulfill in life, whatever that purpose is you know that it is already being fulfilled, because you experience the past, know the future, and live in the present. Your attention is totally involved in whatever is at hand. Such attention becomes an experience in both the object of your attention and an absorption in the field itself in which the object resides so that the dichotomy between you and your world merges into an experience of a unity of all things. Everything is observed as one thing and that one thing is perceived and experienced as being within you, within your control.

Another way of stating this experience of oneness between your and your environment is that at the highest levels of the cognitive awareness you, in effect, transcend the limitations of the ego personality. You become people centered rather than ego centered in your perceptions and feelings toward others and the many events of life. You can thus become more empathetic, understanding, all knowing of others and of the flow of life in general. You sense a unity, a wholeness, between you and them. You experience a transcendent state where self and others merge into a greater whole, where their joy becomes your joy and their pain becomes your pain. However, you do not experience their pain as the pain of the ego self but rather the pain of their unawareness and ignorance of the ways of life and its rules for living. For you know the ways of the world for you become the ways of the world.

At the highest levels of self awareness you experience self

as totality, a completion within self, not needing others or events in the world to complete or validate your self identity. By identifying with the greater whole you know that you are complete just as you are. You have a self acknowledge of your completion, a knowingness of perfection in cognitive congruence, an attainment of self assurance and of self absorption but not of self centeredness. Nothing can be added to you, or taken away from you, to make you a better person. Nothing can be truly taken away from you to make you less than what you already are. You are wholly sufficient and complete just as you are and you have a sense that you were always whole, perfect and complete. It is just that you didn't always experience yourself and your life as such.

You are able to transcend the limitations of mind, body, and emotions. You are able to merge the thoughts of the mind, the physical action of the body and the energy of your emotions into a holistic unity where everyone and everything is experienced as existing in perfect order, perfect order regardless of whatever stage of completion they might currently exist in the greater reality residing in spacetime. Everything is, at some point, already made perfect even though currently it is still incomplete. You, however, can experience both its completion and its incompletion simultaneously in the spacetime of the here and now of your existence. The reason for this is that you have transcended the limitations of ego personality and the limitations of physical time and space while still living in a physical body in present physical time and space. You no longer view people, places and things as separate parts, separate places and separate things, but rather everything as being one thing and that one thing is happening right now. You can see the interconnection and interrelationship of all things throughout the entire universe and you experience self as part of everything happening within the universe. You clearly understand the surface dichotomy of all things and at the same time the deep underlying unity and connectiveness running throughout all of nature.

Normal experiences of everyday life take place in the context of the day, year, and decade in which events take place. History and culture all arose out of the past. We live in current times and the future is either full of hope or full of despair. However, at the highest level of cognitive awareness you transcend time as being in the past or in the future for you take control of time and experience it as happening in the here and now. For as we said, you control the past. The past is no longer behind you and at a distance from you but is essentially within you and within your personal space. It is within your power to change and mold the past in any manner that you wish it to be. So too with the future. It is not something outside of you and outside your space, something and someplace outside of you and in the future where you can only hope good things will happen. Rather the future is within you to be revealed as you direct it to be. There is no out there and no in here. There is no those we oppose and those we champion. There is no those we are against and those we are for. There is no those we hate and those we love. Ultimately there is no longer opposition, diversity or duality in life. Life is composed of only different aspects of the one thing and that one thing is under your direction and control to mold as you will.

As a person moves consciousness up the hierarchy of his cognitive structure he begins to redefine himself in terms of who and what he is for he sees himself reflected in others and in all of life. He becomes more inclusive in his attitudes toward others and less and less judgmental and critical of others for he finally comes to realize that all judgment is ultimately self judgment. In the same manner all love starts with self love and all healing starts with self healing. If all blame is self blame and all criticism is self criticism then how is it that you can blame and criticize others for the seemingly bad things that happened to you? When you become aware, you simply cannot. You can only love yourself and love others and forgive yourself and forgive others and hope that they too will someday become aware. You view all others and life itself from a higher perspective where you now understand that we are not alone and that when one of us suffers we all suffer. Accordingly, you become more compassionate and loving toward all, even those who are critical and harbor ill will toward you. You transcend prejudices of all types whether of race, sex, economical status or education. You begin to be concerned for the welfare of all regardless of their status in life, for you are secure within your own material and spiritual well being.

Such a person is released from the psychodynamics of striving for a better life, of searching for greater meaning in life, of amassing greater material possessions, seeking greater status and power, or seeking more physical pleasures from life. For he comes to know the meaning of his life. He regards simply living day to day to be experienced as the ultimate joy life has to offer. In a very deep spiritual sense consciousness is its own reward and expanded states of awareness yield far greater experiences of true happiness, joy, and knowledge than you could ever conceivably obtain from all of the physical wealth of this world. As you persevere you will ultimately find that inner happiness which life holds for each one of us who seeks in earnest to unlock its mysteries.

As you transcend the limitations of ego personality you no

longer are subject to the dynamics of self asserting itself in the world. You no longer are intent on accumulating material wealth in order to expand the outer definitions of ego. You no longer have to seek power and status in society in order to prove to yourself and to others who and what you are. Being whole, perfect and complete you transcend the characteristics of striving, grasping, taking, and forever trying to gain ground in the daily affairs of life. Rather, you are detached and more of an observer passively watching the myriad activities of others in life.

You simply create what you need and want easily and motionlessly without tension and exertion, creations without the expenditure of needless energy and emotions. You put back into life more than you ever take for yourself for when you know that you can easily possess all there is you have no need to take more then you can reasonably use for your own comforts in this world. All of life becomes easy and enjoyable without striving and grasping after the illusions of happiness. Happiness is never really found in the search for external material things but rather happiness is a natural state of being at all times simply found within. You no longer have to experience the dichotomy of have and have not, pleasure and pain, happiness and sadness, fulfillment and emptiness, purposefulness and loss, or even strive to experience meaning or the lack of meaning in life. For as I said at the beginning of this section, when you reach the highest states of consciousness and expanded states of awareness at the very top of the cognitive structure of your mind you will ultimately experience only "wholeness, perfection, and completion". You now, at last, rest and experience the peace and quietness that comes from the inner knowing that you have truly attained what the ancient mystics refer to as "degrees of enlightenment and beyond".

CHAPTER 10

MetaCreation of Personal Realities in Space and Time

Introduction to the MetaCreation Process

In Chapter 9 we saw that by moving vertically upward from one level of the cognitive structure of ego personality to the next higher level you are able to shift your perceptions and experiences of your inner and outer realities in life. From this higher vantage point in viewing all of life you are able to resolve issues you might have been facing which at the lower level seemingly had no immediate answer. This chapter deals with horizontal shifts at the level at which you are presently creating and experiencing your inner and outer realities in life. In this Chapter we will explore the process by which you create physical manifestations of material wealth, abundance, and prosperity, how you rid yourself of detrimental emotional attachments to possessions, and how to achieve the relationships and career positions you desire at the cognitive level at which you are presently experiencing and creating a life.

Almost everyone seeks greater wealth in the form of greater financial security, more income or cash flow on a regular and continuing basis. We all desire more material possessions such as a better house or car, more savings and money for our children's education. We all could use additional income so that we wouldn't have to work another day in our life if we chose not to do so. Everyone would like to have a retirement program that allows you to enjoy a comfortable retirement without any financial worries whatsoever. Of course, there is more to life than just making money or earning a living. However, having enough money can free you to pursue your life's dreams, make a positive contribution in other people's lives, develop your own unique talents, skills and abilities, and be able to enjoy life more fully.

Keep in mind that financial freedom is not derived from how much money you have in the bank but is found in knowing that you have the mental ability to create whatever wealth you desire. Financial wealth is a state of mind, not the state of your pocket book. Financial wealth is but a reflection of the creative power of your own mind, just as lack and limitation are borne out of the poverty of your own mind. Mental freedom like financial freedom is in the certainty that you, at all times, have the ability to create whatever wealth you desire for yourself. Financial wealth doesn't come from other people, circumstances in life, or anything else outside of you. It comes from knowing and working the processes of the conscious and subconscious mind to create whatever you want in life. Financial freedom is derived from believing in yourself and not just in believing in your pocketbook. Accordingly, in this chapter you will learn the six step MetaCreation process used to create material wealth, abundance and prosperity. You will also explore the method by which you can discreate unwanted realities that are a hindrance to you. Let's get started creating the type of life you have always dreamed for yourself!

Theory of the MetaCreation Process

We saw in Chapter One that at the most quantum level all thought creating personal realities exist as polarized energy and matrix configurations forming an infinitely large variety of realities, both tangible and intangible. All thought has substance consisting of the form of its energy and has definition in the form of the outer boundaries of its matrix configuration. Every thought and every personal reality exists from its point of creation through the span of its existence to its point of dissolution. It is the energy and matrix configuration that gives substance to a created reality existing in fixed time and space which itself is a created reality. All thought and all personal realities will continue to exist as long as their energy and matrix configuration remain intact. They will not fade from your personal awareness until their energy and matrix configuration are dissolved and dissipated. This is a very important fact and is the basis upon which the MetaCreation process works in creating realities.

With respect to purely inner mental realities they do not dissipate and dissolve until their energy and matrix configurations are dissipated by the mental processes of the mind. With respect to our outer physical realities they do not fade out of your personal outer reality until you consciously withdraw all of your mental thought processes and feelings from the object existing as your personal reality. In that event it might still exist as a material object in the physical world, but it will no longer be a part of your own personal reality. It's like trading in your old car for a new one. The new one now becomes part of your personal reality. The old one may be the property of someone else, but it is no longer part of your physical personal reality. However, if you hold fond memories of it, then it is still part of your inner reality. If you were to borrow it from time to time from the new owner it would then remain, to that extent, a part of your outer physical reality.

All realities created by ego personality are created to be experienced by ego personality in one form or another. This is true even with respect to realities which you create inadvertently or unconsciously, such as when you absentmindedly make a wrong turn and end up in a part of town which was not your intended destination or you become so engrossed in what you are doing that you forget the time and arrive home several hours late. You have created these realities by the inattention of conscious mind instead of the intention of conscious mind. The reason for this is that the subconscious mind brings to conscious awareness an expression of the reality as the result of the inattention of conscious thought or by the none intended action or behavior of conscious mind.

We porsonally experience in one fashion or another all realities that we create. This is true whether we are consciously or unconsciously aware of this fact, whether we are consciously aware of their initial creation, or whether we fully experience them, or whether we begin to resist them once we have created them. The problem arises when we are not aware of their point of origin; when we are unaware that we created them as our own personal realities in the first place; or when we recognize that they exist as our personal realities but we would rather avoid or resist them for they only bring us discomfort in one form or another. Like a bad toothache, you know that it is there, but you would rather ignore it in the hope that it will go away and that you will not have to deal with it by going to the dentist and having it fixed. The reality once created will hang around until it is once again acted upon by a further application of the energy of the conscious mind in order to destabilize it and dissolve it as a conscious reality in life. Deliberate or focused attention by ego personality in the creation of the many events and activities making up the realities of life leads to the accumulation of knowledge, broadening the definition of ego personality and higher states of consciousness.

One further point: while we are often almost instantaneously made aware of the end results of the process giving rise to the creation of all thought, we are not aware of the mechanics involved in the process itself taking place on the level of the subconscious mind. Often we are not even aware of the origin of the existing energy and images of our realities for they have long since been buried and forgotten in our subconscious primary programs as we have seen in Chapter 8. With respect to not knowing the mechanics of the process itself, this chapter will give you the method for taking effective control over the MetaCreation process for creating realities. We will also explore in this chapter the origin of existing realities, particularly unpleasant ones, in order to dissipate and remove them from the inner realities of our mind. The purpose of the MetaCreation processes is intended for you to take effective control over all of your realities in life, both inner and outer, and when you are dissatisfied with them you can dissolve them and create new realities more to your liking.

How the Metacreation Process Removes Unwanted Realities in Life

We will first explore the specific steps involved in dissolving both inner and outer realities. As indicated above all realities are created to be experienced by ego personality. However, far too often, particularly when a reality is no longer serving a useful purpose, it is difficult to extinguish and remove it from our life. This is so since it represents an aspect of ego personality which we are not yet ready to deal with in order to remove the unsatisfactory reality. Or we simply don't know how to go about removing if from our life. In either event we simply do not recognize that it is a reality created by us and as such it can be dissolved without negatively impacting or diminishing ego personality.

To start the process you first consciously get an impression of what it is that you want to discreate out of your life. Here we are primarily dealing with things in our life that create emotional attachments which are difficult to give up. This could be a tangible thing such as getting rid of an old house or car, or a sentimental possession such as a piece of art, jewelry or family heirloom, but more likely, it would involve an intangible thing such as getting out of a bad relationship, stressful job or unwanted friendship in a manner that is freeing to you and does not offend another person and where all benefit. You first visualize the situation or thing to be discreated out of your life and shift perceptual positions where you no longer have an emotional or other attachment to the object or thing to be discreated. You get a sense of what your life will be like without the object that is being discreated. If you have trouble experiencing what your life will be like or look like without this undesired reality you will want to explore its meaning for you. How is it affecting your sense of ego personality? Once you have detached a personal sense of self from the physical reality and you have become detached from its energy you can then effectively dissolve the unwanted reality from your life. If you still maintain a conscious or subconscious emotional attachment to the object you want out of your life you need to first explore the purpose of the emotional attachment and discreate it first before effectively discreating the original object itself.

You want to start the process by first delabeling it. Simply describe the undesired reality just the way it is without any positive or negative judgment about it, or how it is affecting your life. For instance, instead of the object being grandfather's old clock it simply becomes an old clock. If you find it difficult to separate your sense of self from its effects in your life you are being controlled by its energy and image in your life. Maintaining a connection or emotional attachment to the object being discreated perpetuates its effect in your life. Remember in Chapter 2 we started out in life by labeling everything we came in contact with, and then we acquired an attitude concerning most of those things. Here we reverse this process and delabel it and remove the attitude from the object. In this way we start the process of discreating the unwanted reality from our life.

You will find that by simply taking the label off of it and removing any judgment about it, this in itself starts the process of reducing its effects in your life. By doing so you have made the perceptual shift from being attached to the thing to be discreated to being detached emotionally, mentally or physically. After delabeling it you can more easily dissolve it in your life for it no longer serves the purpose for which it was created in the first place. Its creation was to reflect ego personality in some fashion or another. Finally, delabeling it and removing any value judgment you held concerning the reality places you in a safer objective perceptual position to dissipate its energy, particularly with respect to highly charged emotional realities. The reason for this is that if you held strong emotional ties to the object to be discreated and you were able to discreate it in spite of these emotional ties you might end up feeling a tremendous loss or void in your life instead of feeling liberated and free of the object or intangible thing to be discreated.

Once its energy tie is disconnected your perceptual position would be viewing the object outside yourself. The object or thing would exist outside the normal parameters of what gives meaning to your life. You are no longer relying on the object to add meaning to your life. You are now in a position to effectively destabilize its energy and matrix configuration in order to remove it from your life. While connected to and perceptually within the experience of the reality you have very little control over it. When you make the perceptual shift and experience the unwanted reality as being or lying totally outside of you, outside of your experience of ego self, you are able to effectively take control and remove it from your life. You do so by removing its submodalities and mentally taking its energy down to zero.

You can start this process by simply relaxing the conscious mind and accessing the subconscious mind where all realities originate. Once in a relaxed state, image in your mind the thing to be discreated. Then simply state silently to yourself something like, "This is not me, this is my reality that I now discreate." When silently stating this to yourself you will want to make a mental energy shift by visualizing a perceptual shift in your mind from your connection to the reality to a perceptual position of viewing the reality as being outside of your view of ego self, that is outside the parameters of your life. As you make the mental perceptual shift from being connected to the reality to being detached from the reality you also shift from an associated state to a dissociated state. You shift your perceptual position from one of seeing the undesired reality through your own eyes to a position of seeing the reality as existing outside of you and disconnected from your sense of ego personality. You are now able to see yourself without any attachment to the object. Once you make this internal perceptual shift in your awareness from being connected to the reality to disconnecting self from the energy and image of the reality you can then take control over the reality instead of being controlled by it. In this manner you have become a detached and objective observer of the reality. You can now delabel it and remove any value judgments from it. By delabeling it the reality no longer represents a value judgment reflecting who or what you are in life

To complete the process simply use a reversal of the creation process. You start to withdrew all modalities and submodalities from the undesired reality on your inner mental screen of your mind. Once you have reduced its energy and image down to zero you release it from your internal visual frame in which you were experiencing it and let it dissolve and disappear from your awareness. You will want to check to see if it is still within your inner experience even if only in a reduced or modified form. If so, do a mental check to see whether there are any opposing thoughts or feelings holding it in awareness. If there are, use this same discreation process on those opposing thoughts and feelings and take their energy and image down to zero and return to destabilizing and dissolving the original unwanted reality from your life.

You simply allow the energy of the analog system of the subconscious mind to dissipate the energy and image of the unwanted reality by extinguishing it in your life while at the same time leaving ego self intact. While ego personality is intact, it is, at the same time, being redefined by the discreation of the unwanted reality in question. Remember, the reality always affects ego personality, whether for good or bad, while it exists in your life. In this manner the analog system of your subconscious mind will immediately start to sort for those people, places and things which will effectively remove the unwanted reality from your outer physical life in the most appropriate way. Remember it took you some time to create the unwanted reality, and, therefore, do not become impatient if it is not immediately dissolved and removed from your outer physical life. However, mentally you will be free of it.

Also, you must support its removal by doing those things

necessary or appropriate as indicated by the conscious mind. As we have seen such direction may come in the form of a specific thought or idea or it could be simply an intuition on your part. It might even come through the actions of a friend or third party or action initiated by the event or person you seek to remove from your life. As you adopt conscious analog thinking this will facilitate the removal of the unwanted event or person.

Remember, that once removed the unwanted physical reality might still exist as part of the physical world, but it is no longer a part of your own personal reality in life. It no longer holds any meaning in your life. In order to be a part of your personal reality you maintain a connection or relationship to it in some fashion or another. This is an important point because it is easy to confuse your personal realities for the physical objects existing in the fixed three dimensional space of your world. The two are not synonymous. *While physical objects make up your personal reality your personality reality does not consist of physical objects.* Your personal reality consist of your perceptions, feeling and thoughts concerning those objects. All objects exist as part of the physical world but not necessarily part of your personal reality.[1] Therefore once you remove all attachments or connections to an object, emotional and otherwise, it no longer exists as part of your personal reality although it might still exist as part of the physical world.

You will notice that by using the silent discreation statements set forth above you have taken control and responsibility for the reality. You have separated self identity from the reality and recognized it simply for what it is, a reality consisting of energy and images. You have taken ownership over it. In this way you are in the perceptual position to effectively dissolve it from your life. When doing so you will notice that you do not defend the reality, justify it in any way, ignore it, resist it by creating a counter reality, modify its size and strength in your life, magnify it out of proportion, make moral judgments about it or about you, make excuses for it being in your life, joke about it in order to minimize its effects on you, or in any other way deal with it in any manner which perpetuates its energy and therefore its existence in your life. You simply delabel it without judgment and destabilize it and dissolve it first as your inner reality and

---

1 Quite often you read in psychology literature that "the map is not the territory." However, in fact, the territory itself is not the territory but rather it is part of the map. The territory lies within, for the territory is your personal reality creating your outer reality. You are creating the territory minute by minute each and every day of your life although you are not aware that you are doing so for we have objectified the territory and perceive it as lying outside of us in an objective world over which we have very little control.

then as your outer reality.

How the Metacreation Process Creates Expanded Realities in Life

In Chapter 8 we have seen how the subconscious primary programs of the mind form your inner thoughts and beliefs. It is these inner thoughts and beliefs which form your beliefs about your outer physical realities. For a diagram showing the sequence in this process of creating outer physical realities refer to Chart 41 at page 376, entitled *Simple Diagram of how Secondary Beliefs Create Outer Physical Realities.* Chart 41-A at page 377 entitled *Dynamics Involved in the Process of Manifesting Your Goals in Life* sets forth the internal mental processes involved in creating outer physical realities in life. The MetaCreation theory of how to create more beneficial realities follows this sequence in the creation process. There are basically 6 major steps in working the MetaCreation process to organize more enjoyable realities. We will review each of these steps and then give you an example of how the process worked in Ricky's life. Chart 42-1 at page 379 sets forth the six major steps in creating a better life for yourself. Chart 42-2 at page 380 sets forth in much greater detail the eighteen steps which expand upon these six major steps. The six major steps are summarized as follows:

1. Belief in yourself, in the process, and your ability to create the object of your desires
2. State the object you desire specifically and in sensory positive terms
3. Take appropriate action to support your manifestations
4. Have faith and trust that the process will always work in your life
5. Eliminate all impeding thoughts and emotions to your creations
6. Allow for the object to manifest in your life and give thanks.

Step # 1 - Belief in Yourself

Many of these characteristics of the creation process we have covered in various ways in the previous chapters. We now look at them as a whole operating system in relation to how they work in creating material realities in life. "Belief in yourself, in the process, and in your ability to create the object of your desires" is listed first for it, more than any of the other steps, is what starts the process of creating personal realities in life. We have seen that once you have acquired the facility to shift mental states at will you have the ability to take total control over your inner mental states. By controlling your inner states you can evoke a positive state under almost all circum-

CHART 41

Simple Diagram of How Secondary Beliefs Create Outer Physical Realities

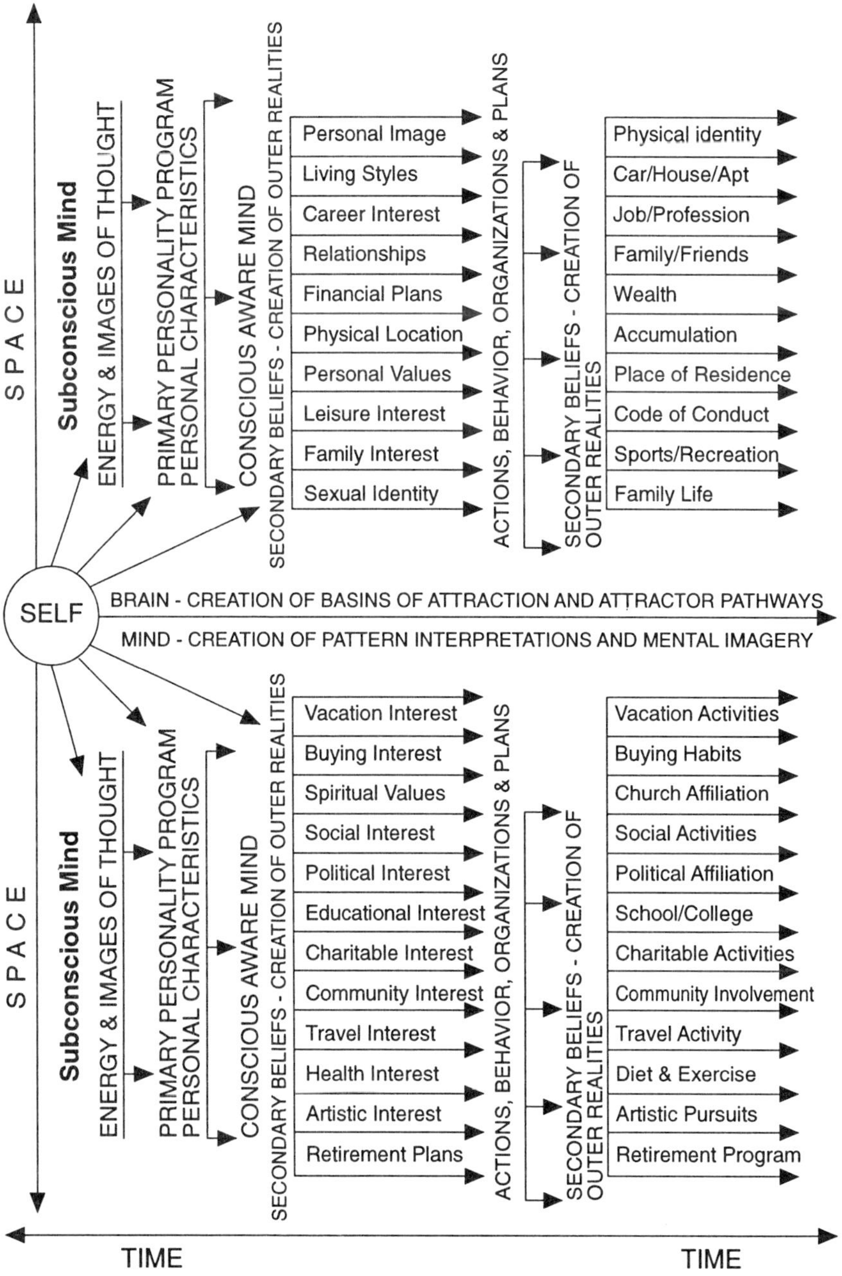

CHART 41-A

Dynamics Involved in the Process of Manifesting Your Goals in Life

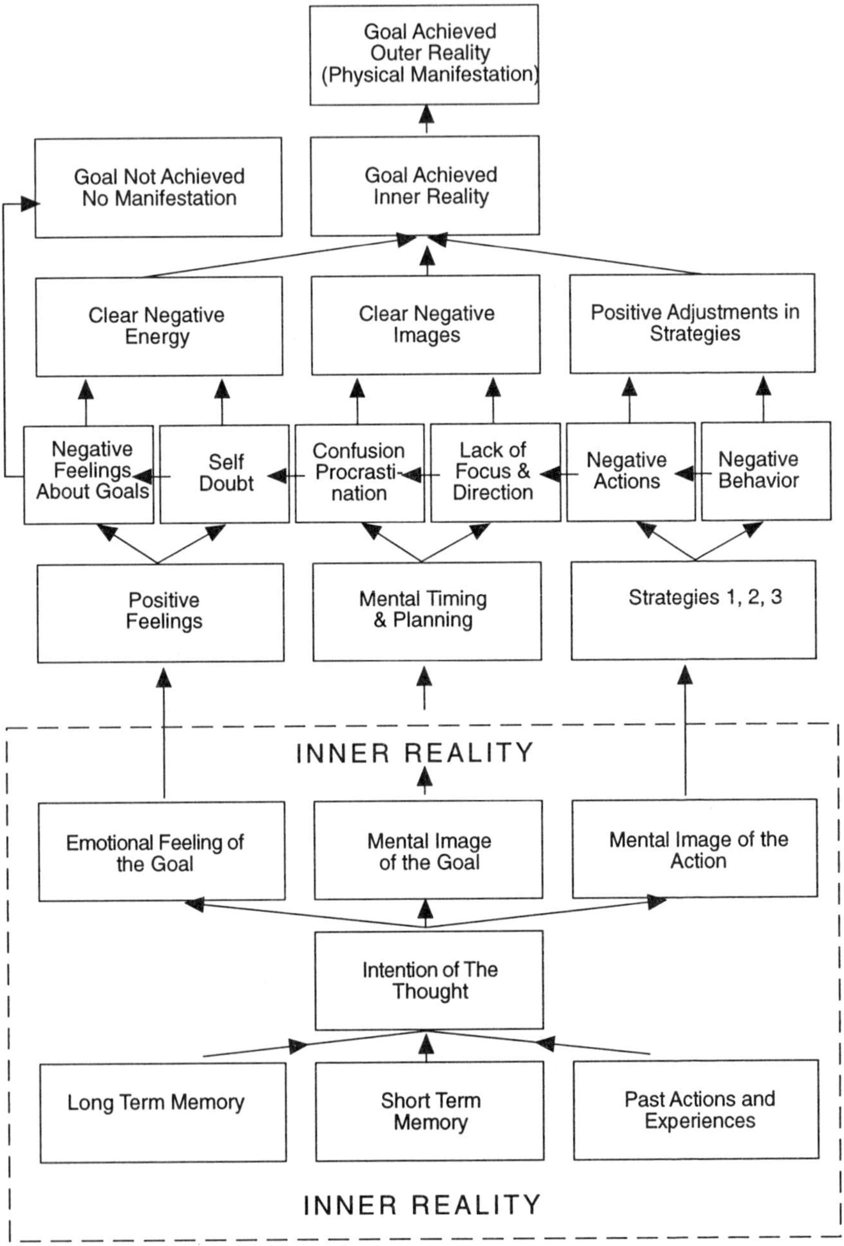

stances in life.With a positive frame of mind you can create positive mental images which correspond to your positive feelings. From a positive frame of mind you can than image your desired realities. By knowing and believing in "your ability to create the object of your desires" you can relax and enjoy the process for you realize that the process itself is what creates realities, not action alone, not desire borne out of lack or frustration, not intent alone, not willpower, not being a good person, not forgiveness and love, not money, not luck, not circumstances in life, not education, not material wealth, not age and experience, not reliance on anyone or anything outside yourself, and not even your fondest hopes and desires. Simply believing in and working the analog MetaCreation processes alone creates realities in your life.

However creating outer physical realities in life is not primarily about taking action to create those realities. This is where the first part of the initial premise "belief in yourself and belief in the process" comes into play. Creating realities in life is about creating a positive sense of well-being which is then reflected in your realities. By forming a positive sense of well-being you can then effectively use the MetaCreation process in creating whatever it is you want in life. It is the act of creating and not the physical creations themselves that is critical to the process of creating a sense of well-being. You need to come to personally adopt, thoroughly understand, and personally know, that you yourself as ego personality are an inherent and essential part of the reality creating process itself. *In fact you are the creation process itself, for you are processing realities all day long.* These realities reflect who and what you are in life. Another way of stating this is that who you are is but a reflection of your life realities. Your life not only expresses who you are, it literally defines who you are. There is no defined ego self image apart from the definition of your life realities. For this reason people continually endeavor to express their self image through the acquisition of more money, status and power in society. However, we are seldom aware of the subconscious processes by which ego self creates a sense of self by creating its realities. The correlation between our self image and our outer physical realities is but the outcome of this process. Therefore, if you are not satisfied with who you are, or with your outer realities in order to effectively change either one of them you need to take control over the processes creating both.

Step # 2 - State the Desired Reality in Sensory Specific and Positive Terms

Identify those things you desire which support and enhance your life style. With respect to your outer reality it would include

CHART 42-1

Summary of the Creation Process

1. Belief in Yourself, Belief in the Process, and Belief in Your Ability to Create the Object of Your Desires

2. State the Object You Desire Specifically and In Sensory Positive Terms

3. Take All Appropriate Action to Support Your Manifestations

4. Have Faith and Trust that the Process Will Always Work in Your Life

5. Eliminate All Impeding Thoughts and Emotions to Your Creations

6. Allow for the Object to Manifest in Your Life and Give Thanks

CHART 42 - 2

The Eighteen Steps Involved in Working the Creation Process
In Order to Physically Manifest the Object of Your Desires
Within the Time Frames of Your Choice

I. Belief in the Process and the Object of Your Desires

1. Believe in the Validity of the Process of Creation.
2. Believe in the Validity of the Object of Your Desires.
3. Believe that You can take control over an objective world and create your own reality.

II. State the Object Specifically and In Positive Terms

4. Define the Object in positive terms.
5. State the Object in specific terms and not in generalities.
6. State the Object in Sensory Specific Terms.
7. Express the Object in present K/V/A Imagery and feel its presence as a present expression of K/V/A emotion.

III. Take Action to Support Your Manifestation

8. The Object desired should be largely within your control and ability to allow for its manifestation.
9. The Object desired should be ecological and should not negatively impact Self or Others.
10. Take all appropriate action to support it's creation in the time frames you desire.
11. The Object should be testable so that you know that you achieved your objective when it manifests in your life.

IV. Have Faith and Trust that the Process will always Work

12. Always have Faith and Trust in the Creation Process. Never doubt your ability to create in accordance with your desire.
13. Create sufficient motivation and desire to support and bring the object into manifestation.
14. Acknowledge that the Process works in your life and trust and release the Object knowing that it will manifest in accordance with your desires.

V. Eliminate All Impeding Thoughts and Emotions

15. Expand your understanding of time and space to allow for the Objects manifestation in the time & space you desire.
16. Eliminate all of the Secondary Thoughts impeding its manifestation in your life.

VI. Allow for the Object to Manifest and Give Thanks

17. Have Fun while working the Process and do not take the process nor life itself to seriously.
18. Enjoy your Manifestations and give thanks by sharing and helping others.

such things as your place of residence, type of job or career of your choice, the type of car and house you want, relationships, and type of family you desire. It would also include all other specific material possessions and financial wealth you desire such as money, property and investments of various kinds. With respect to your inner realities it would include such things as peace of mind, increased confidence about the future, more focus and attention on your priorities in life, harmony and balance in life, optimum health, loving relationships and future plans for a better life and a comfortable retirement. As time goes on you will continue to create for yourself new and different realities and change or eliminate existing ones, for we have seen the MetaCreation process is one of continuous unfoldment rather than one of completion. The MetaCreation process is not intended to be used at specific times only. For its full effectiveness to operate you should consider adopting it as an inherent aspect of your life itself. It should be used on the conscious level throughout the day in order to effectively alter realities you encounter in your daily affairs. After awhile it will become second nature to you and your cognitive structure will almost automatically operate to take control over all aspects of your life. This is called "Conscious Analog Thinking" in cognitively restructuring the mind.

To start the process get an image of what it is that you want to obtain or achieve. You should be able to state your objective in sensory specific terms and not in generalities. The more defined the program parameters are for the subconscious mind the better able it is to manifest your desired realities. You will always want to experience the image of the reality as the reality itself. Visualize it as an actual present reality and not just an idea or pleasant thought of the reality, or just an internal impression of a desired reality, or just something you think that you might like to have sometime in the future. *In order to create it as your personality reality you must experience it as the reality itself.* Visualize the reality you desire in sensory specific terms. Express it in present K/V/A imagery and feel its presence as a present expression of K/V/A emotion. The more vivid the visualization as a present reality the more the subconscious mind is able to bring to conscious awareness those things needed to manifest your desires. Continue the process by getting entirely clear on the desired reality by perceiving it and feeling it completely and in every detail as though you are actually experiencing it as your present reality. Your perceptual position would be seeing the reality through your own eyes rather than seeing yourself experiencing the reality. Once your desired reality is entirely experienced as your inner reality and you have experienced it and know that it exists as your reality you can then release it knowing that it will manifest in accordance with your experience. The subconscious mind will thereafter, unbeknownst to you, continue to sort for its expression in accordance with your mental programming for its creation in the time frames and space frames you intend.

Step # 3 - Take Appropriate Action to Support Your Manifestations

The reality desired should be largely within your control and ability to allow for its manifestation. The manifestation of the reality must be initiated, maintained and controlled by you alone. It is difficult to create those realities you seek if you have to rely on others to assist in their creation. For instance, you might desire that your child grows up to be a rich doctor. To try to force such a reality on your child is a gross misapplication of the process. Rather, you should concentrate on providing a loving and nurturing environment so that the prime directives in your child's life are all positive. In this way he will grow up to be a mature adult contributing to society in a positive way. Further, the object desired should be ecological. It should not negatively impact you or others in your life. To test for its ecology you might ask what will happen if you get your outcome? How will it affect you, your family, others, or your environment?

Once you are clear on your reality and have experienced it as real you release it knowing that it will manifest in your life. It will either manifest immediately in your life, or an opportunity, a pathway, an avenue, or other means or circumstances will open up leading to its manifestation. Just stay open and aware as to how it might manifest. You do not just visualize your reality and sit back and do nothing to support your desire. You are still a cognitive being living in a physical world and you need to take all appropriate action which would naturally occur to you in order to bring about your desired reality. Remember, the subconscious mind has the capacity to sort for people, places, and things in life to bring to conscious awareness the means to manifest the reality as imagined. Therefore, the subconscious mind will also arrange for appropriate avenues to open up leading to your manifestation. You will want to act upon all such appropriate courses of action leading to its manifestation.

The object of your desired reality should be testable so that you know that you achieved your objective when it manifests in your life. This criteria sounds rather obvious. However there are some things which are not so obvious as to whether you have achieved them. For instance, you may want to create as your reality that your parents, who are always unhappy and fighting like Ricky's parents, become a loving and happy married couple. First, such a reality is largely outside your control, and, secondly, you might never know if they are truly happy in the manner you would want them to be.

Step # 4 - Have Faith and Trust that the Process will Always Work

As stated before, you need to believe and have faith and trust in the MetaCreation process. Never doubt your ability to create in accordance with your desires for it is conscious thought that initiates the MetaCreation process by enabling the subconscious mind to sort for the realities you seek in life. You need to create sufficient emotional desire in order to provide clear and vivid program parameters for the subconscious mind to act on in creating your realities. The subconscious mind only acts on the energy of your thoughts. If the energy in the form of emotional feelings is weak, the programmed search by the subconscious mind will correspondingly be weak. Acknowledge that the MetaCreation process is working in your life and trust and release the object desired knowing that it will manifest in accordance with your desires. This is how the conscious and subconscious minds work together in creating a life for you. I am simply inviting you to try this process.

Step # 5 - Eliminate All Impeding Thoughts and Emotions

When experiencing your desired reality experience it in the spacetime of the fourth dimension. By using analog thinking you experience it as already existing as your present reality. Remember space and time are not factors in the initial act of creating realities. They are only factors of where in space and at what time it will manifest in your life. Experience it as your present reality and then see yourself physically experiencing the reality in the time frames and space frames you desire. We have seen if you experience it as a future reality only and not presently existing, the subconscious mind will act on your desires as a future thing only and not as an expression of an existing reality. Feel the reality as already yours, as already existing as your current reality. Experience it fully as though you already have it. In Chapter 7 and 8 it was indicated that you first must experience the object you desire as an inner reality before it manifests as your outer physical reality. Consequently, the experience of the reality as an inner reality is expressed as existing here and now in the spacetime of the inner realities of your mind. If you accomplish this step properly it will then manifest in the time frames you have projected for its physical manifestation in life. Organize only positive thoughts around your desired reality and eliminate all negative thoughts and feelings which do not support your desired reality.

Check for any thoughts or feelings which are in opposition to the original reality and eliminate the energy and images of the opposing thought. You do this by accessing the analog system where the opposing thought originates and destabilizing its energy and dissolving it. Then come back to the experience of the original desired reality and again experience it as presently existing. Continue this process of eliminating any thoughts or feelings which might arise in opposition to the original reality until all such negative thoughts and feelings are eliminated entirely. Otherwise they will continue to inhibit the creation of your desired reality. Stay involved in the MetaCreation process until the desired reality is felt totally without any opposing thoughts surrounding it or distorting it in any way.[2] You want the desired reality to be felt as though it is the only reality of your existence at the very moment of its creation. You do not want anything else impinging on its creation in any way.

It helps in effectively creating positive realities to define your primary mission in life. You need to state your primary purpose or mission in positive sensory specific terms in order to mobilize the subconscious mind to create for you a future that reflects your primary mission. *We are not talking here about just planning your future, although that is very important, but rather actually physically creating that future.* Mobilize the energy of the subconscious mind to create the structure and substance of that future to become your reality. Your primary mission is what provides meaning and fulfillment in life. It helps define who and what you are. It draws you toward completing your future goals. It combines your values, interests and desires in one comprehensive plan for living your life to the fullest. Ultimately it is what defines your future realities in life and it is what the subconscious mind uses as its blueprint for creating that reality. Without a primary mission or purpose your life could become fragmented, shallow and meaningless. You could, in effect, be living from day to day with not much to look forward to in terms of what gives life meaning. It is your primary mission in life which integrates into a cohesive whole the many complex, and often divergent, modular secondary belief systems at the level of the cognitive structure at which you are currently creating and experiencing life.

When you are pursuing your life's mission you will be in the flow of life and you will intuitively know it. You will experience life as an unfoldment of higher states of consciousness and expanded states of awareness which of itself supports happiness, joy, peace, harmony and balance in life. This unfoldment will naturally come about for it follows the evolutionary flow found throughout all of nature. Out of this unfoldment will emerge physical possessions and material wealth for the comfort and enjoyment of life while you are exploring life's journey. The reason for this is that such manifestations are consistent with the conscious unfoldment and enjoyment of life.

You will want to review your secondary beliefs in order to

2 This is simply another way of saying stay consciously involved in the many event taking place in your life.

identify those specific beliefs making up your current life style. You will want to identify those things that support your mission in life and use the MetaCreation process to bring to conscious awareness experiences of them in the manner you envision for yourself. Align all thoughts which move you in the direction of your desired realities and modify or eliminate those thoughts which are either misaligned or in opposition to your desired realities. Finally, eliminate all secondary thoughts, feelings, action and behaviors which do not move you in the direction of your stated mission in life. Of course, at any time you can reevaluate your mission and goals in life and modify them as your circumstances change, for we have seen the creation and discreation processes are ones of continuous unfoldment rather than of completion.

Step # 6 - Allow For the Object to Manifest and Give Thanks

Finally, simply have fun while working the MetaCreation processes and do not take them, ego personality, nor life itself too seriously. Always enjoy your many physical manifestations and give thanks by sharing and helping others who are unaware and do not understand how the mind creates a life for each one of us. Explain to them that if they are unhappy with the way their life is going there are ways of improving their circumstances and that you will be willing to help them. In this manner, after a time, the whole world will become a much better place in which to live.

You will find over time that new possibilities will open up for you. You will experience a new and expanded awareness of possibilities which never before occurred to you. After a while you will become very proficient at the MetaCreation process and you will be able to formulate in your mind those realities you desire. You can just as easily discreate aspects of your old realities based on old beliefs and events by using the MetaCreation processes. The following description of how Ricky Mindful deals with his problems at Primetex Communications will give you a good example of how you can raise your consciousness up the cognitive structure of the mind and create expanded realities in life.

Example of How the Metacreation Process Created New Realities in Ricky's Life

We left Ricky faced with the problem of declining sales in his region and a critical meeting with his boss, Mac the Knife. Following the Thanksgiving weekend with Tommy and his family Ricky went into work early Monday morning, December first, for that fateful sales meeting Mac called to review the Eastern Region's deteriorating sales. Ricky knew that no amount of explaining would ever satisfy Mac since sagging regional sales adversely impacted national sales volume for which Mac was responsible. What made matters worse was that the Western Region's sales volume was up over projections. A.Tad Tiredly, the Western Region V.P. was seven years Ricky's senior. Tad would turn 60 in just 4 months. He had been with the company seven years longer than Ricky and in the early years he was Ricky's mentor. Tad is highly respected within the Company. There is even speculation that Tad might become the next Executive V.P.Sales when it opened up, for Mac probably would become President and CEO when the current President, Carl, retired next year.

Ricky was hoping that he would be chosen over Tad for the Executive V.P. position but with sagging sales his prospects didn't look good right now. By the time Ricky received the third quarter sales reports he realized that he had to take action quickly or matters would just get worse. He studied the daily sales reports closely and saw that young corporate executives as well as the small business person accounted for a large segment of his regions sales. They were buying electronic equipment for both the corporate and home office and were also purchasing items for personal use and as gifts. By early October Ricky quickly devised a sales strategy that if they purchased office and business electronic equipment at full retail price they would receive a certificate for 50% off on their own purchases for personal use and as holiday gifts. Ricky recognized that while this sales strategy might be risky, if he just sat and did nothing at all it could spell disaster.The strategy worked and sales immediately started to climb. At seeing the first few weeks results the Western Region under A.Tad Tiredly quickly adopted the same sales strategy for the remainder of the year.

As Ricky sat at his desk preparing for the 2:00 o'clock meeting Terri, his executive secretary, buzzed him on the intercom and stated she just got off the phone with Mac's secretary, Cheryl, from corporate headquarters in Los Angeles. She said that Mac was rushed to the hospital late Sunday afternoon and was scheduled for emergency surgery that morning for his bleeding ulcer which was acting up again. The meeting was postponed until Monday, January 26, 1998, at which time the year end sales figures would be in, and if they were not good Ricky would have a whole lot of explaining to do. As Terri walked into Ricky's office she could see that he was visibly relieved at hearing this bit of good news. He had a two month reprieve before facing Mac. Mac's secretary called Terri later in the day to report that the surgery went well and that Mac was resting comfortably. He would just have to cut down on the cigars, booze and spicy food.

Later that afternoon as Ricky was contemplating these recent developments Terri came into his office with a cup of coffee for him. Terri mentioned that Ricky should try to cover his flank by shoring up his fences with the old warhorse Tad before the January meeting. In those early years they were close and Ricky was eager to learn all he could from his mentor. But as Ricky rose through the ranks like Tad, Ricky became even more aggressive and he saw the Western Region under Tad as his chief competition. He no longer saw Tad as an ally but as an adversary who could not be trusted. In view of this Ricky thought to himself that Terri's comments were rather curious for there was no secret about the open competition between Ricky and Tad. Terri said that Ricky's aggressive style got him this far in the company but that teamwork was now the company mantra and because of his vulnerable position a little cooperation and teamwork might better serve Ricky's interests.

It was just the week before that Ricky had encountered himself at depth in the north Georgia mountains and finally made a commitment to Betty. The prospects of the January meeting with Mac over Ricky's sagging sales dampened Ricky's enthusiasm somewhat for his upcoming marriage. Tommy's family planned to join them at Ricky's place in Atlanta for the holidays. On the day after Christmas they would get married in a small private ceremony and immediately leave for Palm Beach, Florida. As a wedding present Betty's bother, Neil, a wealthy architect out of New York City, lent them the use of his luxury eight room condo located right on the ocean. The next three weeks before the Christmas holidays went by quickly. Ricky and his staff worked feverishly on their new Christmas sales promotions and to get all of their year end sales reports and next years projections as complete and detailed as possible for Mac's January meeting. Luckily there was a very strong surge in daily sales by the fourth quarter as a result of the new sales strategy. The Christmas crowd was favoring electronic and mobile communications equipment for business as well as their personal use and as holiday gifts.

On Friday, December 19, the week before Christmas, Primetex Communications held its Eastern Region Annual Christmas Party in the grand ballroom of the Ritz-Carlton Hotel on Peachtree Street in Buckhead. While Tad and Mac were not there, Carl, Primetex's President and CEO, was in attendance and gave the State of the Company speech during dinner. Afterward while the orchristra played dance music, Cheryl, Tad's secretary from L.A., come over to Ricky and Betty's table to chat. Ricky often felt that Cheryl would be a total knock out if she would just lose a couple of pounds. Cheryl always had her eye on Ricky and was not pleased to hear that he was going to marry Betty. Everyone was having a good time and you could tell that Cheryl was well into the champagne. Cheryl turned to Betty and said, "Honey if you aren't woman enough to satisfy your man then just step aside and I'll take him". She then turned to Ricky and with her sexiest smile she batted her big blue eyes and asked Ricky for the next dance. As they were dancing she confided in Ricky that after Tad's prostate cancer operation last year Tad has noticeably slowed down and now might be the right time for Ricky to bury the hatchet and make peace with the old warhorse. As they returned to Ricky's table, Cheryl stated to Betty, "Remember, dearie if you don't want him I'll take him in a heart beat." With that she gave Ricky a hug and wished them all a Merry Christmas and walked back to her table across the dance floor.

Saturday morning, the day after the Christmas party, Betty and Ricky slept in and had a late breakfast. Betty then said that she and Louanne had some last minute Christmas shopping to do, and they would shop and have lunch in Buckhead and be back later in the afternoon. Ricky sat at the dinning room table opening Christmas cards. There was a small package from Carl, the President of the Company. Ricky wasn't about to wait until Christmas to open that one. The present was Stephen Covey's book, "Seven Habits of Highly Effective People". The card inside read, "1998 will be a year of teamwork and cooperation. I want each member of my executive staff and all department heads to review one of the seven habits of highly effective people at your monthly staff meetings for the next seven months. Have a Merry Christmas and Happy New Year. Regards, Carl". Ricky thought to himself, Terri was right after all. The Company theme this year was teamwork. These last few months Ricky realized that he was more of a loner than a team player. Over the years the sales staff always admired his brilliant sales strategies and he was liked by most of his associates but he never really felt close to them or a part of the corporate culture. Since that visit with Tommy up in the north Georgia mountains Ricky had become acutely aware of this fact and had started to review his life. He saw that he now wanted to become more of a friend and team player and less of an aggressive adversary with his business associates.

As Ricky continued looking over the Christmas cards, the UPS delivery man dropped off a rather large and heavy carton. The present was from Tad. It was a 24 bottle case of fine California wines from the vineyards of Robert Mondavi. The card read, "From one old Warhorse to another, in the spirit of the good old days when we worked closely together, I send a toast to you and your bride-to-be Betty and wish you and the Eastern Region sales staff of Primetex Communications a prosperous and happy 1998!" Ricky pulled out of the case a rare 1986 vintage bottle of Chardonnay, uncorked it and took in its bouquet before pouring a glass and taking a couple of sips. "Really pretty good stuff," Ricky thought to himself. Tad always sent Ricky a couple of bottles of wine at Christmas but never an entire case, and never of such fine quality. Was it Betty, the upcoming marriage, or that Tad himself had a particularly good year? What was it, Ricky pondered as he poured himself a little more wine.

Then it dawned on him.Tommy always said that if events came in threes, conscious analog thinking would indicate that they probably held a special meaning for you. Ricky recalled that over three weeks ago in his office Terri mentioned that maybe he should adopt a new strategy of cooperation and teamwork. Then at the annual Christmas party Cheryl said to Ricky, while they were dancing, that maybe he should bury the hatchet and make peace with the old warhorse. Now Tad also seemed to be trying to convey a message to Ricky, but what was the message? They had always been adversaries ever since the old days way back when Tad was his mentor. Over the years Ricky became more confident in his sales abilities and relied less and less on Tad's advice and counsel.

As Ricky poured himself a little more wine he thought to himself maybe he should use Tommy's count down relaxation exercise and see what the analogs of his subconscious mind would reveal to him concerning these matters. Ricky didn't know if it was the relaxation exercise or the wine, but whatever it was he was feeling very relaxed. He remembered Tommy saying something about feeling the reality you want to create for yourself fully as a present condition. Then eliminate any opposing thoughts or feelings and come back to the original thought and continue this process until the present thought is experienced as a present reality in something called spacetime. He always dreamed of someday being President of Primetex Communications. So he relaxed and tried to visualize himself sitting in Carl's large office in L.A. as the President of Primetex. He finally got a pretty good impression of what it would feel like to be President. When it came to eliminating any opposing thoughts or feelings he had this slight anxiety of not doing something, or of being something, that was preventing him from raising any further up the corporate ladder. He just couldn't put his finger on whatever it was.

He drifted into a relaxed state which took him back to Lookout Mountain in Chattanooga, Tennessee. Several years ago he stopped by the hang gliding school there to inquire about hang gliding lessons. He recalled the chief instructor stated that it was a piece of cake, that the hang gliders practically flew themselves. Under almost any conceivable scenario the hang glider would automatically catch the wind currents and glide under its own power and land safely in the valley below, with just one exception. With that explanation the instructor turned to help a young student harness up. Ricky remembered saying, "Well?" The instructor said, "Well, What?" Ricky said, "Well what is the one exception?" The instructor replied, "The only way that you could ever crash is if you overcontrolled the glider. You need to relax and catch the air currents and just let the glider do its thing. You need to trust yourself and trust the glider and not panic out of fear, choke up and resist the natural flow of the air currents and literally force the glider into a nose dive. That is the only way you could ever have a problem."

As the scene faded from his mind Ricky thought to himself that he had always been a controller. He believed that if he could control all circumstances in his relationships with women he wouldn't get hurt and if he could control all sales decisions in his region he could stifle all competition and criticism. He now realized that these strategies could only get him so far in life, but ultimately they would fail him in obtaining intimacy in his relationship with Betty and would prevent him from rising any further up in the company for which he worked so hard and was so loyal. But what was the message he needed to learn from all this he thought to himself? Was there a strategy to follow in order to become the President of Primetex within the next five years?

He again drifted into that relaxed state, and this time he visualized himself technical climbing up the steep east face of Long's Peak in Rocky Mountain National Park near Estes, Colorado. As a young army private stationed at Fort Carson in Colorado Springs, Colorado, Ricky remembered climbing most of the Collegiate Peaks and he had climbed Long's Peak a dozen times. However, in his dream he was climbing up the difficult east face of Long's Peak with A. Tad Tiredly. It was late on a cold winter afternoon. They both underestimated the weather and the time in which to make the climb. Both were exhausted and they still had about 45 feet to climb to reach the top. Ricky was in the lead and made his way up to a small ledge. He was able to secure his position by driving an anchor pyton into a small crevice in the rock and locked it in and then he relaxed for a brief moment. Tad was about 20 feet below him and had trouble making the final accent. Tad cried out as he lost his grip and swung out from the face of the cliff and swung back in and hit the surface hard. Tad was dazed but managed to get a hand hold and hang on. Ricky's pyton held the weight of both of them.

Ricky thought to himself that they had maybe a half hour before sunset. In the dark they would not make it to the top and would not survive overnight on the face of the mountain. His philosophy in life had always been when the going gets tough cut your losses and go it alone. Maybe he should cut Tad loose and it would be every man for himself. Then something deep within welled up inside him and revealed to him for the first time in his life that if they worked together they could do better than going it alone. He remembered Terri counseling him, "Teamwork was now the Company's mantra." Carl in his Christmas message said something about "corporate teamwork in 1998," and Tad said something about "in the spirit of the good old days when we worked together." Perhaps he had overlooked it all his life. By going it alone in order to survive he might have missed the greater message of cooperation, help, companionship and sharing which might bring him even greater rewards and success in life. Was teamwork the answer he was looking for? With that thought he shouted down to Tad "if we work together and help one another both of us can make it to the top and succeed." By working together they made it to the top just barely before the sun went down. Ricky recalled seeing himself giving Tad a warm handshake and then a big bear hug and saying "put it there, partner!" At that moment his wine glass slipped from his hand and smashed to the hardwood floor and abruptly brought him out of his reverie.

He had a wonderful time at Christmas with Tommy and his family, and Betty and Louanne. He and Betty were married in a quiet ceremony the next day and their honeymoon went all too quickly in Palm Beach, Florida. Back in the office on Monday, January 12, Ricky stated to Terri that he had given a great deal of thought to what she and others had said before the holidays. He made a News Years Resolution to become a team player and to cooperate and help others attain their goals. By helping others he would be benefiting himself as well as the company. Terri had seen a remarkable change in Ricky these last few months. She said that she felt he was making the right decision and was very pleased for him. The year end sales reports finally were all in and they showed that his region either met their yearly projections or were slightly ahead of them depending on how you interpreted the figures. Christmas sales were extremely heavy as a result of Ricky's new sales strategy and made up for the earlier slump. The sales crisis was over!

Ricky stated to Terri he was thinking about calling Tad in order to try to reconcile things between them. They had grown apart over the years and Ricky said he didn't know just how to go about establishing a better business relationship with Tad without appearing to be coming from a position of weakness. Terri explained to Ricky that it wasn't about coming from a position of either weakness or strength but rather a matter of being true to yourself. If he truly wanted to develop a better business relationship and this was his only motive then Tad would recognize it and accept it at face value. It is simply a matter of coming from a position of honesty and integrity and not from either a position of weakness or strength, for honesty, integrity and openness were themselves strength. It is only when you have some hidden agenda do people sense this and become suspicious and defensive.

However, Terri added that a good entree might be to call Tad and tell him that his Western Region must be faced with the same challenges that we are facing back here in the east in preparing the 1998 sales projections. If the two of you got together maybe you would come up with a joint report covering both regions. One comprehensive national sales projection would probably impress Mac the Knife. Ricky thought to himself, "What an excellent idea. Now I see why I kept Terri around all these years." He immediately called Tad. Tad responded by also agreeing that a joint report, while never done before, was an innovative idea and Tad was certain that it would impress Mac. Further, Tad added that the corporate jet happened to be at Atlanta Hartsfield International Airport after ferrying corporate executives back east following the holidays. Ricky could take the jet back to L.A. on Thursday. By Friday morning they could pick up all of the needed sales reports and fly down to Tad's beach front house in Coronado Beach in southern California. Just before they hung up Tad added that he had something very confidential he had wanted to discuss with Ricky for a long time and now seemed to be the right time. Maybe they could kill two birds with one stone.

Corporate travel was very fast these days Ricky thought to himself. He could always measure the direct flight from Atlanta to L.A. in terms of an on board snack lunch, a couple of hands of poker with whomever else was also on the plane, and maybe a short nap, and before he knew it he was stepping off the plane at the Orange County Airport. That night Ricky had a leisurely dinner with Tad and his wife, Joanne. The next day they gathered up all of the year end reports and by early afternoon they had flown to the San Diego airport. They rented a car and drove out to Tad's beach front house in Coronado Beach.

At dinner that night, after a round of drinks and some leisurely shop talk, Ricky said to Tad that he had given a lot of thought to Carl's teamwork theme and Carl was right that the two of them should work more closely together in the future. Ricky confided in Tad that he recognized that over the years he hadn't been trusting enough. He told Tad he owed a great deal to him for showing him the ropes during those early years and helping him rise up the corporate ladder. Ricky continued that he recognized that early on he became a little too independent and aggressive in his business style and felt that he didn't need tad's advice as a mentor, or even as a friend, any longer. He became too aggressive and a controller and didn't delegate enough to others, particularly the younger men coming up the line, for he didn't trust anyone. But now, somehow, he was able to see everything from a different perspective.

As his brother, Tommy, would put it, Ricky realized that he had shifted to a new higher level up the cognitive structure of his mind, whatever that meant. From this new vantage point everything looked different than before. He really didn't know what he was running from all these years and what he had to hide from. He no longer had to prove himself over and over again. He could finally trust others. Both Betty and Tommy had helped him so much in finding himself, and he would like to, in turn, be able to help others who might have also suffered the disadvantages he did as a child. He stated to Tad that he truly wanted to work with him and not compete on every issue that came along as he had in the past. A trusting, honest and open relationship of working together would strengthen each of them, and both could achieve whatever they wanted and both would be winners. Tad sensed a genuineness in Ricky he had never seen before. Ricky seemed to have truly found himself. Tad saw that Ricky had matured these last few months and knew Ricky could now be trusted and relied upon. Maybe it was Betty and Tommy's influence in his life, but whatever it was, it certainly changed Ricky. He no longer showed that bravado and aggressive style and was more straightforward in his dealings and honest and open about everything which gave him a deeper strength of integrity and character than Tad had ever seen before.

Tad responded that it was now time to also confide in Ricky. Tad stated he felt that he could now trust Ricky and felt the two of them were back working together like the good old days. Tad disclosed that his cancer operation had taken a lot out of him and that he just couldn't keep up the fast pace that the job demanded. He indicated to Ricky that the company had been considering consolidating both regions into one national territory. Nothing had been said about this since it was only in the preliminary planning stage. Ricky's proposal of a combined sales projection was timely. It could work to both of their advantages. If Tad recommended Ricky as the Executive V.P. Sales for the consolidated territory, Mac, Carl and the Board would probably endorse it, for they still held Tad's judgment on these matters in very high regard. In exchange Tad recommended to Ricky that Tad serve as V.P. Sales Administration under Ricky. This way Tad would not be under the intense pressure he had been these last few years. He would not have to travel as much and could spend more time with Joanne, and he could maintain his six figure salary until he reached retirement age in five years.

Tad stated he only now considered such a strategy for he felt he could trust and depend on Ricky fully. Ricky responded that he appreciated the faith Tad had placed in him but Tad would lose those fat annual bonuses he was used to getting. Tad responded that he didn't care about the bonuses as much as keeping intact his six figure salary. Ricky responded that he would personally see to it that the salary base wasn't touched until Tad retired. Tad said, "Then its a done deal." Ricky responded, "Yes" and Tad shook his hand and said "Put it there, partner." Ricky took another sip from his drink and thought quietly to himself that as Executive V.P. Sales of the new consolidated region he would be in line for the Presidency someday since Mac, who was next in line, was already 63 and would be retiring within several years and Tad now was no longer in the running. Tad said, "I know what you are thinking. Some day you could be President." Ricky turned red and said, "What are you, a mind reader?" Tad said, "I would not have made this proposal to you if I didn't think that you now have the maturity, honesty and integrity as well as the competency for the job. I wish you all the best." With that he raised his glass to Ricky in a toast. . . . .

Summary on the Metacreation Process

It helps to keep in mind that you are none of the realities that you experience, even though you experience them as very real in life. They are only creations of the mind and therefore they can be just as easily dissolved as they are created. Once you deeply understand internally and feel the dynamics of this life process itself you finally can rest knowing that you now embody the truth about life, and it is this very truth which will set you free; set you free from a self limiting past and from the fear of an uncertain future. You can effectively change the past and the future. It is not freedom in the sense of creating what you want in life, although this is very important, but freedom to experience fully what you are in life. You are an expression of life itself, that which expresses reality. If your expressions as life are positive, your realities will be positive. As you create positive realities using the Metacreation processes, you will experience yourself as a positive expression of ego personality, positive expression of inner and outer realities, and a positive expression of life itself. This is what I mean by freedom, to create for yourself a positive image of ego personality in order to express self in a meaningful way and to express it fully in life. Once you come to know and embody this truth you are truly free.

The Metacreation process starts you out on the process of creating those things, both material and otherwise, which bring you this freedom in life. Whether it is material and financial wealth, good health and loving relationships, or simply peace of mind, harmony and balance, you will have reached a point where you are finally able to take total control over your very own thoughts and passions and, consequently, exercise control over your inner and outer realities. Whatever path you seek for yourself, by following the MetaCreation processes on a regular basis you will, at long last, come into that knowledge which unlocks the mysteries of life, a personal knowledge and understanding which ultimately leads to mastery over life itself. You will come to finally understand and experience who and what you are and why and what your purpose in life is really all about. As you diligently seek to uncover the mysteries of life, with your newfound knowledge and creative processes of the mind, you will stand in awe at the magnitude of the many revelations life ultimately will reveal to you. I hope that you have found this book exciting, revealing and rewarding.

Conclusion

This book is not intended to be the final definitive word in applied mind science. It is not intended to be the final word on how the human mind and brain work together to process a variety of stimuli to represent our realities in life. We have seen from Chapter 4 alone that there still remains many unknown frontiers to explore in the further reaches of consciousness. However, this book is offered simply as yet another step in describing the process of how we create realities in life. It is offered in the hope that mankind may, one day soon, come to more fully understand and know his true nature while embodied in human form existing on this planet called earth in the fixed time and space of our life. If I have provided you with information, theories and techniques which move you just one step further toward your own truth, your own destiny, and your own reality, I have accomplished my purpose.

For whatever else your life is about, it is certainly a learning experience. As each one of us acquires additional knowledge so that we can comprehend and understand ourselves, our mind's processes, and how it represents our world to us, we can, together, take greater control over those processes as well as over that world. For as I have stated numerous times and in numerous different ways throughout this book, for whatever else life might mean to you it certainly provides you with many opportunities to discover for yourself higher states of consciousness and expanded states of awareness.

For you will, one day soon, connect with your deepest nature in order to experience your highest good.

When that day comes, and it is not too far off, you will personally connect with yourself, with others, with your community at large, our world, and yes even with the energies of our entire universe in the spacetime of the fourth dimension. You will experience yourself as being and existing as both a part of the greater whole, and, at the same time, experience that greater whole made up of many different parts forming the mosaic of your life. Parts, many of which, still lie unknown and unexplored, that are waiting to be discovered by you. This discovery is unique to each one of us and yet it is, at the same time, common to all of mankind. In this manner may you truly encounter and experience the source of these many parts themselves and come to know and understand your true purpose for being here and the destiny of all mankind. For in this way you will ultimately come to know yourself and the source of your being. For isn't this truly what life is really all about? I wish each one of you success on your life's journey with its many new revelations and manifestations waiting to be discovered by you. Always be open to your good. I wish you God speed on your personal journey, a journey of a thousand stars!

-The End -

EPILOGUE

A Description of Seventh Level Consciousness

This chapter is to be read only by those who fully understand the theory of MetaCreation and have mastered the MetaCreation processes spelled out in this book so completely that they have created for themselves all the material wealth and prosperity the world will ever be able to offer. If you have not, then please do not read any further, but rather keep working on your creation techniques until you do.

However, if you do meet this threshold criteria for reading this Chapter, then the next test is, have you created for yourself and experienced all of the beauty the world has to offer. Have you been able to sail the seven seas, seen New York, Paris, London, Rome, Tokyo, Hong Kong and all the exotic places in the world? Have you tasted all of the finest foods and wines the world has to offer? Have you been able to hear all the finest music, and seen all of the finest art, the world has to offer? If not, please do not read this Chapter, but rather continue to work on your creation techniques until you do.

The final test you must meet in order to be invited to read this final Chapter is, have you experienced, truly experienced, unconditional, unrequited love with your true beloved, with the one and only love of your life, a love that is truly so deep, intimate, and pleasurable beyond words that you have truly touched one another's soul on the level of pure being. If you can answer yes to these three questions than you are permitted, yes, even invited, to read on.

If you have accomplished all of these things and have passed these final three tests you have then truly mastered totally all of the techniques the MetaCreation processes have to offer, and, therefore, you have finally graduated and have truly become a Master. You have become a Master over the thought processes of your mind, a Master over all of the emotions of your body and a Master over all that you possess. You have overcome the personality of ego. You have surpassed the operational style of a thousand men. You have transcended both time and space and are eternally present in the here and now of life. You have finally truly internalized an objective world and it is yours to mold and transform into whatever image you desire through your very own thought processes without any effort, doubts, or fears on your part whatsoever. You do not seek anything for yourself for you express all that anyone could ever seek in life.You do not have to count your possessions for you possess all. You do not have to plan for tomorrow, for you are tomorrow, and all of the yesterdays.

You do not have to create the reality of this world for you are the reality of this world!

You have truly become the universe's metaphor expressed as "whole, perfect and complete." Your every thought surpasses any belief system that ever existed and forms the reality of all of the words and thoughts that were ever spoken throughout the ages. You no longer speak, see, or feel with sensory acuity for you have transcended sensory acuity and have become the source of its creation. You now embody all of the submodalities that ever existed in forming the endless realities of life itself. You are the color in the sunset. You are the sound in children's laughter. You are the feelings of love in the newborn child. You are the vibration of every song that was ever sung. You are the passion of the newly wedded couple. You are the security of the elderly. You are all of these things, for you are the hope of the world for you are its destiny!

You are all of these things for you are truly a Master in the sense used by the ancient ones. As Great Eagle and other Shamans and other Masters have so beautifully stated in different ways, at different times and at different places throughout the history of mankind, that once you have mastered your mind and speak only your truth, you become that truth. When you become truth, it becomes impossible to speak anything but truth at all times. When you are truth, if you were to confer a blessing upon a person, even if that person is a sinner, he nevertheless is blessed beyond words and is no longer in a state of sin. If you confer a healing upon a sick person, that person is, at that instant made well, whether or not he deserves the healing. For you do not only believe.You have become belief itself for you have been transformed into the constituent parts of belief. You trust for you truly know, beyond reasoning, beyond cognition, beyond awareness, beyond the many experiences of mankind and beyond the reality of life itself. You have finally become all of the energy and all of the images pervading the entire universe. You are its moons, stars, constellations and galaxies for you have become the universe and the source thereof.

You are the beginning and the end of all that exists. Before this world of ours was ever conceived and brought into manifestation you were there. Wherever a tree may fall in the forest you have already heard it. Before anyone can even think a thought you know of its existence. Before an infant is born you are its life for you are the unity in diversity, the deeper unity in the surface of chaos itself. You already existed long before the particle and the wave of life manifested in all of its many realities. Nothing and no one ever exists without you sustaining them, for you are the breath of life, the love that sustains all of life. You are its visions and its dreams. You are the knowledge gleamed by mankind over the years. You are the scientist in the laboratory. You are the engineer planning the construction of a new and better world for tomorrow's children. You are the professor in the classroom. You are the doctor imparting healing to an ailing patient, the mother nurturing her child.

For you are truly this one. You add meaning where there is no meaning.You add truth where there is no truth. You add life where there is no life. You add hope where there is no hope. You add love where there is no love. One among many and the many within the one. Each one of you is truly this one. You just need to learn how to embrace your rightful inheritance. Such a person lacks all covetous for the things of this world, and is in a state of such pure being and awareness that the world consequently becomes his slave. He that possesses nothing of this world owns everything of this world. He possesses all things and all knowledge and all power is his. All manifestation is under his control and direction. When he gives you a gift of any type, whether or not the gift is of a physical or purely of a mental or emotional nature, the gift is always given unconditionally without thought of receiving anything in return, and the recipient of such a gift has been rewarded more then he will ever know, whether or not he earned or deserved it. You will truly be blessed a thousand times and those blessings in turn will flow out of you to bless others with whom you come in contact for this is the way of the world.

The reason for this is that he who obtains the state of non-convetness, becomes all things, all knowledge and all manifestation. Such a person speaks his truth through love, becomes love, and is love. As love his light shines throughout the world and he lights the way for others. His presence has a calming effect on all those around him. His words are an inspiration to those who are still unaware and as a consequence continue to suffer in darkness and to experience pain, lack and limitation in one form or another in their lives. He is with the person who cries out in anguish and in pain. He is with the mother who lost her son in war. He is with the young man who is slowly dying of AIDS. He is with the elderly man who is fearful and all alone in the nursing home. He is with the small child without a present on Christmas day. He is a source of inspiration. He is a service to all who are in need. He comforts the lonely, heals the sick, enriches the poor, provides for the destitute, and teaches those who are simply unaware and do not understand. He leads by example and never asks anything in return. Such a person does not seek to gain but to give, not to acquire but to provide, not to take advantage of but to help, not to be admired but to encourage, not to oppress but to lift up all mankind.

Such a one does not fear others or anything of this world, not even death, the final arbitrator of the value and meaning of life. He does not fear death for he knows, not only experiences, but truly knows, that death is but a point, a transition to another plane, another world, and finally another dimension. He is finally, at long last, reunited with his true self, pure awareness, unaffected by the things of this world. He can now rest with his true beloved and experience the source of his being, the Source of All. He is at last home, the journey was long and at times arduous, and yet always full of excitement with new adventures at every turn. It was always unfolding in accordance with the grand plan of evolution, whether at times he was aware of it or not. He would not have wished it any other way, and yet at the same time, in a sort of poignant and sad way, he is sorry that it is over so quickly, so soon, for he feels that he has so much more to offer, to learn, to experience, to share, to give and to receive.

Death was never a concern of his, since he long ago realized and experienced himself as immortal even while embodied in human form. He is infinite and death, like life, is but simply a pause in eternity. This is both the end of the line, and a new and exciting beginning. Each one of you is truly this one. You just need to learn how to embrace your rightful inheritance and become that which you already are. That which you seek, you will ultimately find. Rather then looking to the stars or to others you must look within. There you will be able to unlock the mysteries of the universe, the mysteries of this world, the mysteries of life and death. Yes, even the deepest of all mysteries of who and what your are, the meaning of your life, and why you are even here now at this particular time and place on this journey called life. There you will find all of the answers you will ever seek and you will immediately stand in awe at the profound revelations that flow out of the human condition. You will be blessed a thousand times! I, again, wish you God speed on your journey! Your personal journey of a thousand stars!

APPENDIX ONE

CONSCIOUS ANALOG THINKING - Continuation of Chart 9

The following sets forth in comparative fashion the differences between the processes of conscious analytical thinking in fixed space and linear time of an objective three dimensional world and conscious analog thinking which conceptualizes time and space in the en bloc spacetime of the fourth dimension:

Comparative Analysis in terms of the Conscious and Subconscious Mind

1.- 7. Items 1 through 7 are set forth in Chart 9 at page 96.

8. Analytical thinking being logical usually takes place within mental, emotional and environmental constraints. Analog thinking takes place outside those constraints and is often characterized as being in the flow of life where you are supported by the energies in life in all that you undertake to do.

9. Analytical thinking involves rational linear thought. Analog thinking involves spontaneous creative thought.

10. Analytical thinking usually involves intentional rational thought based on deductive and inductive reasoning. Analog thinking involves spontaneous holistic thought where all aspects of a given situation are revealed to conscious mind from the subconscious holistically, that is, all at one time. That means you don't think out all of the aspects of a thing but rather all aspects are revealed spontaneously often when you are in a relaxed state and not concentrating on the thing at hand.

11. Analytical thinking usually arises in daily Beta rhythms of the mind. Analog thinking regularly accesses Alpha rhythms of a relaxed mind to organize creative thought and bring it spontaneously to conscious awareness.

12. Analytical thinking often exists and functions out of alignment with the energy of the mind, body and its environment. Analog thinking is aligned and congruent with the energies of the mind, body, and the environment.

13. Analytical thinking often rationalizes existing negative energy and images arising from the subconscious mind rather than dealing with them at the time that they arise. What is meant by this is that often a very intellectual person using logical rational thought to order his world will often find many arguments and excuses to explain his negative thinking, feelings and behavior. He uses either arguments and excuses rather then dealing with the origins of the negativity in his life. Consequently, such negativity will continue to manifest until it is effectively dealt with for it just has a nasty habit of never going away nor will it be denied expression in one form or another until it is dealt with. Analog thinking, on the other hand, clears negative energy and images as it moves toward greater and

more expanded states of awareness, states of equilibrium and mental homeostasis.

14. Analytical thinking often operates separate from the energy giving rise to and sustaining physical life. Analog thinking experiences self as intimately connected to and one with the greater forces within the mind, body and environment sustaining all of life.

15. Analytical thinking experiences emotions arising from blocked desires as frustration, anger, prejudice, hatred, hostility and other negative forms of emotion. Analog thinking on the other hand experiences emotions arising from expanded states of awareness as peace, harmony, and love. Analog thinking expresses those emotions as help, support and cooperation with others. The reason for analog thinking experiencing positive emotions such as these is that analog thinking seeks to find its primary purpose in life and then endeavors to follow it. When it does so one's life starts to unfold in a natural progression and consequently one encounters far less resistance and negative energies in life. Life always supports, nourishes and sustains the intentions of the mind which is in alignment with its purposes. Such intention and purposes become congruent with the greater energy forces flowing within the mind, the body and the environment in which all of the events in life take place.

16. Analytical thinking often views other people and situations as the cause of its pain. Analog thinking views self as solely responsible for whatever pain self experiences and is solely responsible for its elimination. Analog thinking always takes responsibility and never blames others for any misfortune or pain encountered in life. While it is true that others may have, out of ignorance or even malice, caused you great harm it is still up to each individual to take his or her own responsibility in clearing any negativity arising from such harm and to then move on in their life. Harboring hostility and resentment in life over others or events which may have caused you pain in the past only prolongs the pain and delays its removal.

17. Analytical thinking is based on the premise of competition among your peers for the limited material goods and financial wealth of this world. Analog thinking is based on cooperation among all people and on the premise that there is available within this world unlimited wealth. You need only to relax and access it through the processes of the subconscious mind and allow it to manifest as your very own abundance in life. Quite often the very act of striving for wealth only drives it further away. The reason for this is that wealth in whatever form is a representation of your level of awareness in life. Find something that people want, need and are willing to pay for and wealth will flow to you. Wealth is not something to acquire and amass but something that is the result of a person providing a needed service or product that others need and want in their life.

18. Analytical thinking draws its creative energy from the cognitive structure of the conscious mind. Analog thinking also draws its thoughts from the creative energy and images of the subconscious mind and from its surrounding environment. This creative energy

and images is accessed by the awareness of an expanded cognitive structure. An expanded cognitive structure comes about by becoming knowledgeable as to how the universe and all of life operate on the quantum level of its existence in order to nourish, expand, and grow. Then you will want to emulate these processes for growth in your own life. In this manner you will draw material wealth, prosperity and loving relationships into your personal awareness and be able to claim them as your own.

19. Analytical thinking for the most part only can accommodate rational linear thought images arising within its existing physical reality. Analog thinking also creates its realities from the energy and images arising outside and independent of its experiences of its physical reality. Analog thinking can operate outside its experience of its physical reality for the reason it can expand its cognitive structure to perceive and allow for the assimilation of much greater stimuli from its environment. It can then organize such additional data within the subconscious mind in such a manner as to create for itself expanded realities in life. Remember the subconscious mind can, with proper programming, sort for unlimited possibilities, unlimited potential and unlimited correlation among those potential possibilities to bring to conscious awareness those realities you seek in life.

20. Analytical thinking experiences happiness as physical pleasure arising from the physical senses. Analog thinking experiences happiness from the experience of expanded states of awareness expressed as love, joy, peace and harmony with self, your world and all mankind.

Comparative Analysis in Terms of Time and Space

21. Analytical thinking takes place in linear time and fixed space. Analog thinking takes place holistically which involves the dynamic interplay of both past and future events in life as they relate to manifestations in present spacetime of the fourth dimension.

22. Analytical thinking quite often deals with the events of life from the past or from the future. Analog thinking takes place primarily in the present moment which encompasses both past and future events. Remember we described time as flowing in circular fashion instead of linear fashion. In this way the subconscious mind can allow for the realities you desire in life to emerge on the outward flow of life and allow for negative realities you no longer want in your life to dissipate and dissolve on the inward flow of life.

23. Analytical thinking often arises from the premise that it has little or no control over past or future events in life. Analog thinking arises from the premise that it has total control over past and future events in life for the past and future are perceived as all arising in the present moment in life.

24. Analytical thinking experiences life on a time line from birth, youth, middle age, old age and death. Analog thinking experiences life in the present here and now until you make a transition to another plane of existence.

25. Analytical thinking often views your outer physical world as objectified and largely outside your control. Analog thinking views your outer world as totally subjective and can be accessed, molded and formed in whatever fashion you choose for yourself. The reason for this is that if you truly view your world in this analog fashion you can then program the subconscious mind to sort for unlimited possibilities to bring to conscious awareness those people, places and things needed to manifest your realities in time frames and space frames of your choice.

26. Analytical thinking experiences less control over things and events as they progressively move further out in the space of their environment. Analog thinking experiences total control over things and events in life regardless of where they are located or at what point they occur in the space of their environment. Again as we just explained the subconscious mind can place realities in the space frames of your choice if it is programmed to do so.

28. Analytical thinking endeavors to force creations in time and space frames through its own physical efforts. Analog thinking relaxes and lets go knowing that all of its creations already exist in the spacetime continuum of the fourth dimension. Relaxing and letting go removes whatever constraints exist which might impede the manifestation of your realities in spacetime of the present moment.

29. Analytical thinking relies on the basic premise that if things are to happen you must make them happen through your own physical efforts and talents. Analog thinking views life as existing in spacetime and it is simply a matter of allowing that which is already complete to unfold naturally in life without your limiting it from manifesting in any way through conscious effort on your part.

30. Analytical thinking, that is by definition self-limiting. This is so for the simple reason that all thought regardless of how expansive it might be is ultimately self-limiting. This characteristic of all thought is both good and bad. Good for the reason that it is the outer definitions of thought that give definition to our world and therefore stabilizes our experiences of our world. However, all thought also has its down side in that when it expresses an idea all other options are eliminated at that very moment in time. Analog thinking on the other hand by releasing conscious cognitive thought and relying on the unlimited possibilities of the subconscious mind can readily manifest what is desired in life unimpeded by the definitions created by conscious logical thought. Analog thinking then uses conscious logical thought primarily to support its creations and only secondarily to actually create realities in life.

Comparative Analysis in Terms of Inner and Outer Realities

31. Analytical thinking experiences reality as outside of and independent of oneself. Analog thinking experiences reality as inside and controlled by one's conscious and subconscious mind.

32. Analytical thinking often experiences the world as though you have little direct control over circumstances in life. Analog

thinking experiences the world as though you have total control over all circumstances of your life.

33. Analytical thinking organizes its reality from rational, logical reasoning consistent with the appearance of its inner and outer realities in life. Analog thinking organizes its realities spontaneously from the impulse of the energy and images of subconscious thought independently of outer reality.

34. Analytical thinking employs physical work, effort and labor to manifest creations in outer reality. Analog thinking allows the subconscious mind to bring into conscious awareness those people, places and events in life to manifest creations with minimal personal effort or physical labor on its part.

35. Analytical thinking organizes the things of its world in terms of numerous separate and distinct physical objects. Analog thinking organizes the things of this world in terms of interconnected and interrelated, people, places and events.

36. Analytical thinking views the individual as existing in this physical body living out a life in a given span of years. Analog thinking views the individual as expanding states of consciousness, that life is unfoldment, that the body is a vehicle for expressing life, that we are all interconnected, and that death is a transition to another plane.

37. Analytical thinking places its values on the accumulation of wealth, competition and material success. Analog thinking places its values on the expression of higher states of consciousness and expanded states of awareness.

38. Analytical thinking experiences life as a series of problems and a search for their solutions. Analog thinking experiences life as an unfoldment and advancement of awareness toward completion and fulfillment. Analog thinking experiences realities as energy and its life as process.

39. Analytical thinking attempts to maximize pleasure and minimize pain. Analog thinking seeks to expand awareness and consciousness expressed as love, peace, harmony and balance in life.

40. Analytical thinking experiences self as existing separate and apart from all other living beings. Analog thinking experiences self as connected to, one with, and existing in harmony with all other living beings.

41. Analytical thinking experiences life as an attachment to and identification with material wealth and possessions. Analog thinking experiences life as an unfoldment of higher states of consciousness. Analog thinking does not identify with material possessions in life for the purpose of attaining a personal sense or meaning in life.

How to Personally adopt Conscious Analog Thinking in Life

42. First you need to thoroughly understand and incorporate into your life the theory and processes by which you manifest a life.

You do not just live a life. You are a life. That which you experience as your life is in fact you. You are the realities that you create.

43. In order to create a better life you need to start creating a life from outside the realities you are presently experiencing as your life. You exist as whole, perfect and complete in spacetime and create from this awareness and not from within the realities you are presently experiencing in an objective world.

44. You do this by staying aware at all times. Be aware that what you are presently experiencing is a creation of your mind. In order to create a better reality you need to relax, stop controlling the things in your life and simply let go.

45. Your true identity is not derived from any of your realities including your possessions, physical body or ego personality for you do not exist as any of these aspects of your life. You stop identifying with the things of this world and start to identity with who and what you truly are as self awareness whose existence lies beyond the physical dimensions of your life.

46. You need to mentally and emotionally let go of the things in your life and only allow those things you desire to manifest naturally by imaging them at the subconscious level of the mind and by dissolving any opposing energies to the desired reality. Remember the subconscious mind will always search for unlimited possibilities to bring into manifestation your desired realities.

47. If you remain attached to the realities of your present life you are creating from within those realities and this is very hard to do while you exist as them. In this case you are reacting against your current realities, trying to control them, or striving to force into manifestation newer or better realities in life in opposition to your existing realities. Such an orientation within life creates resistance and tension which is nearly impossible to overcome. *You never create through an existing reality in life.* You extinguish it by taking its energy down to zero and create a better reality in its place. You can only accomplish this by effectively detaching from all of life's realities.

48. In order to create from this higher cognitive level of self awareness without attachment to the things of this world you first need to personally experience what whole, perfect and complete truly means in your own life. To get an experience of this higher cognitive level in life you need to eliminate any negative energy in the primary programs which might be preventing you from directly experiencing your true identity.

49. When you are truly detached you start to experience all of life as completion, for the meaning of life is not found in the physical realities you are presenting experiencing but in the awareness that you have the power within yourself to discreate and create all of life realities as you want them to be.

The Theoretical Basis for Conscious Analog Thinking

50. We have seen that the three dimensional phenomena of linear time and fixed space only arises as the outcome of the creation process of energy into matter which takes place throughout the entire universe. At the absolute speed of light the energy of light is represented as the subatomic wave and particle. As such they exist as dynamic energy in light form and while in light form they do not exist as matter.

51. Another way of expressing this is that light as pure kinetic energy does not have mass. Mass only relates to matter and all matter exists in space and time. Fixed space and linear time only exist as an aspect in the process of converting energy into matter. When the pure kinetic energy of light is not involved in the creation process of matter the aspects of linear time and fixed space associated with the creation process does not exist insofar as light energy itself in concerned.

52. When observing a beam of light what you are witnessing is the energy of light traveling in the spacetime of the fourth dimension as viewed from our relative position in our three dimensional world. While all energy exists within space and time, fixed space and linear time does not exist as an aspect of pure energy. The reason for this is that pure energy without mass only exists within spacetime. However energy manifesting as matter can only exist within fixed space and linear time. Fixed space and linear time, of course, exists within spacetime. Remember at the quantum level and at the universe level the energy of matter exists in spacetime.

53. The implications of this is that conscious analog thinking through the intentionality of the energy of thought takes place in spacetime. The physical manifestations of thought however manifest themselves in fixed time and three dimensional space. Therefore to create the type of life you desire you need to remove yourself from identifying solely with the material manifestations of your life and start creating a life from the dynamic energy of conscious analog thinking in spacetime.

54. Physical, mental or emotional conscious cognitive processes creating the substance of your realities all take place in linear time and fixed space. All of your creations from this conscious level of mind are subject to physical laws governing the operation of physical reality. Creating from the dynamic energy of spacetime at the quantum level of the subconscious mind transcends linear time and fixed space and has the capacity to correlate the energy of all emerging physical manifestations with the intentionality of thought. This is accomplished by the subconscious energy processes of the mind sorting for unlimited possibilities to bring to conscious awareness those things needed to manifest personal physical realities in accordance with the intentionality of your thought.

APPENDIX TWO

Self Organizing Processes of the Cognitive Structure of the Mind

Self Organizing Processes of the Conscious Cognitive Thought

The conscious mind uses self organizing processes to accomplish the myriad tasks assigned to it in the most efficient and economical ways possible. Self organizing processes are process functions of the mind rather than application functions of the mind. Self organizing processes constitute the tools used by the mind in information processing. The following is a summary of the mind's basic self organizing processes governing the creation and operation of all conscious cognitive thought:

1. Converts images into thought
2. Converts thought into experiences of reality
3. Alters realities by viewing them from entirely different perspectives in life
4. Classify experiences
5. Organizes thought analytically
6. Identifies and forms relationships among differing ideas
7. Create abstract thoughts from objective ideas
8. Form generalizations from specific thoughts
9. Create complex relationships from simple relationships
10. Form a hierarchy of cognitive structures from simpler cognitive structures
11. Differentiates among groups of thought with similar characteristics
12. Create higher order process synthesis
13. Integrates groups of seemingly diverse thoughts
14. Makes value discriminations and comparisons.
15. Draws conclusions from incomplete or insufficient data
16. Preserves and enhances ego personality
17. Integrates thoughts into broader generalizations, concepts and principles
18. Uses sorting styles to organize experiences in life.

There are many other self organizing processes at play in facilitating the consolidation, organization, and flow of information and the creation of thought by the mind. Those listed above, however, constitute the self organizing processes most widely used by the mind in organizing thought. All of these self organizing processes taken together form the process functions of the mind and act to facilitate the processing of data into images, and images into thoughts. It is through the application of the personality and operational programs of the mind using these self organizing processes that thought is organized into experiences of inner and outer realities. It is the personality and operational programs of the mind which form the essential application programs for the purpose of creating the content of life's realities.

Self Organizing Processes of the Subconscious Mind and Brain

The subconscious mind also applies basic self organizing processes in the processing of information similar to the self organizing processes of the conscious mind. The self organizing processes of the subconscious mind are also process functions rather than application functions. The following is a summary of the subconscious mind's basic self organizing processes:

1. Conversion of electrochemical impulses of the brain into patterns of imagery which are interpreted by conscious mind as meaningful thought.
2. Processes the brain's basins and pathways into submodelities and modalities for the purpose of creating patterns of imagery.
3. Converts space and time into spatial and sequential analog patterns which can be interpreted by the mind in processing space and time in a three dimensional world.
4. Receives and transforms unorganized stimuli into organized stimuli.
5. Has the ability to carry out the many automatic functions of the physical body
6. Organizes the immense storage capacity of our short and long term memories
7. Processes the subconscious aspects of the self organizing regulation of all conscious cognitive thought
8. Immense capacity to sort almost unlimited possibilities to bring to conscious awareness appropriate realities to give expression to conscious thought
9. Monitors its environment and make internal adjustment in the stimuli it receives through the operation of its cybernetic loops.
10. Mediates stimuli, data, information and thought processes between the conscious and subconscious mind and between the brain and the subconscious mind.

This list describes the most commonly recognized self organizing processes used by the subconscious mind and brain in processing information. Once you understand how the self organizing processes of the conscious and subconscious minds operate to create for itself personality and operational programs to create a life for itself you can begin to appreciate how the mind and brain creates realities in life. By understanding how these processes work you will be in a much better position to take control over these processes and the primary personality and operational programs of the mind and make whatever changes you desire in order to achieve a better life for yourself.

Appendix Three

A Comparative Analysis of MetaPsychology™ With Traditional Psychology

Traditional Psychology

Traditional psychology differs in many respects from the principles underlying MetaPsychology.™ Briefly these differences are summarized as follows:

1. The Mind Creates Realities: Traditional psychology does not emphasize the fact that the primary, if not the sole purpose, of the human mind/brain system is to create inner and outer realities in life, realities from which we create a sense of ego personality and develop life styles for functioning effectively in our world environment. Traditional psychology is based on the premise that the purpose of the human mind is to assist us in more effectively relating to an objective world which exists outside and separate from us. MetaPsychology takes the position that our personal world itself is but a reflection of our mind, and is under our control to change in any fashion we desire in order to make it a better world in which to live.

2. The Environmental Connection: Traditional psychology does not take into account the fact that through the energy and images of our mind/brain we are all interrelated and interconnected within the energy fields of one another and the entire energy field of our world in the creation of our physical realities. Traditional psychology takes the position that essentially we exist and function as separate and independent beings, that our sense of self, our personal identity, and the way in which we relate and experience the world, and one another, ends at the outer layer of skin of our physical bodies.

3. Quantum Reality: Traditional psychology does not take into account the fact that at the quantum level of creation at the primary systems of the mind we develop our life realities out of quantum energy and quantum images which are identical to the energy and matrix configurations of the atoms making up the entire universe, including you and me. Traditional psychology takes the position that mental processes consist primarily of thoughts, beliefs, perceptions, feeling, emotions, actions and behaviors. Through the use of a number of different therapies, traditional psychology takes the position that by changing any one or more of these components making up our life, we can effectively change our inner and external view of ourselves and of our world. MetaPsychology takes the position that traditional psychology deals with changes primarily at the intermediate levels of the mind, and seldom addresses or attempts to affect changes at the much deeper primary levels of origination of reality existing at the quantum energy level of the brain and mind.

4. Spacetime in the Fourth Dimension: Traditional psychology does not take into account the fact that time and space which comprise the temporal and spatial field into which we project our life realities are also creations of mind of both individual consciousness and the collective consciousness of all mankind, nor that human consciousness exists within the energy

forces and fields of the physical world, in which, and from which, we form our physical existence. Traditional psychology takes the position that essentially we exist independently of one another in a fixed three dimensional world and that the time sequencing and physical spacing of our life takes place in linear time and fixed space which exists largely outside and apart from each one us.

5. Human Consciousness: Traditional psychology largely ignores the role human consciousness plays in the creation of life realities. Seldom does traditional psychology endeavor to deal with the process components of human consciousness, namely, self awareness, knowledge, intelligence and volitional will, in order to effectively rearrange the application programs of the mind in creating expanded realities in life. Because of this limitation in not recognizing the central role human consciousness plays in creating life realities, traditional psychology finds it very difficult to deal with characterlogical dysfunctions of ego personality. It is human ego personality and not human consciousness which traditional psychology recognizes as the primary instrumentality giving rise to the human condition we experience as our life.

6. The Psychopathology of Reality: Traditional psychological treatment including therapy, counseling, and guidance should not be geared solely to the removal of mental blocks to living a more effective and productive life. Traditional psychology should also take into consideration developing and using mental processes for creating a better life reality for ourself, including the creation of physical realities, and for developing higher states of consciousness and expanded levels of awareness. Traditional psychology has taken the position that matters concerning bettering one's own life mentally, emotionally and physically, beyond simply removing mental blocks, is better left to other areas of society, such as the self help, personal growth and the human potential movement, as well as social, cultural and religious disciplines and institutions.

7. The Comprehensive Mind/Brain Model, Deep seated characterlogical deficits of the ego personality have been difficult to address and deal with effectively within traditional psychology. The reason for this is that traditional psychology has not developed a comprehensive model that quantifies the structure, functions and processes of the human brain and mind in organizing ego personality and it many realities in life. We have seen that the basic desk reference on clinical psychology referred to as the DSM manual simply describes types of mental dysfunction. It does not provide those practicing in the field with a comprehensive model of how the brain and mind work in creating a life. The MetaCreation Technology effectively fills this gap by providing this needed comprehensive model and therefore provides the foundation for an entirely new psychology we refer to as Metapsychology.™ Basically Metapsychology addresses the origins of characterlogical mental deficits at their very source in order to eliminate them. Metapsychology™ addresses such dysfunction at the quantum level where the atom and molecular energy and matrix configurations creates distortions in the neurological basins of attraction and in the primary analog systems of the subcon-

scious mind. The corresponding clinical therapy of Metapsychology is called "Analog Therapy"™ or "Analog Field Therapy"™.

Summary

The underlying basic differences between traditional psychology and Metapsychology,™ as summarized above, is that traditional psychology is based on the premise that its purpose is to assist the client in developing the skills needed to live effectively in a complex world. MetaPsychology takes the position that we should, in addition to developing skills to live more effectively in our world, assist individuals in developing the skills needed to actually create a better physical world in which to live. Traditional psychology assumes that we live in an objective world which exists separate and apart from us. MetaPsychology assumes that we have created the world in which we live and by becoming aware of the processes by which we created a sense of personal identity and our world we can take effective control over our mental processes and create for ourselves both a better ego personality and a better world!

The primary belief system structures determine how strong a person's ego identity is and how effectively he can function within his world environment. As indicated above, it is these primary belief systems structures where we find what is commonly referred to in the field of psychology as Axis II characterlogical disorders of the mind. These types of disorders are, under current clinical psychological therapies and protocols, difficult to deal with effectively. The reason for this is that the sense of self has nowhere to go conceptually if the ego identity program is substantially altered or eliminated. Notwithstanding this fact traditional psychology has not been able to broaden its working models of the mind in order to allow for the existence of the concept of self awareness as existing behind or beyond its ego identity program. If such a concept of self awareness as a working model does not exist within traditional psychology then there is no effective way for traditional psychology to deal with or eliminate substantial characterlogical dysfunctions arising within the ego identity program of the mind without first overcoming the resistance of fear, and without encountering a considerable amount of pain and suffering in the process.The reason for this is that the sense of self would be lost in the process. It is these Axis II type disorders as well as lesser situational dysfunctions which the MetaCreation processes discussed in this book deal with in order to effectively clear limitations to the full enjoyment of life.

A very close analogy to this concept of self awareness as used herein is found in the AA programs which have developed over the last 50 years, with chapters now existing in almost every major city in this country and abroad. What has made the AA program effective in treating alcoholism is their reliance on a power greater than the individual in helping the individual overcome alcoholism. However, in the AA programs the concept of a power greater then the individual quite often is couched in religious or philosophical terms rather then as an aspect of human consciousness or in other psychological terms, and, therefore, it does not have much carry-over effect in treating characterlogical dysfunctions in traditional therapy protocols.This is not to imply that the concept of a power greater then yourself, or a concept of a loving God involved in a person's life are not very strong symbols in supporting a person's self image, the way he relates to other people, and the manner in which he conducts himself in life in general. The MetaCreation symbol for self awareness as being, "whole, perfect, and complete", is comparable to the AA symbol of a power greater than yourself insofar as mind dynamics are concerned, for they both allow for the organization of personal realities beyond ego personality and the organization of cognitive rational thought within the cognitive structures of the mind in a fixed objective world.

INDEX TO CHARTS

SELECTED READINGS

Quantum Physics

Elliott, David A.,*The Electric Universe* (Inklination Press, 1996)
Lederman, Leon, *The God Particle* (Dell Publishing, 1994)
Brody, David E. & Brody, Arnold R.*The Science Class You Wish You Had* (Berkley Publishing Group, 1997)
Capra, Fritjof, *The Tao of Physics* (Shambhala Publications, 1975)
Capra, Fritjof, *The Turning Point* (Bantam Books, 1982)
Capra, Fritjof, *The Web of Life* (Doubleday, 1996)

Spacetime

Hawking, Stephen W., *A Brief History of Time* (Bantam Books, 1988)
Capra, Fritjof, *The Tao of Physics* (Shambhala Publications, 1975)
Capra, Fritjof, *The Turning Point* (Bantam Books, 1982)

Neuroscience

Pert, Candace B., *Molecules of Emotion* (Simon & Schuster, 1997)
Furman, Mark E. *Neurophysics of Human Behavior* (CRC Press, 2000)
Churchland, Paul M.*The Engine of Reason, the Seat of the Soul* (MIT Press, 1996)
Churchland, Patricia S. & Sejnowski, Terrence J.*The Computational Brain* (MIT Press,1996)
Shore, Rima *Rethinking the Brain* (Families and Work Institute, 1997)

Consciousness

Maslow, Abraham H. *Toward A Psychology of Being* (Chapman & Hall 1968)
Penrose, Roger *Shadows of the Mind* (Oxford University Press, 1994)
Herbert, Nick *Elemental Mind* (Penguin Group, 1993)
Capra, Fritjof *The Web of Life* (Doubleday, 1996)
Bucke, Maurice R.*Cosmic Consciousness* (E.P. Dutton & Co., 1969)

Subconscious Processes

Bechtel, William & Abrahamsen, Adele *Connectionism and The Mind* (Blackwell Publishers, 1991)
Horgan, Terence & Tienson, John *Connectionism and the Philosophy of Psychology* (MIT Press, 1996)
Shore, Bradd *Culture in Mind* (Oxford University Press, 1996)

Subconscious Patterns

Moore, Robert & Gillette, Douglas, *King, Warrior, Magician, Lover* (Harper Collins, 1991)
Rowan, John *Subpersonalities, The People Inside Us* (Routledge, 1990)
Sheldrake, Rupert *The Presence of the Past* (Park Street Press, 1988)
James,Tad & Woodsmall,Wyatt *Time Line Therapy* (Meta Pub. 1988)
Bandler, Richard & MacDonald, Will *An Insider's Guide to Sub-Modalities* (Meta Publications 1988)
James, Tad *The Secrete of Creating Your Future* (Advanced Neuro Dynamics, Inc. 1989)

INDEX